Bose-Einstein Condensation

Proceedings of the
Thirteenth Physics Summer School

Bose-Einstein Condensation

From Atomic Physics to Quantum Fluids

Canberra, Australia
17 – 28 January 2000

Editors

Craig M. Savage

Mukunda P. Das

Department of Theoretical Physics
The Australian National University

World Scientific
Singapore • New Jersey • London • Hong Kong

Published by

World Scientific Publishing Co. Pte. Ltd.
5 Toh Tuck Link, Singapore 596224
USA office: 27 Warren Street, Suite 401-402, Hackensack, NJ 07601
UK office: 57 Shelton Street, Covent Garden, London WC2H 9HE

British Library Cataloguing-in-Publication Data
A catalogue record for this book is available from the British Library.

BOSE-EINSTEIN CONDENSATION
From Atomic Physics to Quantum Fluids

ISBN-13 978-981-02-4484-2
ISBN-10 981-02-4484-3

Contents

Preface vii

Topics in the Theory of the Ultracold Dilute Alkali Gases 1
A. J. Leggett

Statistical Mechanics of Bosonic Systems 43
J. Oitmaa

Condensate Oscillations, Kinetic Equations and Two-Fluid Hydrodynamics in a Bose Gas 65
A. Griffin

Finite-Temperature Effects in Bose-Einstein Condensates 117
K. Burnett

The Theory of Atom Lasers 153
R. J. Ballagh and C. M. Savage

Vortices and Solitons in Bose-Condensates 189
P. D. Drummond

Cooling, Trapping and Manipulation of Neutral Atoms and Bose-Einstein Condensates by Electromagnetic Fields 231
K. Helmerson and W. D. Phillips

Experimental Aspects of Bose-Einstein Condensation 281
A. C. Wilson and C. R. McKenzie

Preface

The thirteenth Physics Summer School on *Bose-Einstein Condensation: atomic physics to quantum liquids* was held for two weeks during January 2000 at the Australian National University, Canberra. This summer school was one of an annual series organised by the Department of Theoretical Physics as part of the program of the ANU Centre for Theoretical Physics. On this occasion the School was jointly organised by Craig Savage (Physics and Theoretical Physics, Faculties) and Mukunda Das (Theoretical Physics, IAS) as convenors.

The program covered basic and advanced topics in the area of Bose-Einstein condensation and atom optics. Tony Leggett discussed several conceptual issues in the physics of BEC dealing with phase coherence and superfluidity and gave a lucid picture of dilute Fermionic condensates at ultra low temperatures. Jaan Oitmaa presented a pedagogic survey of Bosonic condensation for both noninteracting and interacting atoms in two and three dimensions. Allan Griffin dealt with the kinetic equations by deriving collective oscillations in two fluid hydrodynamics for the trapped BEC. Keith Burnett took this further by discussing finite temperature effects on the collective states of BEC in traps.

Atomic lasers are an exciting outcome of recent BEC studies. Robert Ballagh covered this aspect of coherent atom optics clearly and in detail. Nonlinearity is an important aspect of BEC studies in recent years. Peter Drummond analysed solitons and vortices in nonlinear field equations. One can not talk about the BEC in confinement without knowing how the atoms are trapped. Bill Phillips and Kris Helmerson presented a detailed picture of the science and technology of how the atoms are cooled in a variety of electromagnetic traps. This series on cooling and trapping was supplemented by Andrew Wilson on their recent experimental accomplishments on BEC and atom lasers. Finally we had a series of lectures on BEC-BCS crossover by Mohit Randeria. Unfortunately Randeria was unable to provide his manuscript in time to be included in this Proceedings.

We are grateful to Professor Erich Weigold, Director of the RSPhysSE for his keen interest in organising this School. We are also thankful to Professors Robert Dewar and Hans Bachor for their support. We are indebted to all the lecturers, participants and members of our Departments for their helpful cooperation. We are particularly thankful to Heli Jackson for her excellent work in running the Summer School.

Canberra
15 August 200

Craig Savage
Mukunda Das
Editors

1. Philip Brydon
2. Nick Robins
3. Wooshik Shim
4. Craig Savage
5. Bob Ballagh
6. Heung-Ryoul Noh
7. Jay Gambetta
8. Robert Sang
9. James Swansson
10. Peter Drummond
11. Prabasaj Paul
12. Sang-Hoon Kim
13. Jonathan Ashmore
14. Catherine Bellair
15. Michael Fraser
16. Jeanette Fyffe
17. Julie McCabe
18. Nicholas Whitlock
19. Heli Jackson
20. Federica Cattani
21. Sodano Posquale
22. Fred Green
23. Jeroen Koelemeij
24. Jaan Oitmaa
25. Scott Edwards
26. Vanessa Leung
27. Mark Lee
28. Marteen Hoo Gland
29. Jacquez Tempère
30. Björn Hall
31. Joshua Grossman
32. Adam Norrie
33. Jacob Dunningham
34. Robert Dall
35. Mohit Randeria
36. Allan Griffin
37. John Close
38. Ken Baldwin
39. Mukunda Das
40. Michael Walker
41. Tony Leggett
42. Brian Robson
43. Max Colla
44. Andreas Penckwitt
45. Jamie Robbins

TOPICS IN THE THEORY OF THE ULTRACOLD DILUTE ALKALI GASES

A. J. LEGGETT

Department of Physics, University of Illinois at Urbana-Champaign, 1110 West Green Street, Urbana, IL 61801-3080, USA

E-mail: tony@cromwell.physics.uiuc.edu

Web: http://web.physics.uiuc.edu/People/faculty/deansforms/leggett.html

These lectures address Bose and Fermi ultracold dilute alkali gases. For Bose gases I discuss general BEC concepts, superfluidity, the Josephson effect, and issues related to the establishment of a "phase standard". For Fermi gases I discuss prospects for Cooper pairing and its detection. The four lectures on alkali Bose gases which I actually gave at the School were based on a review paper in preparation for Reviews of Modern Physics. In order to avoid extensive duplication of the contents of that review, what I present here, condensed into three lectures, is a melange of informal commentary on the results presented there and some material which for space and other reasons had to be omitted. This material is ideally read in conjunction with the review. The fourth lecture, on the Fermi alkali gases, is presented here in (nearly) the form in which it was actually given.

1 Introduction. History, the experimental systems, comparison with liquid helium. General concepts concerning BEC

Let's start with a bit of potted history (for a much more detailed account, see Griffin (1999)). As we all know, in 1924 Bose proposed that the Planck radiation law could be obtained by regarding photons as indistinguishable particles such that the wave function is symmetric under their interchange (so that in modern terminology they obey "Bose statistics"); in this way he obtained the distribution

$$n_i = \{\exp(\beta\epsilon_i) - 1\}^{-1}, \quad \beta \equiv 1/k_B T, \tag{1}$$

which yields the Planck radiation law. Inspired by this result, Einstein very shortly afterwards proposed that systems of massive particles with integral spin (including zero) should show a similar distribution; however, there is a crucial difference with the photon case in that the total number of atoms is conserved whereas that of the photons is not, and as a result the distribution postulated by Einstein for atoms is

$$n_i = \{\exp[\beta(\epsilon_i - \mu)] - 1\}^{-1} \tag{2}$$

where μ is the chemical potential. It is clear that the distribution described by (2) makes sense only if μ is smaller than the lowest possible value of ϵ_i, and that at sufficiently low temperatures this condition is not compatible with conservation of the total number of particles. In fact, for N particles in a volume Ω there is a critical temperature $T_c \equiv 1/k_B\beta_c$ defined by the condition

$$\sum_i \{\exp(\beta_c \epsilon_i) - 1\}^{-1} = N \tag{3}$$

which leads to the explicit expression

$$T_c = \alpha \frac{\hbar^2}{mk_B} n^{2/3}, \quad \alpha \equiv 2\pi/[\zeta(3/2)]^{2/3} \cong 3.3, \quad n \equiv N/\Omega \tag{4}$$

Einstein postulated that below T_c a finite fraction $N_0 \equiv fN$ of the total number of particles N would occupy a single one-particle state, namely the state corresponding to the lowest value of ϵ_i (with the standard periodic boundary conditions, this state is the zero-momentum state and its energy is zero). The quantity f increases from 0 at T_c to 1 at $T = 0$. This phenomenon (or a generalization of it) is known as Bose-Einstein condensation, or BEC for short.

Einstein's prediction was initially not taken very seriously, since there was a suspicion that it was a pathology of the free Bose gas which was likely to disappear as soon as interactions were properly taken into account. However, in 1938 two laboratories working on liquid helium-4, that of Kapitza in Moscow and that of Allen and Misener in Cambridge, UK, simultaneously reported the discovery that below the so-called λ-temperature T_λ (about 2.17 K at saturated vapor pressure) this liquid appeared to flow through very narrow capillaries without detectable friction. This property (which must, incidentally, have been occurring almost daily in the laboratory for twenty years previously, without ever having been explicitly noticed!) is a spectacular manifestation (though not the conceptually simplest one, see 2.1) of the property, or rather complex of properties, now known as superfluidity. Excited by its observation and by the proximity of the λ-temperature of helium to the value of T_c ($\sim$ 3.13 K) predicted by formula (4) for a free Bose gas of the same mass and density, Fritz London immediately proposed that below T_λ liquid ^{4}He was indeed showing the phenomenon of BEC, and that this was the origin of its superfluid behavior. Over the last sixty years there has been overwhelming experimental and theoretical evidence that London's hypothesis is correct, and it is now almost universally accepted by workers in the field.

However, there is one unsatisfactory feature of this scenario: it has proved extremely difficult to verify directly that BEC is occurring! The best direct evidence comes from high-energy neutron scattering and from the evaporation of atoms from the ^{4}He surface (see Leggett (1999a) for details) but, while certainly entirely consistent with a condensate fraction which is of order 10% at $T = 0$, it cannot be said to rigorously establish it. In the mid-70's it was realized that the phenomenon of BEC might be much easier to "see" directly in a dilute gas of Bose atoms, and this was part of the motivation for a program set up by several laboratories to cool atomic hydrogen into the relevant ultra-low-temperature regime. As you know, BEC was in fact achieved in atomic hydrogen in the fall of 1998, but by that time the phenomenon had already been attained in other systems of atomic alkali Bose gases (^{87}Rb, ^{23}Na and ^{7}Li); it is these latter systems which are the main subjects of these lectures, though many of the considerations apply equally to H.

Just to remind you of the relevant parameter regimes: for a typical Bose alkali gas trapped by magnetic or laser methods, the total number of atoms can range from a few hundred to $\sim 10^8$ (10^{10} for H), the maximum density (usually achieved in the BEC phase) ranges from $\sim 10^{11}$ cm^{-3} to $\sim 5 \times 10^{15}$ cm^{-3} (still several orders of magnitude less than that of air at STP!) and the temperatures required to achieve BEC are typically $\sim 10^{-7}$ to 10^{-5} K, which means that the atomic velocities in the BEC phase can be as low as 1mm/sec. Obviously, in the long term such a gas of atoms is unstable against recombination and they will eventually form a solid, but in practice, by a suitable selection of the hyperfine states, etc., sample lifetimes can be made of the order of minutes.

The alkali gas systems differ from the "paradigmatic" BEC system, liquid ^{4}He, in just about every possible respect: for a detailed comparison, see Leggett (1999). Probably the two most important differences are (a) that the alkali gases are very dilute and weakly interacting, so that perturbation theory in the interatomic interaction is expected to be a very reliable method of calculating their properties, and (b) that, partly as a result of (a), they are extremely sensitive to manipulation of the external potentials in space and in time, something which is moreover quite straightforward to achieve in experimental practice. Two well-known examples of such sensitivity are, in the time-independent case, the manifestation of the onset of BEC by a marked change in the density profile (Anderson et al. (1995)) and, in the time-dependent case, the interference fringes observed when a barrier between two independently formed condensates is suddenly removed (Andrews et al. (1997)); neither of these phenomena have (realistically observable) analogs in liquid ^{4}He.

Let's now turn to some general considerations concerning BEC in an interacting system. The first point to note is that in order to talk sensibly about BEC, one must talk about the atoms as wholes, and in particular not try to analyze the behavior of the nuclei and electrons separately. This is actually rather obvious in the case of alkalis such as ^{87}Rb or ^{23}Na, since the nuclei in these cases when taken by themselves obey Fermi rather than Bose statistics, but it is equally true for ^{4}He: it was long ago shown rigorously by Girardeau (1970) that if one excludes the obviously unphysical case of zero nuclear-electron interaction, the nuclei by themselves can never show BEC. The point is most easily made within the framework of the standard Born-Oppenheimer approximation, in which the complete wave function of a system such as ^{4}He containing N nuclei (labelled by coordinates $\mathbf{R}_i$) and ZN electrons (labelled by coordinates $\mathbf{r}_\nu$) is of the generic form

$$\Psi(\mathbf{r}_\nu; \mathbf{R}_i) = \Psi_n(\mathbf{R}_i)\Psi_{el}(\mathbf{r}_\nu : \mathbf{R}_i) \tag{5}$$

where the "electronic" wave function Ψ_{el} may be regarded as depending parametrically on the nuclear coordinates $\mathbf{R}_i$, since it is obtained by solving the Schrödinger equation for the electrons at fixed $\mathbf{R}_i$. Because of this dependence, it can be shown (Girardeau (1970)) that if we use the complete many-body wave function Ψ the single-nucleus density matrix cannot have an eigenvalue of order N, and thus according to the definition given below BEC cannot occur. (In essence, the atomic electrons are continually "observing" the position of the nucleus, and thereby destroying the coherence between the amplitudes to be at different (mutually distant) positions.) However, this result, while formally correct, is of no great significance. A much more natural and physical approach is to define BEC in terms of the "nuclear" wave function $\Psi_n(\mathbf{R}_i)$, which may for most purposes be regarded as the wave function of the (center-of-mass of the) atom as a whole. There are then no a priori obstacles to achieving BEC, as we now indicate.

Let's then focus for the moment on a gas of atoms without (relevant) internal structure, and note that the single-particle density matrix obtained from $\Psi_n(\mathbf{R}_i)$, namely

$$\rho_1(\mathbf{R}, \mathbf{R}'; t) \equiv N \int d\mathbf{R}_2 ... d\mathbf{R}_N \Psi^*(\mathbf{R} : \mathbf{R}_2 ... R_N : t)\Psi(\mathbf{R}' : \mathbf{R}_2 ... \mathbf{R}_N : t) \tag{6}$$

can, by virtue of its Hermiticity, always be expressed in the form (we now revert to lower case $\mathbf{r}$ for its arguments, since there is no danger of confusion)

$$\rho_1(\mathbf{r}, \mathbf{r}' : t) = \sum_i n_i(t)\chi_i^*(\mathbf{r}t)\chi_i(\mathbf{r}'t) \tag{7}$$

where the $\chi_i(\mathbf{r}t)$ form, at any given time t, a complete orthonormal set of single-particle states. We will say that at any time t the system shows "BEC" if <u>at least one</u> eigenvalue $n_i(t)$ is of order of the total particle number N, and that it shows "simple BEC" if one and only one eigenvalue has this property. These definitions are straightforwardly extended to the case where the atoms have an internal (e.g. Zeeman) degree of freedom, by including the relevant discrete variable α (e.g. the m_F-value) on an equal footing with the continuous variable $\mathbf{r}$: $\rho_1 \equiv \rho_1(\mathbf{r}\alpha, \mathbf{r}'\alpha' : t)$ etc. (see Leggett (2000)). In the existing literature it is very often implicitly assumed that BEC, if it occurs at all, must be of the "simple" variety, but in lecture 3 we shall see that this is sometimes very far from obvious.

If simple BEC does occur in the system of interest, it is convenient to label the single eigenvalue which is of order N as $N_0(t)$ and the associated single-particle eigenfunction (the "wave function of the condensate") as $\chi_0(\mathbf{r} : t)$. The "order parameter" $\Psi(\mathbf{r}t)$ can then be defined simply by the relation

$$\Psi(\mathbf{r}t) \equiv \sqrt{N_o(t)}\chi_0(\mathbf{r}t) \tag{8}$$

and it is also convenient to define the superfluid velocity $\mathbf{v}_s(\mathbf{r}t)$ by writing $\chi_0(\mathbf{r}t) \equiv A(\mathbf{r}t)\exp[i\varphi(\mathbf{r}t)]$, where A and φ are real, and setting

$$\mathbf{v}_s(\mathbf{r}t) \equiv \frac{\hbar}{m}\nabla\varphi(\mathbf{r}t) \tag{9}$$

so that whenever $A(\mathbf{r}t)$ is finite we have

$$\text{curl}\, \mathbf{v}_s(\mathbf{r}t) = 0 \tag{10}$$

The quantity $\mathbf{v}_s$ defined by (9) may be contrasted with two other characteristic "velocities". For a single particle described by Schrödinger's equation, one can separate the Schrödinger wave function into amplitude A and phase φ and define a "velocity" $\mathbf{v}_1(\mathbf{r}t)$ in terms of the latter analogously to Eq.(9) (and satisfying (10)), and this quantity plays an important role in some interpretations of the quantum-mechanical formalism, see e.g. Bohm (1952); however, because one is dealing with only a single particle, the fluctuations in associated physical quantities are quite comparable to the mean values. On the other hand, one can consider a many-body system and define the "hydrodynamic" velocity $\mathbf{v}_h(\mathbf{r}t)$ as the ratio of the current $\mathbf{j}(\mathbf{r}t)$ to density $\rho(\mathbf{r}t)$: using the representation (7) and defining $\chi_i(\mathbf{r}t) \equiv A_i(\mathbf{r}t)\nabla\varphi_i(\mathbf{r}t)$ we find

$$\mathbf{v}_h(\mathbf{r}t) = \frac{\hbar}{m}\left\{\sum_i n_i(t)A_i^2(\mathbf{r}t)\,\nabla\,\varphi_i(\mathbf{r}t)\right\} / \sum_i n_i(t)A_i^2(\mathbf{r}t) \tag{11}$$

The quantity defined by (11) is stable against fluctuations, but in general does not have simple properties like (10). The superfluid velocity (9) has in some sense the advantages of both worlds: it is truly a simple quantum-mechanical object analogous to $\mathbf{v}_1(\mathbf{r}t)$, but at the same time is associated with a macroscopic number of particles and thus stable against fluctuations. In the simple case of a very weakly interacting gas, the particle current $\mathbf{j}(\mathbf{r}t)$ in a state corresponding to a given value of $\mathbf{v}_s(\mathbf{r}t)$ is under suitable circumstances given by the intuitively plausible expression

$$\mathbf{j}(\mathbf{r}t) = |\Psi(rt)|^2\mathbf{v}_s(\mathbf{r}t) \tag{12}$$

But it should be emphasized that this statement is in general not correct for an interacting system, and the definition of the "superfluid density" which relates $\mathbf{j}$ to $\mathbf{v}_s$ is actually quite a delicate matter, cf. e.g. Huang (1987), ch. 13.

We finally enquire briefly why BEC should occur at all, and why it is so often (though not always, cf. lecture 3) of the "simple" variety. An obvious remark is that Bose statistics and low temperatures are not by themselves sufficient conditions for the occurrence of BEC, as evidenced inter alia by the existence of a solid (crystalline) phase of ^{4}He which certainly does not exhibit this property! If we confine ourselves to the dilute-gas limit relevant to the alkalis, we can say that the same "statistical" effects which lead to BEC in the non-interacting system will persist for weak interaction. As to the effect of the interaction itself, this depends essentially on its sign; it is believed that an attractive (effective) interaction will in general lead to an instability, while weak repulsion actually leads to a small increase in the transition temperature (cf. e.g. Grüter *et al.* (1997)). The basic physics behind this somewhat counter intuitive result is that weak repulsion actually favors multiple occupation of the same (orbital) state, thereby not only reinforcing the statistical effect but favoring "simple" BEC, at least in spinless systems, over the more general kind. For details see Leggett (2000), where the few rigorous results on the existence or not of BEC known to the author are also listed.

2 Aspects of Superfluidity

2.1 Criteria for superfluidity

"Superfluidity" is not a single phenomenon but a complex of phenomena which, at least up to now, have generally been observed to occur together (though see below); this complex of phenomena is not peculiar to BEC systems in the literal sense but is observed (at least to an extent) also in liquid ^{3}He below 3 mK and, in light disguise, in various metals where it is known as superconductivity. (These systems may be regarded as undergoing "pseudo-BEC", since pairs of fermions form molecular-like objects (Cooper pairs) which can be treated as (something like) bosons.) Phenomena which are related to superfluidity and sometimes taken to be part of its definition include:

(1) The ability to flow through narrow pores without apparent friction.

(2) The Hess-Fairbank effect (refusal of the system, at sufficiently low angular velocity, to rotate with its container).

(3) The ability of the system, when placed in a toroidal geometry, to sustain a persistent circulating current even when the container is at rest.

(4) "Quantized circulation".

(5) Zero or much reduced friction on objects moving in the system with velocity less than some critical velocity.

(6) Certain peculiar types of collective excitation such as "second sound".

(7) The Josephson effect.

Although these various phenomena are related, the exact nature of the relationship is quite a delicate issue, and I will discuss it only fairly briefly here. My own view is that the most basic manifestation of superfluidity, and thus the one perhaps most suitable as a definition, is property (2), which is the neutral version of the Meissner effect in superconductivity; alone of the phenomena listed, this is unambiguously a property of the thermodynamic equilibrium state of the system. To see its connection to BEC, we argue qualitatively as follows (see Leggett (2000) for details): As we shall see in the next section, for a spinless system placed in a container rotating uniformly with angular velocity $\boldsymbol{\omega}$, the thermodynamic equilibrium state is obtained by minimizing the free energy $\langle \hat{H}' \rangle - TS$, where the "effective" Hamiltonian $\hat{H}'$ is given by

$$\hat{H}' \equiv \hat{H} - \boldsymbol{\omega} \cdot \hat{\mathbf{L}} \tag{13}$$

$\hat{H}$ being the Hamiltonian of the system at rest and $\hat{\mathbf{L}}$ the angular momentum around the axis of rotation. Let us for definiteness consider a geometry such as a thin torus of radius R which possesses nearly exact cylindrical symmetry around the rotation axis, so that we can characterize the single-particle states by their angular momentum quantum number $\ell \equiv L_z/\hbar$ around this axis. Then in a normal (non-BEC) system with $\omega = 0$ the number of particles $\langle n \rangle$ with quantum number ℓ has a wide distribution ($\Delta\ell \sim (mR^2 k_B T/\hbar^2)^{1/2}$ where R is a typical radius of the torus), and the system accommodates to a finite value of ω by shifting the distribution slightly, thereby producing an angular momentum $\langle L_z \rangle$ which is essentially the classical value $NmR^2\omega$ (For a quantitative implementation of this argument for the case of the free Bose gas, see e.g. Leggett (2000), and for the more generic case, Kohn and Sherrington (1964)). By contrast, let us consider the case of a weakly interacting Bose gas at $T = 0$, for which we may suppose that to a first approximation all atoms are in a unique condensate ("simple BEC"). Since this condensate is characterized by a unique value of ℓ, the only possible values of $\langle L_z \rangle$ are $N\ell\hbar, \ell = 0, \pm1, \pm2...$, and it is easy to convince oneself, in particular, that for sufficiently small ω ($|\,\omega\,| < \frac{1}{2}\omega_c$, $\omega_c \equiv \hbar/mR^2$) the effective Hamiltonian in (13) is minimized by the choice $\ell = 0$. Thus the system has zero angular momentum, i.e. appears to remain at rest in the lab frame despite the rotation of the walls – the Hess-Fairbank effect, which I re-emphasize is a property of the thermodynamic equilibrium state.

By contrast, property (3) is an indication of a high degree of metastability of certain excited states. In fact, for a toroidal container at rest in the laboratory frame, a rigorous variational argument shows that the angular momentum $\langle L_z \rangle$ in the (global) thermodynamic equilibrium state cannot be greater than $\alpha N\hbar$, where for a sufficiently "thin" torus α is less than 1; thus, any states which carry angular momentum (i.e. circulating current) greater than this, such as are certainly observed in ^{4}He, can be only metastable. The qualitative explanation of this metastability in a BEC system is that any kind of deformation of the condensate state so as to change ℓ from its initial finite value to zero (or to a smaller value) involves a free energy barrier which is, at least prima facie, [a] extensive and thus cannot be overcome by thermal (or quantum) fluctuations; this consideration is generally believed to give an adequate account of the persistence of circulating currents in ^{4}He in a toroidal geometry. However, in the context of the alkali gases it is crucial to appreciate that the validity of the argument requires (a) that the effective interatomic interaction be repulsive (and finite), and (b) that the geometry have an adequate degree of cylindrical symmetry. In fact, in Leggett (2000) it is shown by explicit study of a "toy" model that a p-state is metastable against decay to an s-state only to the right of a "separatrix" in the parameter plane defined by interaction strength and (lack of) cylindrical symmetry. In other words, it is possible for a system to show property (2) but not property (3).

The relation of property (1) (which is the originally and most readily observed manifestation of "superfluidity" in ^{4}He) to properties (2) and (3) is

[a] The situation in real life is complicated because of the possibility of nucleating "local" defects such as vortex rings (see Langer and Fisher (1967)); however, the conclusion remains qualitatively valid except very close to T_c.

quite amusing. To examine it, it is convenient to re-express the problem of thermodynamic equilibrium in a rotating container by working in the frame of the container; then the Hamiltonian is just H (up to an irrelevant centrifugal term, cf. Eq.(18) below) but the "single-valuedness" boundary condition on the many-particle wave function becomes schematically

$$\Psi(\theta_1 ... \theta_i + 2\pi ... \theta_N) = \exp(2\pi i m\omega R^2/\hbar)\Psi(\theta_1 ... \theta_i ... \theta_N) \tag{14}$$

where θ_i is the angular coordinate, rather than the more familiar special case obtained for $\omega = 0$. The Hess-Fairbank effect is then manifested, in this frame, by a finite value of the angular momentum (circulating current) which, for $\omega < \frac{1}{2}\omega_c$, is proportional to ω (whereas a normal system would show zero current in this frame). Thus, the Hess-Fairbank effect may be regarded, in a more general sense, as the tendency of a (small) "twist" in the boundary conditions to induce flow. Suppose now we consider a narrow tube (pore), and imagine that we "twist" the boundary conditions at the ends of the pore relative to one another (this is automatically achieved, for example, by applying a gravitational or pressure difference across the tube, either continuously or for a finite time interval). If the induced phase difference $\Delta\varphi$ is less than π (corresponds to $\omega < \frac{1}{2}\omega_c$ in the toroidal geometry) then the state of the liquid in the tube corresponds to its true equilibrium state subject to the twisted boundary conditions, and carries a current proportional to the phase drop. The latter property persists even when the phase drop is greater than π, but it is clear that the state is no longer the equilibrium one but rather a metastable state, and corresponds more nearly to the "persistent currents" observed in the toroidal geometry with the container at rest (for which $\Delta\varphi = 2\pi\ell$). Thus, depending on the details of the geometry and the experimental procedure, phenomenon (1) may be regarded as a special case either of (2) or of (3)!

Property (4) can also be regarded as in some sense derivative from (2) and (3), although in the case of the alkali gases it is more easily observable than either (Matthews et al. (1999); Madison et al. (2000)). It is clear from the definition (lecture 1) of the superfluid velocity $\mathbf{v}_s(\mathbf{r}t)$ that it must satisfy the Onsager-Feynman quantization condition

$$\oint \mathbf{v}_s(\mathbf{r}t) \cdot d\boldsymbol{\ell} = nh/m, \quad n = 0, \pm 1, \pm 2 ... \tag{15}$$

For a simply connected container with the order parameter $\Psi(\mathbf{r}t)$ finite everywhere, the irrotationality condition Eq.(10) of course implies Eq.(15) with $n = 0$, but in (e.g.) a toroidal geometry nonzero values of n are possible, and in fact in a nearly cylindrical geometry correspond to the nonzero (but integral) values of ℓ in the current-carrying states. In a simply connected container it is possible to form a "vortex" by constraining $\Psi(\mathbf{r})$ to vanish on a given line and treating the remainder of the volume as effectively toroidal; thus (15) applies, with a nonzero value of n, provided that the circuit is taken once around the line in question. (In practice, $|n|$ is almost invariably 1, since a vortex with $|n| > 1$ tends to be unstable against dissociation into a set of $|n| = 1$ ones). It is needless to remark that the mere existence of vortices is in no way peculiar to a superfluid system, and in fact it would almost certainly be possible, by an appropriate stirring process, to produce a vortex with a (hydrodynamic) velocity field $\mathbf{v}(\mathbf{r})$ satisfying the quantization condition (15). What is special about superfluid (BEC or pseudo-BEC) systems is that (a) if only the condensate is considered, the only vortices which can be produced satisfy (15), and (b) vortices which satisfy (15) tend to have a high degree of metastability once the mechanism producing them has been switched off (whereas for a normal liquid the vorticity would rapidly dissipate). Actually, the theoretical expectation is that a vortex produced by rotating an alkali gas in an approximately cylindrical trap will, once the rotation is switched off, eventually migrate to the boundaries of the condensate and disappear (Rokhsar (1999); Fedichev et al. (1999)), and this seems qualitatively consistent with the experiments to date (Matthews et al. (1999); Madison et al. (2000)).

The exact relation between properties (5) and (1-4) is slightly problematic. At first sight it is very plausible that if the liquid or gas in question can flow through their capillaries at velocities up to some critical velocity v_c without apparent friction, then by Galilean invariance an impurity such as an ion should be able to move through the liquid at velocities up to v_c also without apparent friction, and vice versa; indeed, it is effectively the "vice versa" form of the argument which is at the heart of Landau's celebrated formula for the critical velocity, v_c, namely

$$v_c = \min_p(\partial\epsilon(p)/\partial p) \tag{16}$$

where $\epsilon(p)$ is the spectrum of the elementary excitations of the system. (For the case of the ion, $v < v_c$ with v_c defined by (16) is simply the condition for the absence of "Cerenkov" radiation.) For a single-component system in thermal equilibrium at $T = 0$ the argument indeed looks convincing, and experiments with ions in liquid ^{4}He (Allum et al. (1977)) appear to confirm (16) in this case. For any case more complicated than this, however, a number of qualifications have to be made. For example, in a system of pure ^{4}He containing a dilute concentration of ^{3}He at $T = 0$, the minimum velocity of an "excitation" (in this case a fermion particle-hole pair) is zero, but it is almost certain that such a system would show frictionless flow in a narrow capillary. The point is that it is only the "normal component" (in this case, crudely speaking, the ^{3}He component) which comes into equilibrium with the walls of the capillary, or with which conversely the ion comes into equilibrium (thus experiencing reduced but not zero friction); the superfluid component associated with the condensate continues to flow past the walls (or past the ion) without apparent friction. This consideration makes it clear that in the general case the $\epsilon(p)$ which appears in the Landau criterion should not refer to all possible excitations of the system, but only to those which couple to the condensate (crudely speaking, to its density fluctuations). Similar considerations apply, at finite T, to the noncondensed atoms (or more strictly the "normal component").

Property (6) may be regarded in some sense as a special case of (5), since the existence of the "second sound" mode is a direct consequence of the ability of the normal and superfluid components of the bulk liquid to flow relative to one another without appreciable damping. I postpone discussion of property (7), the Josephson effect, until lecture 3, merely noting here that in the "external" case it is in fact possible (by incorporating the Josephson junction in a ring) to regard it as an extreme case of the "toy model" of rotation treated in Leggett (2000), with a very large value of the cylindrical asymmetry.

2.2 Rotating frames of reference

In order to formulate a quantitative account of superfluidity, whether in liquid ^{4}He or in the BEC alkali gases, we need to know how to do statistical mechanics when the potential in which the system is confined[b] (be it due to a physical container or to magnetic or laser-induced trap, etc.) is rotating. In the following I address until further notice the "spinless" case; with this reservation, the argument of the present subsection is quite general (in particular, it does not assume that the "container" is necessarily multiply-connected). If the container is stationary in the laboratory frame, then in that frame the Hamiltonian has the generic form

$$H_{lab} = \sum_i (\mathbf{p}_i^2/2m + V_{ext}(\mathbf{r}_i)) + \frac{1}{2}\sum_{ij} U(|\mathbf{r}_i - \mathbf{r}_j|) \tag{17}$$

where $V_{ext}(\mathbf{r})$ represents an external potential due to the container walls, (or magnetic fields etc.): it is not necessary to assume that this possesses exact or

[b]In the following I will sometimes use the phrase "container" to indicate the external potential, irrespective of its physical nature.

approximate cylindrical symmetry (though typically in the case of experiments on He it would do so). Note that the interatomic potential is taken to be a function only of the magnitude of $|\mathbf{r}_i - \mathbf{r}_j|$. Now suppose the container is rotating around its axis with angular velocity ω. In order to employ the usual reasoning of thermodynamics, it is necessary to use a frame of reference in which the Hamiltonian is *not time dependent*, i.e. the frame of the rotating walls. [c] That is, we must minimize the energy E (or at finite temperature the free energy $E - TS$) as it would appear to a rotating observer. What is the form of this Hamiltonian? At first sight, it is just the sum of a kinetic energy term $\sum_i \mathbf{p}_i'^2/2\text{m}$ and a time-independent potential energy $V(\mathbf{r}_i')$, where $V(\mathbf{r}_i')$ and $\mathbf{p}_i' \equiv \mathbf{m}\dot{\mathbf{r}}_i'$ are respectively the coordinate and momentum of the i-th atom as seen by the observer. However, if applied to a classical liquid like water this would imply that there is no way the observer would detect, by observing the behavior of the liquid, that he is rotating; in particular, it would not give the meniscus which is observed to occur. So we must exercise a little care. The correct procedure is to start from the (time-dependent) Lagrangian in the laboratory frame, express this Lagrangian in terms of the rotating-frame coordinates $\mathbf{r}_i'$ and their time derivatives using the prescription $\dot{\mathbf{r}}_i' = \dot{\mathbf{r}}_i - \omega \times \mathbf{r}_i'$, define the canonical rotating-frame momentum $\mathbf{p}_i' \equiv \delta\mathbf{L}/\delta\dot{\mathbf{r}}_i$ and use the standard definition $\sum_i \mathbf{p}_i' \cdot \dot{\mathbf{r}}_i' - \mathbf{L}$ for the rotating Hamiltonian H'. This procedure gives

$$H' = \sum_i (\mathbf{p}_i'^2/2m + V_{ext}(\mathbf{r}_i')) + \frac{1}{2}\sum_{ij} U(|\mathbf{r}_i' - \mathbf{r}_j'|) - \frac{1}{2}\sum_i m_i(\omega \times \mathbf{r}_i')^2 \qquad (18)$$

and it may be verified that for a classical liquid the last term gives the observed meniscus. In writing down the expression (18) we have used the fact that since the primed and unprimed coordinates are related by a simple rotation, the distance $|\mathbf{r}_i' - \mathbf{r}_j'|$ is identically equal to $|\mathbf{r}_i - \mathbf{r}_j|$.

The problem then reduces to finding the state which minimizes $\langle H' \rangle$ (or at finite temperature the free energy $\langle H' \rangle - TS$). In the classical case this is straightforward, since it is identical to the stationary problem except for the replacement $V_{ext}(\mathbf{r}) \to V_{ext} - \frac{1}{2}m(\omega \times \mathbf{r})^2$.

In the quantum-mechanical case, however, there is a subtlety: The same

---

[c] There is a delicate point here: we need enough "rugosity" (lack of exact cylindrical symmetry) in V that the system can "tell the difference" between the case where the walls are stationary ($V = V(\mathbf{r})$) and when they are rotating ($V = V(\mathbf{r}, t)$). I assume in the following that this condition is fulfilled.

quantum-mechanical many-body state will be assigned one wave function ψ by a stationary observer and a different one ψ' by a rotating one, and in general the "single-valuedness boundary condition" obeyed by ψ' is different from that satisfied by ψ; cf. Eq.(14) above. Although it is quite straightforward to take this into account, a simpler method is to transform the Hamiltonian (18) back to the lab-frame coordinates. That is, we choose a particular instant t_0 (let us say for definiteness, one in which the configuration of the rotating container happens to coincide with its original one) and express $\hat{H}'$ in terms of the lab-frame quantities $\mathbf{r}_i(t_0) \equiv \mathbf{r}_i$, $\mathbf{p}_i(t_0) \equiv \mathbf{p}'_i + m(\boldsymbol{\omega} \times \mathbf{r}'_i)$. Using the vector triple identity $\mathbf{a} \cdot \mathbf{b} \times \mathbf{c} \equiv \mathbf{b} \cdot \mathbf{c} \times \mathbf{a}$ and the definition of the (lab-frame) angular momentum $\hat{\mathbf{L}}$, and comparing with Eq.(17), we see that the final result is

$$\hat{H}_{eff} \equiv \hat{H}' = \hat{H}_{lab} - \boldsymbol{\omega} \cdot \hat{\mathbf{L}} \tag{19}$$

Thus, at time t_0 the equilibrium state of the system in the rotating container is given by minimizing the quantity $\hat{H}_{lab} - \boldsymbol{\omega} \cdot \hat{\mathbf{L}}$ (or the associated free energy). Since we know that as viewed from the rotating frame the state must be stationary, it is given at general times simply by the state at t_0 rotated through an angle ωT around the axis of rotation. I emphasize that the prescription (19) is quite general, irrespective of the geometry of the container and the strength of the interaction. It is clear that in the (nearly) cylindrically symmetric case (19) is equivalent to the replacement (where ϵ_n is the eigenvalue of the one-particle (kinetic and external potential) energy in the single-particle state n)

$$\epsilon_n \to \epsilon'_n \equiv \epsilon_n - \hbar\omega\ell_n \tag{20}$$

where ℓ_n is the angular momentum (in this case approximately a good quantum number) of the state n in units of $\hbar$.

It should be carefully noted that the derivation of Eq.(19) relies essentially on the fact that the interatomic potential is a function only of $|\mathbf{r}_i - \mathbf{r}_j|$ (or more precisely is invariant under simultaneous rotation of all atomic coordinates around the axis of $\boldsymbol{\omega}$); it is this condition which guarantees the existence of a frame of reference (the frame rotating with the container) in which the Hamiltonian is independent of time. If the condition is not fulfilled (e.g. due to the effect of dipole forces) then there will in general be no such frame and the concept of a thermodynamic "equilibrium" (as distinct from a "steady") state becomes meaningless: cf. Leggett (2000).

Note added in proof: While Eq.(19) and the technique used to derive it are (I believe) correct, the details of some of the intermediate formulae in this section are wrong. They can be formally corrected by the replacement, in Eq.(18) and the $\mathbf{p} - \mathbf{p}'$ relation three lines above Eq.(19) (only!), of $\mathbf{p}'_i$ by the kinematic momentum $m\dot{\mathbf{r}}'_i$.

2.3 Effect of hyperfine degree of freedom

Since until recently our only example of a (true) boson system exhibiting superfluidity has been liquid ^{4}He, it is not particularly surprising that most discussion of the rotational properties has assumed that we are dealing with a system of "spinless" atoms, i.e. atoms with no internal (e.g. hyperfine) degree of freedom. However, it was already some time ago realized that in the case of the pseudo-BEC of Cooper pairs in superfluid ^{3}He the internal degrees of freedom of the pairs might play an important role, and thus it is natural to ask whether the hyperfine degree of freedom present in the BEC alkali gases could modify any of the conclusions obtained for the "spinless" case. The ensuing discussion may be more understandable if read in conjunction with ch. 6 of Leggett (2000).

At first sight one might suspect that the introduction of a hyperfine degree of freedom is unlikely to change the situation qualitatively,[d] and indeed in the case of the Hess-Fairbank effect this is so. However, as originally shown by Ho (1982) in the context of spin-polarized hydrogen, the standard considerations reviewed, e.g. in Leggett (2000), concerning metastability are qualitatively modified.

We quote from section 6.4 of Leggett (2000) the result that in the case of most practical interest (strong repulsive interatomic interaction) there is a large free energy barrier associated with deforming our Gross-Pitaevskii (GP) wave function so as to go from (e.g.) a p-state to an s-state. The reason is essentially topological: so long as the GP order parameter $\Psi(r)$ remains finite in magnitude so that its phase φ can be defined, the winding number $\oint \nabla\varphi \cdot d\mathbf{l} \equiv l$ is a topological invariant, so any l-changing process must involve forming a node, and this costs energy because of the repulsive interaction term.

Why does the existence of a hyperfine degree of freedom change the situation fundamentally? The easiest way to see this is to consider the very simplest possible case, namely the transition of an F=1/2 Bose gas in zero field[e] from a p-state to an s-state. Suppose the initial GP state is, schematically

$$\Psi_{in} = \exp(i\chi)|\uparrow\rangle \tag{21}$$

(where χ is the angle around the annulus). The final state is required to be an s-state, but we put no particular requirements on the spin orientation, since it is clear that this can if necessary be rotated back to the original $|\uparrow\rangle$

[d]Except perhaps insofar as it effects the effective geometry, cf. Matthews et al. (1999).

[e]In principle we might achieve this situation by applying a strong magnetic field to H and working in a frame rotating with $\omega_L = \mu_n B_{eff}$ to remove the hyperfine splitting of the nuclear Zeeman states.

configuration uniformly without any cost in energy. Thus, we write the final state as

$$\Psi_f = \mathrm{const}(\chi)\ |?\rangle \tag{22}$$

Now, following Ho, consider the family of states, parametrized by a real angle θ such that $0 \leq \theta \leq \pi$

$$\Psi(\theta) \equiv \cos\left(\frac{\theta}{2}\right) |\Psi_{in}\rangle + \sin\left(\frac{\theta}{2}\right) |\Psi_f\rangle,$$
$$|\Psi_f\rangle = \mathrm{const}(\chi) |\downarrow\rangle \tag{23}$$

It is clear that this family of states is normalized and interpolates smoothly between Ψ_{in} for θ=0 and Ψ_f for $\theta = \pi$. The crunch, however, comes when we examine the explicit expression for Ψ as a function of the angle χ. Apart from an overall factor of $\exp(i\chi/2)$ (which we must certainly eventually keep when evaluating the energy explicitly)[f] this is:

$$\Psi(\chi : \theta) = \cos\left(\frac{\theta}{2}\right) \exp(i\chi/2) |\uparrow\rangle + \sin\left(\frac{\theta}{2}\right) \exp(-i\chi/2) |\downarrow\rangle \tag{24}$$

But this is simply the wave function of the original state rotated through an angle θ around an axis in the xy-plane making an angle χ with the x-axis! Since the wave function (24) is simply a rotation of the original one, it immediately follows not only that $|\Psi(\theta)|^2$ is constant and equal to its original value, but that the interaction term is unchanged from its original value. Thus the sequence of states corresponding to (23) is "downhill all the way," in the sense that it involves crossing no energy barrier associated with the interaction terms. Since the kinetic energy also goes "downhill" (it is simply $(N\hbar^2/2mR^2) \cdot \cos^2\theta/2$), it follows that provided only the "small" stabilizing field B satisfies the condition $\mu B < N\hbar^2/2mR^2$ the total energy also decreases monotonically as θ rises from 0 to π. Thus there is no metastability in the usual sense (though it must of course be emphasized that in a real-life alkali gas the kinetics associated with the process may be slow enough to give the rotating state a macroscopic lifetime; cf. subsection 6.5 of Leggett (2000)).

It is fairly clear that this argument can be generalized to the case of arbitrary F (and in fact to arbitrary ℓ): however, there is a further curious feature which we might not necessarily have guessed from the simple F=1/2 case. To see it, let's explicitly consider the case F=1, and express the Zeeman substates as explicit Cartesian functions x±iy, z. In analogy to the operation conducted on the F=1/2 system, we should expect that a procedure to relax

[f] And also to ensure single-valuedness.

superflow would be a rotation of the m_L=+1 state by π around some axis in the xy-plane to make it m_L=-1. (Note that any rotation through an angle other than π will leave some m_L=+1 component, which we do not want). We would further expect that the axis in the xy-plane around which we rotate would itself rotate around the z-axis as a function of θ so as to execute an integral number ($\neq 0$) of turns as θ goes from 0 to 2π. But it is easy to demonstrate that the phase with which the $m_L = -1$ state is created by this operation is twice the angle made by the rotation axis with the x-axis: thus if the rotation axis itself rotates, as θ goes from 0 to 2π, n times around the axis, then the phase of the $m_L = -1$ state created corresponds to $2n$ rotations, i.e. $4n\pi$. It follows that for a maximally polarized F=1 state the winding number can decrease only by multiples of 2. As shown by Ho, this result can be generalized to higher F and to non-maximally polarized ($m_F \neq F$) states, the generic result is that any system initially in a state with given F and m_F can decrease its winding number only by multiples of $2m_F$, (irrespective of F). The above is an informal statement of a theorem of homotopy theory, namely $\Pi_1(SU(2))=Z_{2m}$: cf. Mermin (1979). Thus, for example, a maximally polarized state of (hypothetical) BE-condensed ^{133}Cs in the upper hyperfine state (F=4) could change its winding number (circulation) only in multiples of 8, and in particular flow states with circulation of 3 or less are evidently metastable (at least as far as the above argument goes).

3 Some aspects of the Josephson effect and related topics

3.1 Realizations of the Josephson effect in the alkali gases

The term "Josephson effect" is now generally understood not to be restricted to its original manifestation in metallic superconductors but to refer more generically to any system of $N \gg 1$ bosons in which each boson is restricted to occupy a 2D one-particle Hilbert space. The generic statics and dynamics of such a system is discussed in detail in Leggett (2000); in this section I discuss its possible realization in the BEC alkali gases. As in Leggett (2000) I refer to the Josephson effect as "external" when the two basic states 1, 2 correspond to spatially separated configurations and "internal" when they differ in some internal quantum number such as the hyperfine index. The Hamiltonian of any "Josephson" system can always be expressed in terms of the three operators

$$\hat{M} \equiv \frac{1}{2}(a_1^+ a_1 - a_2^+ a_2), \quad \hat{Q}_1 \equiv \frac{1}{2}(a_1^+ a + a_2^+ a_1),$$
$$Q_2 \equiv -\frac{1}{2} i(a_1^+ a_2 - a_2^+ a_1) \tag{25}$$

which turn out to be isomorphic to the angular momentum components J_2, J_x, J_y of a "particle" of spin $N/2$. A common and particularly simple case, which we will call "canonical", is obtained when the Hamiltonian is of the specific form

$$\hat{H} \equiv \hat{H}_{canon} = \frac{K}{8}(a_1^+ a_1 - a_2^+ a_2)^2 - \frac{\Delta\mu}{2}(a_1^+ a_1 - a_2^+ a_2) - \frac{\epsilon_J}{2}(a_1^+ a_2 + a_2^+ a_1)$$
$$\equiv \frac{1}{2} K \hat{M}^2 - \Delta\mu \hat{M} - \epsilon_J \hat{Q}_1 \tag{26}$$

A realization of the "internal" Josephson effect in a BEC gas may in principle be effected by stabilizing two different hyperfine species in the same trap and deriving transitions between them with electromagnetic radiation.[g] To simplify the notation I shall specialize at once to the case closest to the existing experiments of most interest in this context (Hall et al. (1998)), that is two states with different F-values (actually the F=2, m_F=1 and F=1, $m_F = -1$ states of ^{87}Rb), and such that the transition between them is a two-step one via an intermediate state $|I\rangle$ (actually the F=2, m_F=0 one) which itself lies in the electronic ground state multiplet, so that the two lasers operate at frequencies $\lesssim$10GHz. Under these conditions spontaneous transitions are completely negligible and we may treat the laser fields as purely classical.[h]

Consider first the situation in the absence of the coupling lasers (but with both species present). If we neglect recombination processes, conservation of angular momentum means that the number N_i ($i = 1, 2$) of atoms of each species is separately conserved, i.e. M=const. The total energy is now a function of M, both through the single-atom splitting $\Delta\mu_0$ and because the intra- and interspecies scattering lengths are in general different. Assuming complete condensation of both species in the same ground state wave function

[g]For brevity I shall refer to the sources of radiation as "lasers," although in existing experiments this is not technically correct.

[h]This would not necessarily be so in a "Raman" setup where the immediate state involves electronic excitation, and might then lead to interesting effects (Javanainen (1997)).

χ_0 (cf. below), we have

$$\hat{H} = \text{const.} - \frac{1}{2}\Delta\mu_0(N_1 - N_2) + \frac{1}{2}\cdot\frac{4\pi\hbar^2}{m}\cdot\int|\chi_0(r)|^4 dr \cdot \sum_{i,j=1,2} a_{ij}N_iN_j$$

$$\equiv \text{const.} + \frac{1}{2}K\hat{M}^2 - \tilde{\Delta\mu}\cdot\hat{M} \tag{27}$$

where

$$\tilde{\Delta\mu} \equiv \Delta\mu_0 + \frac{2N\pi\hbar^2}{m}\int|\chi_0(r)|^4 dr\cdot(a_{11} - a_{22}) \tag{28}$$

$$K \equiv \frac{1}{2}\cdot\frac{4\pi\hbar^2}{m}\cdot\int|\chi_0(r)|^4 dr\cdot(a_{11} + a_{22} - 2a_{12}) \tag{29}$$

Note that K is of the order of ϵ times the total mean-field energy per particle, where $\epsilon = (a_{11} + a_{22} - 2a_{12})$ divided by the "typical" value of a (say a_{11}) and can be considerably less than unity.

The effect of the lasers is two-fold. First, they will provide a "light shift" (dipole potential) which is in general different for states $|1\rangle$ and $|2\rangle$. This effect just changes the value of the constant $\Delta\mu_0$ and is of no great interest. A more important effect is to provide an effective coherent coupling between the states $|1\rangle$ and $|2\rangle$ of the form

$$\hat{H}_J = -\frac{\Delta\epsilon_{las}(t)}{2}(a_1^+ a_2 + a_2^+ a_1) \tag{30}$$

For simplicity I consider here the (realistic) case where the sum δ of the detunings of the two lasers is small compared to the intermediate-state detuning δ_i. It is then convenient to go into the frame rotating with the sum frequency $\omega_1 + \omega_2$ of the two lasers and make the usual "rotating-wave" approximation. The effective coupling $\Delta\epsilon_{las}$ provided by the lasers is then time-independent and given by the expression

$$\Delta\epsilon_{las} = d_{1i}d_{i2}\epsilon_{1i}\epsilon_{i2}/\hbar\delta_i \equiv \hbar\Omega_R \tag{31}$$

where $\epsilon_{1i}(\epsilon_{i2})$ and $d_{1i}(d_{i2})$ are respectively the rms electric field strengths and dipole matrix elements for the $1\to i$ $(i\to 2)$ transition, and I have introduced the standard quantum-optical notation Ω_R for the two-photon Rabi frequency $\Delta\epsilon_{las}/\hbar$. Putting together Eqs.(27-31) and bearing in mind that the effect of the transition to the rotating frame is to subtract from $\Delta\mu$ the quantity $\hbar(\omega_1 + \omega_2)$, we see that in this frame the Hamiltonian is time-independent and given by the "canonical" form (26) with the identifications

$$\Delta\mu \equiv \Delta\mu_0 - \hbar(\omega_1 + \omega_2) + \frac{2N\pi\hbar^2}{m}\eta\cdot(a_{11} - a_{22}) \tag{32}$$

$$K \equiv \frac{1}{2} \cdot \frac{4\pi\hbar^2}{m} \eta \cdot (a_{11} + a_{22} - 2a_{12}) \tag{33}$$

$$\epsilon_J \equiv \hbar\Omega_R \tag{34}$$

where

$$\eta \equiv \int |\chi_0(r)|^4 dr \tag{35}$$

The above schematic description is unfortunately too simple to apply to most existing experiments. In real life not only is the external potential (and hence the condensate wave function, even in the absence of interactions) different for the two species, but the condensate wave functions will themselves be functions of M. In the "adiabatic" limit, where the Rabi frequency Ω_R is small compared to the characteristic inverse "rearrangement time" of the condensates, it should still be possible to use the generic two-state Hamiltonian (but not the "canonical" special case (26)), with the dependence on M given simply by that of the global ground state energy for that M-value. However, in existing experiments (Hall et al. (1998)) we are quite far away from the adiabatic limit. [i]

We now turn to the "external" Josephson effect. Typically, as in the case of superconductors or liquid ^{4}He, we would expect this to occur in a situation where the potential seen by the system consists of two potential wells, typically of similar shape and equal or nearly equal depth, separated by a potential barrier which, while high enough that its interior is classically inaccessible to the atoms in their (many-body) ground states, is not so high as to make tunneling through it negligible. The simplest way to create the necessary two-well potential would seem to be simply to use the setup of the MIT interference experiment (Andrews et al. (1997)), with a Ioffe-Pritchard magnetic trap split down the center by a strongly detuned Ar^+ laser beam, and to turn the power down to a value substantially less than that explored in the published experiments. [j] The result, as in the interference experiment, would be a potential which has two symmetric minima and a single pair of saddlepoints in the xy-plane. I will denote the coordinate which connects the two minima by z. A point which will have some significance later is that given the strong anisotropy of the original magnetic trap and the configuration of the laser beam, the final effective potential is close to being additive in these

[i] This is at least partly due to the anomalously small value of the quantity $(a_{11} + a_{22} - 2a_{12})$ for ^{87}Rb, and is not necessarily a generic characteristic of such experiments.

[j] Subsequent experiments may in fact have attained this regime (see Rohrl et al. (1997)).

coordinates, at least in the region where the probability of finding an atom is appreciable:

$$V(xyz) \sim V_1(x) + V_2(y) + V_3(z) \tag{36}$$

I will assume in what follows that the effect of the laser beam is indeed simply to contribute a "light-shift" term to $V(xyz)$ and nothing more, and in particular that effects associated with real absorption and radiation of photons in the barrier region are negligible. Although this assumption seems plausible for strong detuning, it should really be examined.

It should be observed that one possible ansatz for the many-body wave function in this potential is just the standard Gross-Pitaevskii form

$$\Psi_N = \Pi_{i=1}^{N} \chi_0(\mathbf{r}_i, t) \tag{37}$$

where $\chi_0(\mathbf{r}, t)$, generally speaking, will have appreciable weight in each of the two wells but little weight under the barrier. It would then be natural to describe the time evolution of the order parameter $\Psi(\mathbf{r}t) \equiv \chi_0(\mathbf{r}, t)$ by the TDGP equation. I show in Leggett (2000) that there are some circumstances where this is a good approximation, but for the present we do not wish to assume it. On the other hand, it is convenient to restrict ourselves to the case where, *within a single well*, the form of the wave function is GP-like and moreover corresponds to the "local" ground state. This suggests that we proceed as follows: We first make a GP ansatz for the many-body *ground state* and identify the condensate wave function $\chi_0(\mathbf{r})$. We then write the latter in the approximate form $a\chi_0(\mathbf{r})+b\chi_R(\mathbf{r})$, where $\chi_L(\mathbf{r})$ and $\chi_R(\mathbf{r})$ are mutually orthogonal normalized wave functions which are predominantly confined to the left (right) wells respectively. Then we take $\chi_L(\mathbf{r})$ and $\chi_R(\mathbf{r})$ as the basis of our 2D one-particle Hilbert space. Note that this procedure does *not* assume that the ground state within the GP ansatz is the true many-body ground state.

There is a technical difficulty here, which may be illuminated by the special case of a symmetric double-well potential. In this case it is obvious from symmetry that $\chi_L(\mathbf{r})$ is the mirror image of $\chi_R(\mathbf{r})$ and that in the GP ground state the average number of particles in the left and right wells is equal. The precise form of (e.g.) χ_L must be determined by a quantitative solution of the GP equation, but it is intuitively clear that for weak tunneling it is very similar to the GP ground state which one would get if the R well were absent and L were occupied by $N/2$ particles. Now, the problem is that we would like to be able to use our 2D basis to discuss, *inter alia*, states in which the number of particles in the two wells is far from equal. But if we change the number N_L in (e.g.) L away from $N/2$, the form of the "best" GP wave function

$\chi_L(\mathbf{r})$ within that well will itself change! I will assume nevertheless for the moment that we can proceed "as if" χ_L and χ_R are fixed states independent of $M \equiv \frac{1}{2}(N_L - N_R)$ (compare the situation in the internal effect in the adiabatic limit). With this understanding it is straightforward to calculate the term in the Hamiltonian which depends only on M; it comes purely from the N-dependence of the single-well energies, including the interaction term. For simplicity I give the result only for the case of two identical near-harmonic wells, under the assumption that neither N_L nor N_R is so small that the TF approximation breaks down: from the standard results for a single well (see e.g. Dalfovo et al. (1999))

$$f_1(M) = \lambda\{(\frac{N+M}{2})^{7/5} + (\frac{N-M}{2})^{7/5}\}, \tag{38}$$

$$\lambda \equiv \frac{15}{4}(\frac{15a_s}{\bar{d}})^{2/5} \times \beta\hbar\bar{\omega}$$

where $\bar{\omega}$ is the geometric mean of the trap frequencies and $\bar{d}$ similarly that of the zero-point widths, and β is a numerical factor of order unity which depends on the ratios of the trap frequencies and is 1 for an isotropic trap. For small M we may expand (38) to give

$$f_1(M) = \frac{1}{2}KM^2, \quad K \equiv 2^{-3/5}\frac{7}{25}N^{-3/5}\lambda \tag{39}$$

The calculation of the tunneling (Josephson) term involving the operators $\hat{Q}_{1,2}$ is a bit more tricky. Since the GP ansatz (37) approximately diagonalizes these operators, it is convenient to make this ansatz, with the (time-independent) form of χ_0 given by an expression which is schematically of the form

$$\chi_0(r) = \cos\theta \cdot e^{i\varphi/2}\chi_L(r) + \sin\theta e^{-i\varphi/2}\chi_R(r) \tag{40}$$

i.e., it reduces to the first term within the L well and close to it, and to the second term within the R well. Deep under the barrier we do not know a priori that $\chi_0(r)$ is well approximated by (40). With this ansatz we can attempt to solve the time-independent GP equation and obtain the dependence of the energy on θ and φ; in particular, if it should turn out to be of the form $A_0 \sin\theta\cos\theta\cos\varphi$, where A_0 is a constant, then this justifies a description of the tunneling energy by the last term in (26), with $\epsilon_J = A_0/N$. Note incidentally that for the ansatz (40) we have $\langle M\rangle = \cos 2\theta \cdot (N/2)$. In the following I will assume for simplicity, in view of (37) that deep under the barrier the wave function χ_0 for ϕ=0 (which is real) is factorizable into a transverse part and a function of z which I denote $\psi_0(z : \theta)$; however, the result can be generalized (see Zapata et al. (1998)).

Omitting the derivation, which is straightforward though not entirely trivial (Leggett (1999b)), I quote the following result: Provided we are well in the WKB limit, then the effect of interactions deep under the barrier is negligible and irrespective of the value of θ we can write

$$E(\theta, \varphi) - E(\theta, 0) = A(\theta)(1 - \cos\varphi) \tag{41}$$

where

$$A(\theta) \equiv \frac{N\hbar^2}{m}\left[\int \frac{dz}{\psi_0^2(z:\theta)}\right]^{-1} \tag{42}$$

However, it should be strongly emphasized that the dependence of $A(\theta)$ on θ (i.e. on M) is, except possibly in very special cases, <u>not</u> given by the expression $\sin\theta\cos\theta$, and hence it is not justified to describe the tunneling term by the last term in (26) for general values of M; the physical reason is that in contrast to the case of a single particle, the main way in which a finite asymmetry of the wave function affects the quantity $\psi_0(z:\theta)$ is not through the boundary conditions as specified in (40), but rather through the effects of the interactions on the "local" (in-well) chemical potentials. This point is currently under investigation. The only limit, then, in which we can justify the "canonical" Hamiltonian (26) for the external Josephson effect in the BEC alkali gases is the limit of weak asymmetry[k] ($M \ll N$), where the M-dependence of the tunneling term is negligible.

Finally, one might raise the question whether the above results for ϵ_J, which are based on a GP ansatz for the many-body wave function, might not be importantly affected by "Bogoliubov-level" corrections. In particular, if the ϵ_J calculated using the GP ansatz should turn out to be negligibly small, could these connections provide a larger φ-dependent term in the energy (presumably varying as $\cos 2\varphi$ rather than $\cos\varphi$)? An attack on this problem using a particular model has been made by Meier and Zwerger (1999), with results which are somewhat negative, and recent calculations (Kohler et al. (2000)) appear to indicate that the coefficient of the $\cos 2\varphi$ term is indeed of higher order in the WKB tunneling amplitude than that of the single "GP" term (41).

3.2 Interference and "phase standards"

In section 7 of Leggett (2000) I deal with a two-state system which started off in a fully coherent superposition state and gradually loses the coherence.

[k]Or more generally when the excursions of M, i.e. θ, from some (possibly finite) reference value θ_0 are small then we set $A(\theta)$ equal to its value at $\theta = \theta_0$.

In this subsection I want to address very briefly the inverse process, in which a system which starts off with no phase relation between its two states acquires one as result of "observation." The best-known and most spectacular example of this process is the MIT interference experiment (Andrews et al. (1997) which has been analyzed in many theoretical papers in the literature (e.g. Javanainen and Yoo (1997), Rohrl et al. (1997)). The analysis of this subsection, which is more generic, is in essence a very stripped-down version of that of Castin and Dalibard (1997) and Dunningham and Burnett (1999).

Consider a "toy" model of the MIT interference experiment along the lines of the discussion of Castin and Dalibard (1997). We start with two condensates which are initially in a "relative Fock state," that is, each with a fixed number of particles, say N_1 and N_2, in physically separated states ψ_1 and ψ_2 respectively and thus no phase coherence between them. To examine how "measurement" induces a definite value of relative phase, we imagine that they are subjected to a setup in which the combination of (states evolved from) the linear combination $2^{1/2}(\psi_1+\psi_2)$ induces a particular macroscopic state Φ_+ of some measuring apparatus, while the orthogonal combination $2^{-1/2}(\psi_1-\psi_2)$ induces a state Φ_- which is orthogonal to Φ_+ and distinguishable by direct macroscopic inspection from the latter. (An approximate physical realization of such a scheme would permit weak "leakage" of a well-collimated beam from each of the condensates and combine the two beams at an appropriate beamsplitter: the states Φ_+ and Φ_- would then correspond to the registration of an atom by the counter in the upper (lower) output channel respectively). Let us imagine that we set up such a device and wait until just one atom has been detected[l] (we do not yet know by which counter). This corresponds, schematically, to the transition (where Φ_0 is the initial untriggered state of the counters)

$$\begin{aligned}|N_1,N_2\rangle|\Phi_0\rangle \to\ & \frac{1}{2}\{(|N_1-1,N_2\rangle+|N_1,N_2-1\rangle)\Phi_+ \\ & +(|N_1-1,N_2\rangle-|N_1,N_2-1\rangle)\Phi_-\}\end{aligned} \qquad (43)$$

Now suppose we inspect the counters to determine which of the states Φ_+, Φ_- is realized, and suppose (for example) that we find Φ_+. Then we can immediately conclude that the correct quantum state to assign to the combined condensates as an initial condition for any subsequent measurement is

[l]We could provide an auxiliary device which would give a signal whenever either of the counters is triggered, without indicating which.

the (properly normalized) state

$$\Psi^{(+)}_{cond} = \frac{1}{\sqrt{2}}(| N_1 - 1, N_2\rangle + | N_1, N_2 - 1\rangle) \tag{44}$$

It is clear that measurement of the operator $\hat{Q}_1 \equiv \frac{1}{2}(a_1^\dagger a_2 + \text{H.c})$ on the combined condensates in the state (44) would give the result $\frac{1}{2}\sqrt{N_1 N_2}$ (for large N_1, N_2), which is equivalent to the statement that the two condensates now have (prima facie) a well-defined relative phase of zero. Similarly, had we on inspection of the counters obtained the result Φ_- we could conclude then that the appropriate description of the coupled condensates is the combination orthogonal to (44) namely

$$\Psi^{(-)}_{cond} = \frac{1}{\sqrt{2}}(| N_1 - 1, N_2\rangle - | N_1, N_2 - 1\rangle) \tag{45}$$

which again corresponds to a well-defined relative phase, this time of π. The crucial point is that while the fact that (say) counter 1 has registered tells us that an atom has been detected, it does not tell us from which of the two condensates that atom came, and our description must thus be by a superposition rather than a mixture of the two possibilities. This point may perhaps be further clarified by imagining that we time-reverse the experiment: an atom incident on the beamsplitter from (what is actually) output beam 1 will actually end up in a linear superposition of states corresponding to the (input) beams leading to condensates 1 and 2 respectively. In this very simple toy model complete phase coherence between the two condensates is, at least at first sight, achieved already after a single detection [m]; of course, it is necessary to ensure that subsequent detections do not "upset" this result. and this is the content of the Castin-Dalibard calculation. However, for our purposes the above schematic discussion is sufficient.

One point about the above discussion needs to be emphasized: While we can inspect the final state of the counters and from it infer the final state of the combined condensates, including their relative phase, there is no way, in this setup, that we could predetermine the condensates to have, say, a relative phase of 0 rather than π; we are stuck with whatever nature happens to give us on a given run! Of course, there may well be other procedures we would use so as to prepare, without "selection," a given final relative phase; cf. Leggett (1995a).

[m] In fact the phase coherence is not "complete". The expectation value $(1/2\sqrt{N_1 N_2})$ of the operator $\hat{Q}_1$ is only $\frac{1}{2}$ of the maximum achievable: as shown by Castin & Dalibard, the maximum value is rapidly achieved in subsequent detections.

So far, so good: now, following Dunningham and Burnett (1999), let's set about the (schematic) construction of a "phase standard." Let's label the system which will set us the standard S and two possible systems to be tested 1 and 2, and set the initial values of the number of atoms in systems 1. S and 2 to be K, L, and M respectively. As above, we imagine that we prepare a device which will respond to the combinations, involving 1 and S,

$$\Psi_{\pm} = \frac{1}{\sqrt{2}}(| K-1, L\rangle \pm | K, L-1\rangle) \tag{46}$$

respectively with output $\Phi_{\pm}$, which (and this is crucial!) do not depend in any substantial way on the precise values of K and L. (Since both K and L will be assumed macroscopically large, this condition is extremely natural). We also prepare a similar but physically different [n] device which will respond to the symmetric and antisymmetric combinations involving S and 2, namely

$$\Psi'_{\pm} = \frac{1}{\sqrt{2}}(| L-1, M\rangle \pm | L, M-1\rangle) \tag{47}$$

with outputs $X_{\pm}$ respectively. We now imagine a sequence in which we first put 1 and S "in contact," i.e. subject them to the first (Φ_-) device, while system 2 is still distant. Then we remove system 1, and subsequently put S and 2 "in contact" (i.e. subject them to the second (X_-) device). We then remove system 2. Finally, we imagine that we bring 1 and 2 into contact (e.g. by a third device of a similar nature) and ask whether, as a result of the "standardization" process they have undergone, they possess a definite relative phase?

It is straightforward to follow through the evolution of the complete wave function of the "universe," i.e. the systems plus the counters. At stage 1 we have

$$\begin{aligned} |K, L, M\rangle|\Phi_0 X_0\rangle \rightarrow \; & \frac{1}{2}\{(|K-1, L, M\rangle + |K, L-1, M\rangle)\Phi_+ \\ & +(| K-1, L, M\rangle - |K, L-1, M\rangle)\Phi_-\}|X_0\rangle \end{aligned} \tag{48}$$

Let us suppose for definiteness that inspection of the counters at this point reveals the state Φ_+. Then the projected wave function which is input to stage 2 is

$$\frac{1}{\sqrt{2}}(| K-1, L, M\rangle + | K, L-1, M\rangle) \mid \Phi_+ X_0\rangle \tag{49}$$

[n] In principle we could allow the two devices to be physically identical, but this introduces some irrelevant technical problems.

and the output of stage 2 is (omitting the Φ_+ which plays no further role)

$$\frac{1}{2\sqrt{2}} \{(|K-1,L,M-1\rangle\rangle + |K-1,L-1,M\rangle$$
$$+|K,L-2,M\rangle + |K,L-1,M-1\rangle\,)X_+$$
$$+(\,|K-1,L,M-1\rangle - |K-1,L-1,M\rangle$$
$$+|K,L-2,M\rangle - |K,L-1,M-1\rangle\,)X_-\} \tag{50}$$

Suppose again for definiteness that inspection of the X-device yields the result X_+; then the projected wave function of the condensates (which, of course, simply multiplies the counter state Φ_+X_+) is

$$\Psi_{cond} = \frac{1}{2}\ (\,|\ K-1,L,M-1\rangle +\ |\ K-1,L-1,M\rangle$$
$$+\ |\ K,L-2,M\rangle +\ |\ K,L-1,M-1\rangle\,) \tag{51}$$

We now enquire whether or not a measurement of the relative phase of systems 1 and 2 will yield a definite value. Such a measurement would involve some operator which is of the general form $\hat{Q}_1 \equiv \frac{1}{2}(a_1^\dagger a_2 + \text{H.c})$ (or $\hat{Q}_2 \equiv -i(a_1^\dagger a_2 - \text{H.c.})$, etc.) but is a unit operator with respect to the "standard" S; thus it is clear that there can be no interference between any pairs of states in (51) which differ in their values of the second argument (the number of particles in the "standard"). The only terms that can show interference, therefore, are the second and last: taking out the common argument $L-1$, we have for this component

$$\Psi \sim \frac{1}{2}(|\ K-1,M\rangle +\ |\ K,M-1\rangle) \tag{52}$$

which indeed corresponds to a relative phase of zero. However, the phase coherence is less than "complete", in the sense that the actual expectation value of the operator $\hat{Q}_1$ is only 50% of its maximum possible value. It is plausible that a thought-experiment which allows multiple detections on both the pair 1-S and the pair S-2 would improve this result, but the combinatorics rapidly get rather messy.

What this argument shows (which can evidently be straightforwardly generalized to the case where "measurement" can yield arbitrary relative phases $\Delta\varphi$) is that it is in principle possible to set up a "phase standard" such that if we know that the phase difference between test system 1 and the standard S is $\Delta\varphi_{1s}$, and that between test system 2 and S is $\Delta\varphi_{2s}$, then we can legitimately infer that systems 1 and 2, which may never have been in direct contact, nevertheless process a definite relative phase, namely $\Delta\varphi_{12} \equiv \Delta\varphi_{1s} - \Delta\varphi_{2s}$.

Dunningham and Burnett (1999) argue that this "standardization" procedure justifies the use of the concept of "absolute" phase; however, as they emphasize, the standardization needs an explicit procedure (it does not occur automatically in Nature!). In this respect I believe that for the reasons given in Leggett (1995b) the popular analogy (e.g. Anderson (1984)) between the "absolute" phase of a Bose condensate and the "absolute" direction of (e.g.) the magnetization of a ferromagnet may still involve some subtle difficulties.

4 Prospects for Cooper pairing and its detection in the Fermi alkali gases

Most of the known even-A isotopes of the alkali gases are unstable, with lifetimes short enough that it would probably be difficult or impossible to cool them into the degenerate regime, but there are three exceptions: ^{2}H (deuterium), ^{6}Li and the long-lived ($\sim 10^9$ years) isotope ^{40}K. Most of the early experimental and theoretical work related to Fermi degeneracy and Cooper pairing was done on ^{2}H, but the light mass and high recoil energy of this isotope has meant that, like its bosonic cousin ^{1}H, it is difficult to cool to microK temperatures, and it has not been to my knowledge the subject of much work in recent years. A few months ago, de Marco and Jin (1999) reported evidence for the cooling of ^{40}K into the degenerate regime; however, as we shall see, the most promising candidate for the observation of Cooper pairing is ^{6}Li, and it is on this isotope that most of the recent theoretical work in this field has been done. In Table 1, I list some important properties of these three isotopes:

Table 1. Properties of some fermionic alkali isotopes.

Isotope	^{2}H	^{6}Li	^{40}K
Nuclear spin	1	1	4
Nuclear moment (μ_N)	0.88	0.82	-1.3
s-wave scattering length (Å)	-3.7	-1150	($\sim +100$)

Let's first briefly discuss the atomic energy diagram in a magnetic field, and the expected behavior in the degenerate normal state. Because, as mentioned, it is ^{6}Li which is the prime candidate for Cooper pairing, I will specialize to this case; the case of ^{2}H is very similar, but that of ^{40}K somewhat different because the negative value of the nuclear dipole moment means that the higher-F (9/2) hyperfine multiplet lies lower.

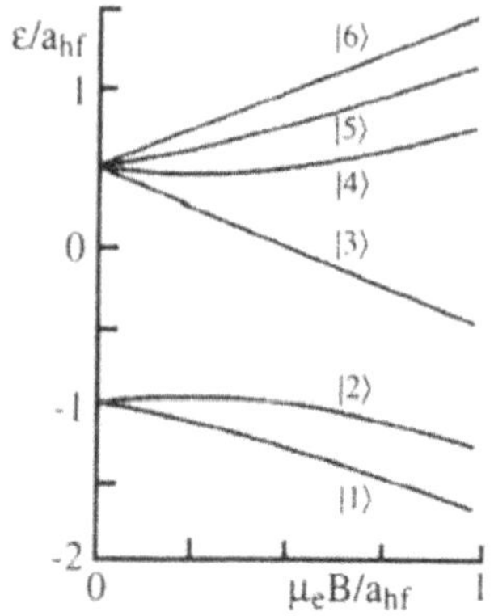

Figure 1. Energies of the hyperfine states of ^{6}Li as a function of magnetic field.

The atomic energy diagram of ^{6}Li is of the general form familiar from similar graphs for the boson isotopes: I show it in Fig. 1 with the conventional labeling of the six states. The state $|6\rangle$ ($F=m_F = 3/2$) is the analog of the $F=m_F=2$ state of (e.g.) ^{23}Na, and the state $|2\rangle$ ($F=1/2, m_F = -1/2$) is the analog of the $F=1, m_F = -1$ state (it is weakly low-field-seeking for small B). For reasons we shall see, the most promising combination of states for Cooper pairing is that of two low-field seekers, say $|6\rangle$ and $|5\rangle$, in fields large compared to the "hyperfine crossover field" B (or, in a laser trap, their high-field-seeking twins $|1\rangle$ and $|2\rangle$ which can be discussed in identical terms); for this reason most theoretical research has focussed on this combination, and I shall do the same.

Consider two ^{6}Li atoms both in the upper hyperfine multiplet (at high field, so that the electron spin projection m_S is effectively fixed and it is the nuclear spin projection m_I which is the "free" variable). Each atom has $I=1$, so the possible values of the total nuclear spin K of the pair are 2, 1 and 0. It is clear that the $K=2$, $m_K=2$ state (and hence all the $K=2$ states) are symmetric in the spin coordinates and hence, from the Fermi statistics, must be antisymmetric in the spatial coordinates; thus the relative orbital states must have odd orbital angular momentum, i.e. ℓ must be at least 1. By orthogonality to the $K=2$, $m_K=1$ state the $K=1$, $m_K=1$ state must be antisymmetric in the spin indices, and hence so must be all the $K=1$ states: thus they must be symmetric in the orbital coordinates, and in particular a relative s-wave state is allowed. It is to this state (for $B \gg 100$ G) that the huge negative value of the s-wave scattering length quoted in Table I refers (there is no particular reason to think that (e.g.) the $|6\rangle - |2\rangle$ scattering length is of anything like this order of magnitude). Finally, the (unique) K=0 state by the same token is spin symmetric, hence orbitally antisymmetric. The upshot is that the only spin states which tolerate a relative s-wave orbital configuration are the K=1 states, in particular the antisymmetric combination $|6\rangle|5\rangle - |5\rangle|6\rangle$. Our discussion of Cooper pairing will refer exclusively to this combination (although it is not difficult to generalize it to arbitrary K=1 states, cf. Modawi and Leggett (1997)).

Let's now consider the normal state of one and of two coexisting species, first in free space and then in a trap. A generic point is that even in the absence of appreciable interaction effects, the whole physics is very different for a fermion system, whether single- or multiple-species, than for a bosonic one; whereas for a bosonic system, in the absence of interactions, there is nothing to prevent the "collapse" of the condensate into the lowest single-particle state, in a fermionic system not only is there no one-particle condensation in the first place, but the degeneracy pressure is always finite and is sufficient to prevent any qualitative change in the density profile even on the onset of the "pseudo-Bose condensation" due to Cooper pairing (see below).

The case of a single hyperfine species in free space is rather boring. The Fermi momentum in this case is related to the density n by

$$k_F = (6\pi^2 n)^{1/3} \tag{53}$$

and the Fermi temperature $T_F \equiv \frac{\hbar^2}{2mk_B} k_F^2$ is therefore given by

$$T_F = k_B^{-1}(\hbar^2/2m)(6\pi^2 n)^{2/3} \tag{54}$$

Not surprisingly, apart from a numerical factor of order unity this is just the formula for the Bose condensation temperature of a Bose gas of the same mass and density. Putting in the atomic mass of ^{6}Li, we find

$$T_F \simeq 10^{-14} n^{2/3} \mathrm{Kcm}^2 \tag{55}$$

so that for $n = 10^{12}\mathrm{cm}^{-3}$ we get a Fermi temperature of about 1μK.

The above calculation assumes that we can treat the single-species system as a noninteracting gas. Actually, this is likely to be true to an excellent approximation, at least at all currently attainable temperatures, since as pointed out above the relative orbital state of any two atoms must be (at least) p-wave, and the effective interaction is therefore of order $U_p(k_F a_p)^2$ where a_p is a characteristic length associated with the p-wave scattering, and U_p can be taken to an order of magnitude as given by $4\pi\hbar^2 a_p/m$. Thus relative to the characteristic (Fermi) energy the interaction is of order na_p^3; if a_p is of the order of atomic dimensions or even longer (say 50-100 Å) this quantity is still only of order $\lesssim 10^{-6}$.

As orientation for the more interesting multi-species problem considered below, we now discuss briefly the problem of a single fermionic species confined in a harmonic trap, which for simplicity we take to be isotropic. As in the Bose case for $T > T_c$, provided that the characteristic wavelengths (which are always $\lesssim k_F^{-1}$) are short compared to the zero-point radius d (a condition almost invariably fulfilled in practice), we can approximate the distribution as semiclassical, i.e. coarse-grain so as to introduce an approximate momentum $\mathbf{p}$ and coordinate $\mathbf{r}$ and write

$$n(\mathbf{p},\mathbf{r}) = \left\{\exp[\beta(p^2/2m + V(\mathbf{r}) - \mu)] + 1\right\}^{-1} \tag{56}$$

which differs from the Bose case by the sign of the 1 in the denominator. The density profile $n(\mathbf{r})$ is obtained simply by integrating (56) over $\mathbf{p}$. At high temperatures ($T \gg T_F$) the Fermi distribution (56) reduces to the classical one, and the integration over $\mathbf{p}$ then leads to the simple classical formula

$$n(\mathbf{r})_{T>>T_F} = \text{const.} \exp\left(-\beta V(\mathbf{r})\right) \tag{57}$$

This is equally true for the Bose case, so that the shape of the distribution gives no information on the statistics. As T falls to values $\lesssim T_F$, the distribution in coordinate space becomes <u>broader</u> than formula (57) would imply; however, there is no singularity at any value of T. At $T = 0$ the distribution in phase space tends to a theta-function:

$$n(\mathbf{p},\mathbf{r}) = \theta\left[\mu - (\frac{p^2}{2m} + V(\mathbf{r}))\right] \tag{58}$$

and the density profile is correspondingly given by a simple Thomas-Fermi equation (obtained by integrating (58) over $\mathbf{p}$):

$$An^{2/3} + V(r) = \mu = \text{const.}, \qquad A \equiv (6\pi^2)^{2/3}\hbar^2/2m \tag{59}$$

In the case of a harmonic trap ($V(\mathbf{r}) = \frac{1}{2}m\omega_0^2 r^2$) the resulting profile is of the form

$$n(r) = n(0)(1 - r^2/r_0^2)^{3/2}, \quad n(0) = (\mu/A)^{3/2}, \quad \mu = \frac{1}{2}m\omega_0^2 r_0^2 \tag{60}$$

where r_0 is the classical cloud edge, determined by setting $\mu = \frac{1}{2}m\omega_0^2 r_0^2$. Note that the shape of the profile (60) is more concave upwards than in the Bose TF limit. The parameters μ and r_0 can be found by equating the integral of $n(r)$ to the total number N of fermions present; in this way we obtain up to numerical constants of order unity

$$r_0 \sim N^{1/6} d, \qquad \mu \sim N^{1/3}\hbar\omega_0 \tag{61}$$

Although these formulae are different in their specific N-dependence from those for the interacting (TF-limit) Bose case, it is amusing that the formula for the "effective" Fermi temperature in the trap, which is clearly of order $\mu(0)/k_B$, is of the same form as that for T_c in the Bose case, namely

$$k_B T_F \sim N^{1/3}\hbar\omega_0 \tag{62}$$

We see that the condition $k_F^{-1} \ll d$ (which is equivalent to $k_B T_F \gg \hbar\omega_0$) is automatically fulfilled in the limit $N \gg 1$.

Let's now turn to the more interesting case of two co-existing species, again starting with the free-space case. The first question is, what are the independent variables? In view of the very long inter-species relaxation times observed by de Marco and Jin (1999), it is at least prima facie plausible to treat the numbers $N_\uparrow$ and $N_\downarrow$ (or the densities $n_\uparrow$ and $n_\downarrow$) as independent and externally fixed variables. A second complication is that since two opposite-spin atoms can now be in a relative s-state, the interactions are not necessarily negligible. In the Hartree-Fock approximation the groundstate energy is given by the sum of the kinetic energy, which for each species separately is just $3n\epsilon_F/5$, and the interaction energy $n_\uparrow n_\downarrow U_0$, where U_0 is given as in the Bose case by $4\hbar^2 a_s/m$. Thus we have

$$E = \frac{3}{5}A(n_\uparrow^{5/3} + n_\downarrow^{5/3}) + n_\uparrow n_\downarrow U_o \tag{63}$$

with A given by Eq.(59) above (note that both A and U_0, when expressed in terms of a_s, contain factors of $\hbar^2/m$).

We now consider, following Houbiers et al. (1997), the stability of the system described by (63) against phase separation or collapse. It is rather intuitively obvious that if U_0 (i.e. a_s) is positive then any phase separation is likely to be into regions with $n_\uparrow$ and $n_\downarrow$ more "unequal" than their average values, while if a_s is negative as in real-life ^{6}Li, then the system may spontaneously collapse so as to increase both $n_\uparrow$ and $n_\downarrow$ in part of the volume and leave voids elsewhere. The condition for either instability to occur is that one eigenvalue of the 2×2 matrix $d^2E/dn_\alpha dn_\beta$ becomes negative (or equivalently that the determinant, which is a product of the eigenvalues, does). From Eq.(63) we see that this condition is equivalent to

$$\frac{4}{9}A^2(n_\uparrow n_\downarrow)^{-1/3} \leq U_o^2 \tag{64}$$

or in terms of the scattering length a_s,

$$(n_\uparrow n_\downarrow)^{1/2}|a_s|^3 \geq \pi/48 \tag{65}$$

in agreement with Houbiers et al. Attainment of this condition in ^{6}Li requires densities of the order of 3×10^{13} cm^{-3}, so is not obviously unachievable. However, it should be strongly emphasized that the whole idea of describing the true inter-atomic interaction by a pseudopotential is unlikely to be valid when $k_F|a_s|$ becomes of order 1 (cf. the situation in the Bose case, and below).

The generalization to a trap geometry is relatively straightforward and has been carried out explicitly by Houbiers et al. (1997). For the moment let us assume, following them, that the trap potential is the same for the two species up to a constant offset (which is irrelevant under the assumption which we are making of no cross-relaxation); this is true to a good approximation in the experimentally relevant limit $B \gg B_{hf}$. To the extent that interactions are neglected, the two species behave independently and have separate cloud edges determined by $r_{0\uparrow,\downarrow} \sim N_{\uparrow,\downarrow}^{1/6} d$. However, interactions, though less qualitatively crucial than in the Bose case, still need to be taken into account. Doing this by the usual Hartree-Fock scheme, we get the appropriate generalization of Eq.(59):

$$An_\alpha^{2/3}(r) + V(r) + U_0 n_{-\alpha}(r) = \mu_\alpha = \text{const.}\,(\alpha = \uparrow, \downarrow) \tag{66}$$

Qualitatively, the effect of a negative value of U_0 is to increase the density of both species at the trap center above the value it would take in the absence of interactions, i.e. in some sense to offset (partially) the effects of the Fermi statistics. We should expect a mechanical instability to occur when the condition (65) is met (at least within a factor $\sim$1) at the center of the trap. Let's see what this implies under "symmetric" conditions. Since $n(0) \sim N r_0^{-3}$, we have from Eqs.(61) and (65) that to an order of magnitude the condition is

$$N \gtrsim (d/a_s)^6 \tag{67}$$

a condition which for the type of trap used in the Rice experiments would require a value of N of the order of 10^8, greater than currently achieved but not obviously unattainable.

Suppose now that the s-wave scattering length for two different hyperfine states is negative, but not so large that the "spinodal" instability discussed above occurs. We still have the possibility of the onset of a "Cooper-paired" state such as occurs for the electrons in metallic superconductors. While such a phase occurs in the neutral system ^{3}He below 3 mK, and is also believed to occur in nuclear matter (e.g. in neutron stars), both of these systems are dense and strongly interacting, and the only dilute system of fermions which has until now been a candidate, namely dilute solutions of ^{3}He in superfluid ^{4}He, has so far shown no evidence of superfluidity down to $\sim$0.1mK. Thus, it would be very valuable for quantitative tests of some aspects of the phenomenon to have it occur in a dilute gas. As noted, so far ^{6}Li in a high ($\gg$100G) field seems to be the best candidate, but it is possible that suitable conditions may be achieved in ^{40}K by tuning close to a Feshbach resonance. For present purposes I concentrate on ^{6}Li.

Let me first briefly review the theory of Cooper pairing in a bulk system, assuming a certain background knowledge of this subject. The original observation of Cooper was that if we start with equal numbers of fermions of spin up and spin down filling up a Fermi sea, and then add two more in opposite-spin states with a weak attractive interaction between them and require that they occupy only states outside the pre-existing Fermi sea, then there exists a solution corresponding to a bound state, that is, one which has a bounded mean-square value of the relative coordinate $\mathbf{r}_1$-$\mathbf{r}_2$ and moreover an energy less than the minimum energy $2\epsilon_F$ needed to add two free particles. The binding energy $B \equiv 2\epsilon_F - E$ is exponentially dependent on the strength of the attraction $-V_0$:

$$B \sim \epsilon_C \exp[-2/(N(0)V_0)] \tag{68}$$

where $N(0)$ is the density of states of one spin at the Fermi surface, and ϵ_C is the phenomenological cutoff energy (taken to be $\ll \epsilon_F$) on the states coupled by the interaction (for a realistic model of electrons in a metal coupled by the exchange of virtual phonons, ϵ_C would be of the order of the Debye energy). The lowest-energy state has the center of mass of the two added electrons at rest, and is thus of the form of a spin singlet function multiplied by the orbital state

$$\varphi(|\mathbf{r}_1 - \mathbf{r}_2|) = \sum_k c_k \exp(i\mathbf{k} \cdot (\mathbf{r}_1 - \mathbf{r}_2)) \tag{69}$$

where the coefficients c_k are proportional to $(\epsilon_k + B)$ where ϵ_k is the energy relative to the Fermi energy; it then turns out, crudely speaking, that at large distances the relative wave function falls off as $\exp(-|\mathbf{r}_1 - \mathbf{r}_2|/\xi)$, where the "pair radius" ξ is of the order of $\hbar v_F/B$ ($v_F \equiv \hbar k_F/m$ = Fermi velocity), and thus is very large if the attraction is weak.

An essential point of Cooper's argument is that the density of states for the excitation of the pair $(\mathbf{k}, -\mathbf{k})$ does not vanish as $\epsilon_k \to 0$ but rather tends to a constant value (so that the problem is effectively that of a particle moving in free space in two dimensions; it is known that in this case the turning on of an arbitrarily weak attractive potential leads to a bound state). It is this feature, and this alone, which leads to the celebrated result that a bound state exists for <u>any</u> attractive potential, however weak. In the present context it is essential to appreciate that a necessary condition for this constant density of states is the equality of the Fermi surfaces, which in turn is a consequence of the exact equality of the numbers of spin-up and spin-down particles. Let's consider for a moment the opposite case, in which the number of particles is unequal and as a result there is a splitting $\Delta k_F \sim k_F^{(0)}(N_\uparrow - N_\downarrow)/(N_\uparrow + N_\downarrow)$ of the Fermi wave vectors of the up- and down-spin particles. Then, if we take our zero of energy as the energy $\epsilon_{F\uparrow} + \epsilon_{F\downarrow}$ required to add two free particles, the minimum energy required to add a pair with total momentum zero is clearly $\epsilon_{F\uparrow} - \epsilon_{F\downarrow}$, i.e. $v_F \Delta k_F$. Thus the relevant density of states is cut off at this value of ϵ, and the result is that the paired state is now bound only if $\epsilon_{F\uparrow} - \epsilon_{F\downarrow}$ is $\lesssim$ the original binding energy B. (We might ask whether it is really necessary to form the pair with center-of-mass momentum zero; the answer is actually no, and this leads to an intriguing possibility which will be discussed further below; but for the moment I will assume we are interested only in the original simple Cooper scheme.)

As is well-known, the BCS theory of superconductivity in metals is essentially based on Cooper's observation, but handles all the electrons on the same footing, without making the artificial separation into a fixed Fermi sea of $N-2$ electrons and the last added pair. The result is a many-body collective state: the N-particle wave function is schematically given by an expression of the general form

$$\Psi_N = \mathcal{A}\varphi(\mathbf{r}_1\mathbf{r}_2\sigma_1\sigma_2)\varphi(\mathbf{r}_3\mathbf{r}_4\sigma_3\sigma_4)\ldots\varphi(\mathbf{r}_{N-1}\mathbf{r}_N\sigma_{N-1}\sigma_N) \tag{70}$$

where $\mathcal{A}$ is the antisymmetrization operator and the "molecular" (two-particle) wave function is of the general form (for the relevant case of s-wave pairing)

$$\varphi(\mathbf{r}_1\mathbf{r}_2\sigma_1\sigma_2) = \frac{1}{\sqrt{2}}(\uparrow_1\downarrow_2 - \downarrow_1\uparrow_2)\cdot\chi(|\mathbf{r}_1 - \mathbf{r}_2|) \tag{71}$$

where the first factor indicates the spin singlet state of two spin-1/2 particles. The crucial point is that the "molecular wave function" φ is common to all the $N/2$ pairs of particles: violation of this condition immediately leads to a substantial decrease in binding energy. The binding energy of the state (70) is

$$E_0 = \frac{1}{2}N(0)\Delta^2 \tag{72}$$

where the quantity Δ, usually called the energy gap (cf. below) is given by the formula

$$\Delta \sim \epsilon_C \exp[-1/(N(0)V_0)] \tag{73}$$

where ϵ_C is, as above, a phenomenological energy cutoff on the states coupled by the (constant) attraction $-V_0$.

The description in terms of a "molecular" wave function φ is formally correct but very inconvenient, because of the need to explicitly antisymmetrize the many-body wave function as in (70). It turns out that physical quantities are expressed much more conveniently in terms of a quantity which is related to φ and is usually called the "pair wave function" and denoted $F(|\mathbf{r}_1 - \mathbf{r}_2|)$. For large distances $F(r)$ falls off exponentially:

$$F(r) \sim \exp(-r/\xi), \qquad \xi \sim \hbar v_F/\Delta \tag{74}$$

Thus the quantity ξ is a measure of the radius (size) of a Cooper pair. The energy of a single-particle excitation from the groundstate, which in the normal state would be ϵ_k, is given by the well-known formula

$$E_k = (\epsilon_k^2 + |\Delta|^2)^{1/2} \tag{75}$$

so that $|\Delta|$ is indeed the "energy gap", i.e. the minimum possible excitation energy of a single fermion.

The theory can be generalized to finite temperature, and one then finds that the gap is temperature-dependent and disappears, along with the collective bound state, at a temperature T_c given in terms of the zero-temperature gap Δ_0 by

$$kT_c = \Delta_0/(1.75) \tag{76}$$

Let's now consider the application to the fermion alkali gases. I will concentrate on the case of ^{6}Li in a high magnetic field (since this is the case in which the s-wave scattering length is believed to be large and negative) and assume (for the moment) that we have only the low-field-seeking states $|6\rangle$ and $|5\rangle$ present. They then play the role of the $\uparrow$ and $\downarrow$ states in standard BCS theory, and if we could assume they are present in equal numbers, the general form of the pair wave function would be

$$F(r_1, r_2)\frac{1}{\sqrt{2}}(|6\rangle_1|5\rangle_2 - |5\rangle_1|6\rangle_2) \tag{77}$$

For the moment we neglect the trap potential and imagine we are in free space.

We cannot take over the BCS calculation exactly as it stands, because it assumes the "bare" inter-fermion potential is (a) weak and (b) has a cutoff in energy space; neither assumption is natural in our problem. Our problem is in effect that of a dilute Fermi gas, possibly in a magnetic field $\mathcal{H} \equiv (\epsilon_{F\uparrow} - \epsilon_{F\downarrow})/2\mu$ which mimics the effect of a difference in $n_\uparrow$ and $n_\downarrow$, and subject to a weak s-wave attraction which is parametrized by the scattering length $a_s < 0$. For the case $n_\uparrow = n_\downarrow$, ("$\mathcal{H} = 0$") this problem was solved long ago by Gor'kov and Melik-Barkhudarov (1962) by a minor modification of the original approach of BCS. The calculation of the transition temperature T_c is straightforward and yields the result (for $a_s < 0$)

$$T_c = K_2 \epsilon_F \exp\left[-(\pi/(2k_F|a_s|)\right] \tag{78}$$

where the numerical constant K_2 is given by [o]

$$K_2 = \frac{8}{\pi}\exp(\gamma - 2) \cdot \exp(-\frac{1}{3}(1 + 2\ln 2)) \approx 0.31 \tag{79}$$

where γ is Euler's constant (the subscript "2" indicates that the expression is valid for a two-species system). Because of the large value of $|a_s|$, T_c need not be unattainably low: e.g. for $n_\uparrow = n_\downarrow = 10^{12}$ it is about 11nK, rising to 1.7μK at the value of n corresponding to mechanical instability (though see below).

In their original paper Gor'kov and Melik-Barkhudarov work only to the lowest nonvanishing order in the "small parameter" $\lambda \equiv 2k_F|a_s|/\pi$ and it might seem rather dangerous to apply formula (78) for values of this parameter of order unity. In a very recent paper Combescot (1999) reconsiders this problem, taking account of what may be the principal new physical effect which comes in at stronger coupling, namely the effect of virtual interactions of the fermions via the collective density fluctuations of the medium. He concludes that while the GMB formula (78) severely overestimates T_c for all but very small λ, the attractive effect of this indirect interaction leads, for λ approaching 1, to a considerable enhancement of T_c; he concludes that the maximum value of the ratio T_c/ϵ_F, attained for $\lambda \sim 0.96$, is about 0.025, putting T_c still within reasonable experimental reach. However, it should be cautioned that values of λ of order 1 correspond to values of na_s^3 also of this order, and at these high densities it is not at all clear that the parametrization of the atom-atom interaction in terms of the single variable a_s will remain quantitatively, or indeed perhaps even qualitatively, valid.

The most detailed paper to date on the possible occurrence of Cooper pairing in real-life ^{6}Li is that of Houbiers et al. (1997) who inter alia take account of the effect of unequal numbers of the populations of the states $|6\rangle$ and $|5\rangle$. As we have seen, any asymmetry between the Fermi surfaces, such

[o]The exponent and most of the prefactor can be correctly obtained by a straightforward adaptation of the BCS calculation. However, the last factor in K_2 comes from a more sophisticated effect, namely virtual polarization of the medium by the attractive interaction, cf. below.

as would occur for $n_\uparrow$ and $n_\downarrow$ unequal, will tend to spoil the simple Cooper picture, because now if one partner of the pair is just above its Fermi surface, the other will be distant by a finite amount from its own one. Crudely speaking we expect that the Cooper instability will be removed as soon as the splitting of the Fermi surfaces $\delta\epsilon_F$ becomes comparable to $k_B T_c^{(0)}$ (where $T_c^{(0)}$ is the transition temperature for $n_\uparrow = n_\downarrow$). Since $\delta\epsilon_F/\epsilon_F^{(0)} \sim \delta n/n^{(0)}$, this gives a critical value of the asymmetry $\delta n/n^{(0)}$ of order $k_B T_c^{(0)}/\epsilon_F$, i.e. according to Eq.(78), of order $\exp(-\pi/2k_F|a_s|)$. Thus the lower the density, the smaller the asymmetry needed to destroy superfluidity; this effect can be seen clearly in fig. 4 of Houbiers et al., who confirm the above qualitative argument by a quantitative calculation of T_c.

What of the possibility for superfluidity in a trap geometry? Houbiers et al. assume (reasonably, under their assumption that $V(r)$ is spin-independent) that one can treat each region of the trap locally, i.e. solve the gap equation for the local gap $\Delta(r)$ in terms of the local number densities $n_{\uparrow\downarrow}(r)$. The critical temperature of the system as a whole is clearly that corresponding to the value of n at the trap center, or more precisely to the value of k_F there: since we have from Eq.(60) that $k_F(0) \sim N^{1/6}d^{-1}$, and thus from (78)

$$T_c \sim \exp\left\{-\text{const.}(d/N^{1/6}|a_s|)\right\} \tag{80}$$

The numerical calculation of Houbiers et al. confirms this, with the constant given by 1.214 (and an extra, next-order term added). It is intuitively clear that we expect the boundary of the superfluid (paired) region to propagate outwards from the center, filling the whole cloud as $T \to 0$ (since T_c is finite for any finite value of n).

Now we come to what at first sight is a serious difficulty with the program of seeking Cooper pairing in a real-life system such as ^{6}Li. We argued earlier that the simple BCS state (with center-of-mass momentum of the pair zero) would become unstable as soon as the energy splitting of the two Fermi surfaces ($\Delta\epsilon$) became comparable to the binding energy of the original (zero-splitting) state, and this is confirmed by the detailed calculation of Houbiers et al. for a trap geometry. In effect, to keep this state stable we need the relative asymmetry $(n_\uparrow - n_\downarrow)/(n_\uparrow + n_\downarrow)$ to be of the order of $k_0 T_c/\epsilon_F$ or smaller. Bearing in mind that even under optimum conditions the latter ratio, even at the trap center, is unlikely to be more than 0.1 at most and may well be much smaller, we see that this is a stringent condition. In the literature various ways of attaining it have been considered, for example based on various pulse sequences, but it still seems problematical from a practical point of view.

Oddly enough, there is one possible way out of this dilemma which has not to my knowledge been discussed in the alkali-gas literature: Long ago, in the context of alloy superconductivity, it was appreciated that under very extreme "type-II" conditions, where the usual "Meissner" suppression of superconductivity was weak, an upper limit on the magnetic field in which ordinary BCS pairing was stable was given by the Pauli paramagnetic depairing effect discussed above; this limit is of order $k_B T_c/|\mu_B|$ as above and is known as the Chandrasekhar-Clogston limit. However, in 1964 two groups (Fulde and Ferrell (1964), Larkin and Ovchinnikov (1964)) pointed out that a paired state might still be possible for fields beyond the CC limit, provided we allow the center of mass momentum of the pair to be non-zero. This "FFLO state" has never been seen in any existing Fermi superfluid, possibly because it is (believed to be) extremely fragile against impurity scattering. Nevertheless, there seems to be a real possibility it could occur in a Fermi alkali gas, where such scattering is very weak.

In the original papers, which considered systems moving in free space, the pairs were taken to have a definite center-of-mass momentum **K** which, while fixed in magnitude by the energetics, was arbitrary in direction, implying a spontaneous breaking of the rotational symmetry. Evidently, in a trap this will not work, and the obvious adaptation of the FFLO ideas to this geometry is to pair in terms of the relevant single-particle quantum numbers n, l, m:

$$(n, \ell, m, \uparrow) \leftrightarrow (n + \Delta n, \ell, -m, \downarrow) \tag{81}$$

where the quantity Δn is chosen so that $\mathcal{E}(n + \Delta n) - \mathcal{E}(n) = \Delta\epsilon$. (Note that both for the case of a flat potential with a boundary and for the isotropic harmonic oscillator, $\mathcal{E}(n, \ell, m)$ is a function only of n). However, there is a delicate point which arises here: If one does the most obvious thing and indeed pairs single-particle energy eigenfunctions, these are real and it then turns out that although the pairing lowers the energy below that of the normal groundstate, the expression for T_c (and for the energy gap) has a crucial extra factor of 2 in the exponent (i.e. the latter corresponds to $2/(N(0)V)$ rather than $1/(N(0)V)$. Since the exponent is likely to have been already large, this would correspond to a drastic suppression of T_c although the latter is still finite for any $\Delta\epsilon$. It does not seem, alas, that more sophisticated pairing schemes are likely to get around this difficulty.

I finally turn to the question of how we might detect such a paired state, should it indeed occur. The literature is full of ingenious proposals for this, most of them in my opinion ranging from optimistic to wildly optimistic. Let me try to approach the problem in a relatively systematic way, by considering successively the static real-space distribution, the single-particle properties and the two-particle properties and asking how each is likely to be affected by the formation of Cooper pairs. In this discussion I shall assume that the ratio of T_c to the Fermi temperature T_F will never in real life be greater than 0.1, an estimate which I actually believe to be optimistic; should this limit be exceeded by a substantial amount, then some of my criticisms of various proposals are likely to be rendered invalid.

In a Bose alkali gas, the most spectacular signature of BEC is the sharp change in the real-space density distribution. Will something similar happen in a Fermi gas at the onset of pairing? In principle yes, but the actual magnitude of the effect turns out to be of order $(T_c/T_F)^2$, so it is at most a 1% effect. Moreover, it will happen gradually, since as we have seen the pairing is initially localized near the trap center and then with falling temperature spreads outwards. Thus it seems unlikely to form a practical method of detection.

The one-particle properties, e.g. the fermionic energy spectrum and the associated specific heat, will undergo "order-1" changes at and below the transition. But how to measure the spectrum on the scale required? An optical method (the analog of angle-resolved photoemission in a solid) looks attractive–until one remembers that the linewidth of the principal transition is several orders of magnitude larger than the looked-for gap in the spectrum. In principle, the modified density of states should be reflected in the transport and collision properties, but the problem here is that in the degenerate normal phase, because of the effect of the Pauli principle, collisions are already very rare.

Let's finally turn to the two-particle properties. Since the pairs in ^{6}Li, unlike those in ^{3}He, are not expected to possess internal degrees of freedom, the only quantities whose behavior is likely to be substantially modified by pairing are the (number) density, the spin density and the transverse current density. As regards the spin density correlations, there should be an "order-1" effect, and indeed the true thermal equilibrium value of the long-wavelength static spin susceptibility is predicted to fall precipitously below T_c and to tend to zero for $T \to 0$; however, in view of the very long cross-relaxation times it seems probably hopeless to measure this "true" value. I suspect that the same problem may arise with other aspects of the spin dynamics, though in the absence of specific proposals we may have to reserve judgment.

Most of the proposals in the literature for the detection of pairing essentially boil down to proposals to look for the changes it induces in the static or dynamic density correlation functions; in particular, most of the proposals for optical detection in the last resort reduce to this. If indeed one is interested in the density correlations, one does not have to do the calculation from scratch, as the effect of pairing on these has been well studied in the context of existing Fermi superfluids (metallic superconductors and ^{3}He). The bottom line is that the static correlations are modified, but only to order $(T_c/T_F)^2$, making the effect very difficult to observe in practice. As to the dynamic susceptibility, there is a qualitative modification in the long-wavelength limit ($q\xi \ll 1$, where ξ as above is the pair radius): the spectrum changes over from a "Landau particle-hole continuum" to a delta-function peak at the so-called Anderson-Bogoliubov mode. For shorter wavelengths the effect is much smaller (at most of order $1/q\xi$). Thus in principle an examination of the propagation of sound in the long-wavelength limit should be able to detect the onset of pairing. However, in real life, by the time one reaches the condition $q\xi \ll 1$ effects associated with the trap geometry are likely to start coming into play, so we might as well go directly to the limit $q = 0$, i.e. examine the fundamental trap modes; see below.

We are left with the transverse current correlations. These are indeed modified drastically by pairing, and the result is just the phenomenon of superfluidity as described in lecture 2 (but now in a Fermi system). Thus, in the absence of other possibilities a viable means of detection should be to look for one of the fundamental manifestations of superfluidity, the "Hess-Fairbank effect" or persistence of circulating currents. However, this may be unnecessary, for the following reason.

Let's consider how the fundamental trap modes are modified by pairing. I will as usual assume an isotropic trap, so that the modes can be classified as monopole (ℓ=0), dipole (ℓ=1), and quadrupole (ℓ=2). The frequency of the dipole mode is completely determined by the relevant f-sum rule and is thus impervious to the onset of pairing. The frequency of the monopole mode is affected, but only to the same order as the bulk compressibility, namely $(T_c/T_F)^2$. However, the quadrupole mode is drastically affected. The easiest way to see this is to note that the frequency of this mode is determined by the ratio of two sum rules, of which one is essentially the familiar f-sum rule in disguise and thus depends only on the trap frequency. The second sum rule is that for the "shear modulus" of the system at frequencies of the order of the trap frequency. Now, in the normal phase, by the time the temperature has fallen to something of order T_c, the Pauli principle has suppressed collisions so effectively that $\omega_o\tau \gg 1$, i.e. we are in the so-called "collisionless" limit. In this limit there is no time for the energy to flow between different (Cartesian) degrees of freedom, so the susceptibility to shear (inverse shear modulus) is low. Now consider the paired state at $T = 0$. Since the only ways in which the pairs (superfluid component) can deform is either to expand (or contract) uniformly or to flow, the different Cartesian degrees of freedom are coupled together and the effect is, perhaps somewhat counter-intuitively, to increase the susceptibility to shear, as it turns out by a factor of 2. Thus, one is led to the prediction (Vichi and Stringari (1999)) that while the monopole and dipole modes of the trap should be virtually unaffected by the onset of pairing, the frequency of the quadrupole mode should decrease from $2\omega_o$ to $\sqrt{2}\omega_0$ as T falls from T_c to zero (a "100%" effect). To my mind this is almost certainly the most practical suggestion to date for detecting the onset of pairing.

References

D. Allum,, P. V. E. McClintock, A. Phillips and R. M. Bowley, Phil. Trans. Roy. Soc. (London) A **284**, 179 (1977).

M. H. Anderson, J. R. Ensher, M. R. Matthews, C. E. Wieman and E. A. Cornell, Science **269**, 198 (1995).

P. W. Anderson, *Basic Notions of Condensed Matter Physics*, Benjamin, Menlo Park, CA (1984).

D. Bohm, Phys. Rev. Lett. **83**, 3766 (1999).

Y. Castin, and J. Dalibard, Phys. Rev. A **55**, 4330 (1997).

R. Combescot, Phys. Rev. Lett. **83**, 3766 (1999).

F. Dalfovo, S. Giorgini, L. P. Pitaevskii and S. Stringari, Rev. Mod. Phys. **71**, 463 (1999).

B. De Marco and D. S. Jin, Science **285**, 1703 (1999).

J. A. Dunningham and K. Burnett, Phys. Rev. Lett. **82**, 3729 (1999).

P. O. Fedichev and G. V. Shlyapnikov, Phys. Rev. A **60**, R 1779 (1999).

P. Fulde and R. A. Ferrell, Phys. Rev. A **550**, 135 (1964).

M. Girardeau, J. Math. Phys. 11, **684** (1970).

L. P. Gor'kov and T. K. Melik-Barkhudarov, Zh. Eksp. Teor. Fiz. 40, 1452 (1961); translation, Sov. Phy. JETP **13**, 1018 (1961).

A. Griffin, in *Proc. International School "Enrico Fermi"*, Course CXL, eds. M. Inguscio, S. Stringari and C. E. Wieman, IOS Press, Amsterdam (1999).

M. Grüter, D. M. Ceperley and F. Laloe, Phys. Rev. Lett. **79**, 3549 (1997).

D. S. Hall, M. R. Matthews, C. E. Wieman and E. A. Cornell, Phys. Rev. Lett. **81**, 1542 (1998).

T-L. Ho, Phys. Rev. Lett. **49**, 1837 (1982).

M. Houbiers, R. Ferwerda, H. T. C. Stoof, W. I. McAlexander, C. A. Sackett and R. G. Hulet, Phys. Rev. A **56**, 4864 (1997).

K. Huang, *Statistical Mechanics*, 2nd ed., John Wiley and Sons, NY (1987).

J. Javanainen, Phys. Rev. A **54**, R4629 (1996).

J. Javanainen and S-M. Yoo, Phys. Rev. Lett. **76**, 161 (1996).

S. Kohler, S. Paraoanu, F. Sols and A. J. Leggett, (2000), in preparation.

W. Kohn and D. Sherrington, Rev. Mod. Phys. **42**, 1 (1970).

A. I. Larkin and Yu. Ovchinnikov, Zh. Eksp. Teor. Fiz. **47**, 1136 (1964); translation, Soviet Physics JETP **20**, 762 (1965).

A. J. Leggett, Found. Phys. **25**, 113 (1995a).

A. J. Leggett, in *Bose Einstein Condensation*, eds. A. Griffin, D. W. Snoke and S. Stringari, Cambridge University Press (1995b).

A. J. Leggett, in *Proc. 16th International Conference on Atomic Physics*, Windsor, Ontario, Canada, Aug. 1998, eds. W. E. Baylis and G. F. Drake, AIP Conference Proceedings, No. 477 (1999a).

A. J. Leggett, Lectures at University of Illinois, unpublished (1999b).

A. J. Leggett, in preparation for submission to Rev. Mod. Phys. (2000).

K. W. Madison, F. Chevy, W. Wohlleben and J. Dalibard, Phys. Rev. Lett. **84**, 806 (2000).

M. R. Matthews, B. S. Anderson, P. C. Haljan, D. S. Hall, C. E. Wieman and E. A. Cornell, Phys. Rev. Lett. **83**, 2998 (1999).

F. Meier and W, Zwerger, cond-mat/9904, 147 (2000).

N. D. Mermin, Rev. Mod. Phys. **51**, 591 (1979).

A. G. K. Modawi and A. J. Leggett, J. Low Temp. Phys. **109**, 625 (1997).

A. Rohrl, M. Naraschewski, A. Schenzle and H. Wallis, Phys. Rev. Lett. **78**, 4143 (1997).

D. S. Rokhsar, Phys. Rev. Lett. **79**, 2164 (1997).

L. Vichi and S. Stringari, private communication, 1999.

I. Zapata, F. Sols and A. J. Leggett, Phys. Rev. A **57**, R28 (1998).

STATISTICAL MECHANICS OF BOSONIC SYSTEMS

J. OITMAA
*School of Physics, The University of New South Wales,
Sydney, NSW 2052, Australia
E-mail: j.oitmaa@unsw.edu.au
Web: www.phys.unsw.edu.au/STAFF/ACADEMIC/oitmaa.html*

1 Introduction

These lectures are intended to give an overview of various aspects of the theory of Bose systems and of Bose-Einstein condensation (BEC) and will, I hope, be useful in their own right as well as providing an introduction, and a basis, for the more advanced lectures to follow. The emphasis is pedagogical. I assume a knowledge of undergraduate statistical mechanics, but not much more. Of course, in these lectures it is impossible to do justice to this fascinating subject, and many topics are either mentioned too briefly or not at all. Roughly speaking my goal is as follows:

Lecture 1 Ideal Bose gas, effect of reduced dimensionality and confining potentials.

Lecture 2 Weakly interacting Bose gas/liquid.

Lecture 3 Strongly interacting systems.

The first two lectures provide a basis for understanding BEC in trapped atomic gases, while the third lecture considers mainly lattice systems where one can go beyond mean field theory.

It is worth giving a brief historical overview. In 1924 Bose and Einstein independently formulated the statistical mechanics of non-interacting quantum particles with integral spin and hence symmetric wavefunctions (bosons). The BEC phenomenon, in a non-interacting system of bosons, was predicted by Einstein in 1925. Immediately after the discovery of superfluidity in liquid ^{4}He by Kapitza in 1937, London and others proposed that the superfluid transition was the first experimental realization of BEC, albeit in a system where interactions could not be neglected. Tisza's "2-fluid model" was the first introduction of the concept of a "condensate".

The 1940's and 1950's saw much theoretical work on the interacting bose gas, although the results were perturbative and restricted to low density and/or weak interactions. The end of this period also saw the development of more general theoretical ideas for treating BEC, such as broken gauge symmetry and off-diagonal long-range-order (ODLRO), as well as a quasi-macroscopic energy functional approach to interacting Bose systems by Gross and Pitaevskii, independently. In 1963 Lieb and Liniger found an exact solution for bosons in one-dimension interacting through repulsive δ-function potentials, which remains to this day the only non-trivial exact result known.

In the late 80's a number of experimental groups began to seek to achieve BEC in a gas of trapped atoms. This stimulated a new period of theoretical work in a new direction, where the BEC phenomenon was more naturally thought of in real space and where the effect of the external trapping potential needed to be included. Experimental BEC was achieved in 1995, and the interest in this area and the number of published papers has increased exponentially since then.

Some of the material of these lectures can be found in standard texts and monographs [G1-G5]. The Proceedings of the International Workshop BEC93, [G6] provides a comprehensive account of the field at that time. Three review articles [G7-G9] contain much useful material. These lectures have been influenced by the most recent review, Dalfovo *et al.* [G7].

As the introductory article by Snoke and Baym [G6] points out, BEC is a rather common phenomenon, occurring in a wide variety of systems from condensed matter, through atomic systems, nuclear physics, particle physics and astrophysics. Table 1 in [G6] lists 14 bosonic systems of current interest. The theme for this Summer School is thus particularly exciting and timely.

2 The Ideal Bose Gas

I will start by considering a uniform system of non-interacting bosons, in 3-dimensions. As is well known, the partition function in the grand ensemble is easily shown to be

$$\ln \mathcal{Z}_g = -\sum_k \ln(1 - e^{-\beta(\epsilon_k - \mu)}) \tag{1}$$

where the single particle quantum states are labelled by wavevector $\mathbf{k}$, with energies $\epsilon_k = \hbar^2 k^2/(2m)$, μ is the chemical potential and $\beta = 1/(k_B T)$, as usual.

The criterion for importance of quantum effects is that the degeneracy parameters $n\lambda^3 \gg 1$, where $n = N/V$ is the number density and $\lambda = \hbar/\sqrt{2\pi m k_B T}$ is the "thermal wavelength". For liquid ^{4}He at 2 K, $n\lambda^3 \approx 4$, while for ^{87}Rb at 170 nK, in the trapped atom experiment, $n\lambda^3 \approx 0.2$.

Quantum effects are dominant in the former case, less so in the trapped atomic gas.

In a large system the single particle states form a continuum, and the summation can be replaced by an integral over energy

$$\sum_k f_k \to \int f(\epsilon)\rho(\epsilon)d\epsilon$$

where $\rho(\epsilon)$ is the density of states. For non-relativistic particles in 3-dimensions

$$\rho(\epsilon) = \frac{2\pi V}{h^3}(2m)^{\frac{3}{2}}\epsilon^{\frac{1}{2}}. \tag{2}$$

It then follows from (1), using standard thermodynamic relations [G1] that

$$\frac{P}{k_B T} = -\frac{2\pi}{h^3}(2m)^{\frac{3}{2}}\int_0^\infty \epsilon^{\frac{1}{2}}\ln(1 - ze^{-\beta\epsilon})d\epsilon - \frac{1}{V}\ln(1-z) \tag{3}$$

$$\frac{N}{V} = \frac{2\pi}{h^3}(2m)^{\frac{3}{2}}\int_0^\infty \frac{\epsilon^{\frac{1}{2}}d\epsilon}{z^{-1}e^{\beta\epsilon}-1} + \frac{1}{V}\frac{z}{1-z} \tag{4}$$

where the term with zero energy has been explicitly separated out, and the fugacity is defined as $z = e^{\beta\mu}$. The zero energy term gives negligible contribution unless $z \to 1$. Note that z is directly related to the (average) number of particles in the lowest single particle state, *viz.*

$$N_0 = \frac{z}{1-z} \quad \Rightarrow \quad z = \frac{N_0}{N_0+1}. \tag{5}$$

The integrals can be evaluated with a simple change of variables, giving

$$\frac{N - N_0}{V} = \frac{2\pi(2mk_BT)^{\frac{3}{2}}}{h^2}\int_0^\infty \frac{x^{\frac{1}{2}}dx}{z^{-1}e^x - 1} = \frac{1}{\lambda^3}g_{\frac{3}{2}}(z) \tag{6}$$

where $g_n(z)$ is the standard Bose integral

$$g_n(z) = \frac{1}{\Gamma(n)}\int_0^\infty \frac{x^{n-1}dx}{z^{-1}e^x - 1}. \tag{7}$$

These integrals are ubiquitous in Bose systems and, for completeness, we give some of their properties in Appendix A.

The number of particles in excited states is, from (6)

$$\frac{N_e}{V} = \frac{1}{\lambda^3}g_{\frac{3}{2}}(z)$$

and this implicitly determines z as a function of n, T. Now $0 < z < 1$ and, since $g_{\frac{3}{2}}(z)$ is a monotonically increasing function of z with maximum value

$g_{\frac{3}{2}}(1) = \zeta(\frac{3}{2}) = 2.612\cdots$, the number of particles in excited states is bounded by

$$N_e \leq V\frac{(2\pi mk_BT)^{\frac{3}{2}}}{h^3}\zeta(\frac{3}{2}). \tag{8}$$

If the total number of particles N exceeds this value, then the lowest state $\epsilon = 0$ will have a **macroscopic occupancy**. This is **Bose-Einstein Condensation** (BEC), as originally formulated by Einstein [1]. The transition temperature, for fixed density, follows immediately from (8)

$$k_BT_C = \frac{h^2}{2\pi m}\left(\frac{n}{\zeta(\frac{3}{2})}\right)^{\frac{2}{3}} \tag{9}$$

and the particle density in the condensate is therefore (for $T < T_C$)

$$n_0 = n\left[1 - \left(\frac{T}{T_C}\right)^{\frac{3}{2}}\right]$$

while

$$n_e = n - n_0 = n\left(\frac{T}{T_C}\right)^{\frac{3}{2}}. \tag{10}$$

Figure 1 shows the particle densities, fugacity, and specific heat as functions of T/T_C. The specific heat has a cusp at T_C.

2.1 The Effect of Reduced Dimensionality

The above development is the standard one for 3-dimensions. The effect of dimension enters through the density-of-states formula (2). It is well known that the ideal uniform Bose gas in 2-dimensions does not exhibit BEC. The density of states is $\rho(\epsilon) = 2\pi mA/\hbar^2$, where A is the area of the system. An explicit calculation for 2-dimensions gives

$$N = \frac{z}{1-z} - \frac{2\pi mk_BT}{h^2}\ln(1-z). \tag{11}$$

Both terms diverge as $z \to 1$, and there is no macroscopic occupancy of the lowest state.

Hohenberg [2] has shown that, even in the presence of interactions, there is no BEC for $d \leq 2$. This general result is based on the inequality

$$\langle n_k\rangle \geq -\frac{1}{2} + \frac{mk_BT}{\hbar^2k^2}\frac{n_0}{n} \tag{12}$$

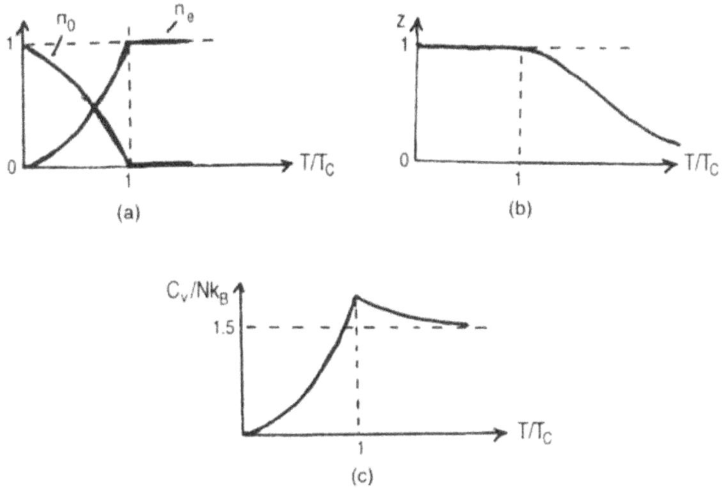

Figure 1. Properties of an ideal bose gas in 3-dimensions: (a) densities of condensate and excited states; (b) the fugacity and (c) the specific heat, as functions of temperature.

where $\langle n_k \rangle$ is the average occupancy of the single-particle state k. The density of excited particles is then

$$\begin{aligned} n - n_0 &= \frac{1}{V} \sum_{k \neq 0} \langle n_k \rangle \\ &= \frac{1}{(2\pi)^d} \int d^d k \langle n_k \rangle \end{aligned} \tag{13}$$

and the integral diverges for $d \leq 2$, unless $T = 0$ or $n_0 = 0$. Thus there is no macroscopic occupancy of the lowest state, except at $T = 0$. There can however be a superfluid transition at finite T in 2 dimensions [3], as is observed in helium films.

2.2 *Inhomogeneous Systems*

The above considerations must be modified if the system cannot be considered to be translationally invariant. This is the case, to a greater or lesser extent, in all real systems where the presence of boundaries, a gravitational field, impurities etc. break the translational symmetry. Hohenberg's theorem no longer applies and BEC can, in principle, occur in 2 and even 1 spatial dimension. In such situations the question of the thermodynamic limit can be subtle, and the condensate density is spatially non-uniform and may diverge at some points. References to work in this area can be found in [G9].

An important example of such a situation occurs when a system of atoms is confined by a trapping potential, and may form a **spatial Bose-Einstein condensate**. This is, of course, the case in the trapped atom experiments [4-6]. I will repeat the published calculation for a 3-dimensional ideal Bose-gas in an isotropic harmonic trap $V(\mathbf{r}) = \frac{1}{2}m\omega^2 r^2$ [7].

The number of particles is given by

$$N = \sum_{\epsilon} \frac{1}{e^{\beta(\epsilon-\mu)} - 1}$$

where $\epsilon = (n_x + n_y + n_z + \frac{3}{2})\hbar\omega$. As before we separate out the ground state

$$N = N_0 + \sum_{\epsilon}{}' \frac{1}{e^{\beta(\epsilon-\mu)} - 1}. \tag{14}$$

The proper thermodynamic limit is obtained by letting $N \to \infty$ and $\omega \to 0$ with $N\omega^3$ finite. In this limit the level spacing becomes small and we can use the **semiclassical approximation**. Replacing the sum by an integral, and defining the fugacity as $z = e^{-\beta(\epsilon_0-\mu)}$ gives

$$N = N_0 + \int_0^{\infty} \frac{\rho(\epsilon)d\epsilon}{z^{-1}e^{\beta\epsilon} - 1}, \tag{15}$$

where the density of states is

$$\rho(\epsilon) = \frac{2\pi(2m)^{\frac{3}{2}}}{h^3} \int_{V^*} \sqrt{\epsilon - V(r)}\, d^3r \tag{16}$$

with V^* the accessible volume *i.e.* $V^* = \frac{4}{3}\pi R^3$ with $\frac{1}{2}m\omega^2 R^2 = \epsilon$. The integrals can be evaluated to give

$$N = N_0 + \left(\frac{k_B T}{\hbar\omega}\right)^3 g_3(z). \tag{17}$$

As before the Bose integral $g_3(z)$ is bounded by $g_3(1) = \zeta(3) = 1.202$. The condition for BEC is therefore

$$N = \left(\frac{k_B T_C}{\hbar\omega}\right)^3 \zeta(3) \tag{18}$$

which gives

$$k_B T_C = 0.94\hbar\omega N^{\frac{1}{3}}. \tag{19}$$

The condensate function is

$$\frac{N_0}{N} = 1 - \left(\frac{T}{T_C}\right)^3 \tag{20}$$

different to the $\frac{3}{2}$ power law for the uniform case.

There are a number of other interesting differences between this system and the uniform Bose gas. The specific heat, for example, has a discontinuity of $6.75k_B$ at the transition point, rather than a cusp [8].

Bagnato and Kleppner [9] have also investigated the possibility of BEC in lower dimensions for a trapping potential of the form $V(r) \equiv r^\eta$. They find that BEC occurs in 2-dimensions for any η, and in 1-dimension for $\eta < 2$. These results are of relevance to trapped atom experiments where the confined gas is non-spherical.

2.3 The Effect of a Finite Number of Particles

Real systems contain a finite number of particles ($10^3 - 10^6$ atoms in recent experiments). It is well known [10] that for finite N there can be no singularity in the thermodynamic functions and, strictly speaking, no phase transition. Furthermore the thermodynamic limit $N \to \infty$, $\omega \to 0$ with $N\omega^3$ finite, of the previous section is somewhat unrealistic. A number of authors [8,11] have considered a finite number of particles in a harmonic trap, and some results are shown in Fig.2. The sharp peaks or discontinuities in the condensate fraction and specific heat become rounded near T_C, and T_C itself (appropriately defined) is lowered

$$\frac{\Delta T_C}{T_C} = -0.73N^{-\frac{1}{3}}.$$

However, apart from the rounding, the curves are very similar to the limiting case $N \to \infty$, even for as few as 100 particles, and it is still appropriate to describe the behaviour as BEC, subject to the *caveat* above.

3 Weakly Interacting Bose Systems

Once interactions, which occur in all real systems, are included we have a many-body problem where there are (essentially) no exact results. If the interactions are weak then, one hopes, the physics will not change too much. Thus it is meaningful to ask whether, and in what ways, the ideal picture of BEC must be modified. Interactions also allow the possibility of collective excitations, which have been known in liquid He for a long time and are now

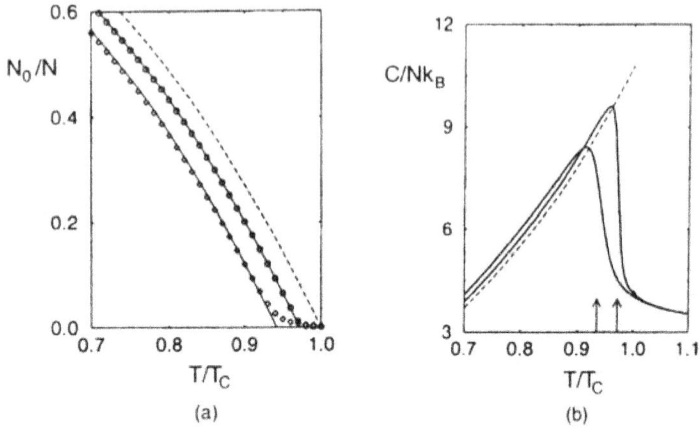

Figure 2. Condensate fraction (a) and specific heat (b) versus temperature for 2,000 and 20,000 particles. The dashed line shows the limiting case of $N \to \infty$ (after Grossman and Holthaus [8]).

being actively studied in the trapped atomic systems. This important topic is not discussed in these lectures (see [G5], [G8]).

3.1 Bogoliubov Theory of Interacting Bosons

Historically, the first successful theory of interacting bosons was due to Bogoliubov [12] (see also [G1] for details). Starting from the Hamiltonian

$$H = \sum_k \epsilon_k a_k^\dagger a_k + \frac{1}{2} \sum_{k,k',q} v(q) a_{k+q}^\dagger a_{k'-q}^\dagger a_{k'} a_k \tag{21}$$

where $v(q)$ is the interaction potential in k space and $\epsilon_k = \hbar^2 k^2/(2m)$, the terms corresponding to the zero momentum states are separated out, and treated as classical variables. The remaining quartic terms are discarded and the resulting quadratic Hamiltonian diagonalized through a canonical transformation (Bogoliubov transformation) to new quasiparticle operators. The Hamiltonian becomes

$$H = E_0 + \sum_k{}' E_k b_k^\dagger b_k + \cdots \tag{22}$$

where the excitation energies E_k are

$$E_k = [2Nv(k)\epsilon_k + \epsilon_k^2]^{\frac{1}{2}} \tag{23}$$

and the ground state energy E_0 is

$$E_0 = \frac{1}{2}v(0)N^2 + \frac{1}{2}\sum_k{}' [E_k - \epsilon_k - Nv(k)]. \tag{24}$$

The following approximation is now usually made

$$v(k) \approx v(0) = \frac{4\pi a\hbar^2}{mV}(1 + \frac{4\pi a}{V}\sum_k{}' \frac{1}{k^2}) \tag{25}$$

where V is the volume and a is the s-wave scattering length, which is positive for repulsive potentials. The parameterization of the interaction potential in this way will be accurate at low energies. It is exact for hard spheres and this model is often termed the "hard sphere Bose gas".

The excitation energies are then

$$E_k = \left[\frac{4\pi na\hbar^4 k^2}{m^2} + \left(\frac{\hbar^2 k^2}{2m}\right)^2\right]^{\frac{1}{2}}. \tag{26}$$

For small k (long wavelengths) the first term dominates and $E_k \equiv \hbar ck$ where $c = \sqrt{4\pi na\hbar^2}/m$ is the speed of sound. These collective excitations are sound waves or phonons. For larger k the free particle term becomes dominant. The dispersion curve does not show the "roton dip" characteristic of liquid ^{4}He, which can be obtained through a more complex treatment [13].

The ground state energy can be evaluated and expressed in the form

$$\frac{E_0}{N} = \frac{2\pi na\hbar^2}{m}\left\{1 + \frac{128}{15\sqrt{\pi}}(na^3)^{\frac{1}{2}} + \frac{8(4\pi - 3\sqrt{3})}{3}(na^3)\ln(na^3) + C(na^3) + \cdots\right\}, \tag{27}$$

an expansion in the parameter (na^3). The Bogoliubov theory gives the first two terms only, which were obtained in the 1950's by Lee, Huang and Yang [14] via several different methods. The $(na^3)\ln(na^3)$ term, which occurs for the strict hard-sphere gas, was obtained by Wu [15], and the last term only recently by Braaten and Nieto [16]. This term is the first which depends not only on the s-wave scattering length a but on other details of the interaction potential.

The Bogoliubov theory also gives the condensate fraction, to lowest order in a, as

$$\frac{N_0}{N} = 1 - \frac{8}{3\sqrt{\pi}}(na^3)^{\frac{1}{2}} + \cdots \tag{28}$$

which shows that the condensate is partially depleted (even at $T = 0$) by the repulsive interactions.

Calculation of finite temperature thermodynamic properties is more problematic. Low temperature properties are determined by the linear dispersion for small k, which gives, for example, a T^3 specific heat (*cf.* $T^{\frac{3}{2}}$ for the ideal Bose gas). At high temperature accurate results can be obtained through the virial expansion. At intermediate temperatures all the difficulties of a fully quantum mechanical many-body problem are manifest, and this remains an active area of research. An illustration of this difficulty is the controversy, which remains unresolved, about the change $\Delta T_C = T_C - T_{C0}$ in the transition temperature [17-19].

3.2 General Considerations

I have avoided, until now, any discussion of the precise meaning or criterion for BEC in an interacting system. In an interacting system the single particle states are no longer eigenstates of the system, and hence the criterion for BEC in an ideal gas has no meaning. This question was addressed in the 1950's by Penrose and Onsager [20], Yang [21] and others.

We define the one-particle density matrix

$$\rho(\mathbf{r}, \mathbf{r}') = \langle \hat{\psi}^\dagger(\mathbf{r}') \hat{\psi}(\mathbf{r}) \rangle \tag{29}$$

where the $\hat{\psi}(\mathbf{r})$ are single particle operators and the average is over either the ground state or a thermal average for the interacting system. Then BEC exists if

$$\lim_{|\mathbf{r}-\mathbf{r}'|\to\infty} \rho(\mathbf{r}, \mathbf{r}') = n_0 \ (\neq 0), \tag{30}$$

a condition generally referred to as "off-diagonal long-range order (ODLRO)". The quantity n_0 is the density of the condensate. Alternatively we write

$$\lim_{|\mathbf{r}-\mathbf{r}'|\to\infty} \rho(\mathbf{r}, \mathbf{r}') = \Phi^*(\mathbf{r}')\Phi(\mathbf{r}) \tag{31}$$

where $\Phi(\mathbf{r})$ is called the "**condensate wave function**". This is a complex field, with an amplitude and a phase.

The existence of a condensed state implies a state of **broken gauge symmetry**, where the phase of the condensate wave function becomes locked at a specific value, despite the Hamiltonian being gauge invariant. Anderson [22] showed that **phase coherence** exists throughout the condensate. His elegant argument goes as follows. Suppose the system has N particles all in the condensate, and divide the system into M subsystems. If a_0 is the destruction operator for the condensate, then

$$a_0 = \frac{1}{\sqrt{M}} \sum_{i=1}^{M} A_i$$

where A_i is the corresponding destruction operator for the ith subsystem. Then

$$\begin{aligned} \langle a_0^\dagger a_0 \rangle = N &= \frac{1}{M} \sum_{i,j} \langle A_i^\dagger A_j \rangle \\ &= \frac{1}{M} \sum_{i} \langle A_i^\dagger A_i \rangle + \frac{1}{M} \sum_{i \neq j} \langle A_i^\dagger A_j \rangle \\ &= \frac{N}{M} + \frac{1}{M} \sum_{i \neq j} \langle A_i^\dagger A_j \rangle. \end{aligned}$$

As $M \to \infty$ the 2nd term dominates and hence $\langle A_i^\dagger A_j \rangle \neq 0$, which implies a correlation between phases in different parts of the system.

Another interesting general issue involves the question of whether more than one quantum state can be macroscopically occupied. At first sight, in a large system where the energy levels are quasi-continuous it is not obvious why the condensate only involves the lowest state. The answer was given by Nozières [23], who showed that it is the exchange interaction energy which prevents fragmentation of the condensate. In this sense "genuine" BEC is a consequence of particle interactions: it is not an ideal gas effect.

3.3 The Gross-Pitaevskii Formalism

The fully microscopic approach, discussed in §3.1, while of fundamental importance, is not easily able to treat systems which are of finite extent or which are inhomogeneous in other ways. An alternative approach, in which the condensate wavefunction $\Phi(\mathbf{r})$ is the primary quantity, was developed independently by Gross [24] and Pitaevskii [25]. Their formalism, which is closely analogous to the Ginzburg-Landau theory of superconductivity, can be developed in a number of ways. One is to start from the equation of motion for the quantum field $\Psi(\mathbf{r},t)$

$$i\hbar\frac{\partial\Psi(\mathbf{r},t)}{\partial t} = \left[-\frac{\hbar^2}{2m}\nabla^2 + V_{\text{ext}}(\mathbf{r}) + \int d\mathbf{r}'\Psi^\dagger(\mathbf{r}')V(\mathbf{r}-\mathbf{r}')\Psi(\mathbf{r}')\right]\Psi(\mathbf{r},t) \tag{32}$$

and to write

$$\Psi(\mathbf{r},t) = \Phi(\mathbf{r},t) + \Psi'(\mathbf{r},t) \tag{33}$$

where $\Phi(\mathbf{r},t) = \langle\Psi(\mathbf{r},t)\rangle$ is the condensate wavefunction and $V(\mathbf{r}-\mathbf{r}')$ is the interaction potential, assumed to have the form $V(\mathbf{r}-\mathbf{r}') = g\delta(\mathbf{r}-\mathbf{r}')$. An alternative approach is to postulate an energy functional

$$E[\Phi] = \int d\mathbf{r}\left[\frac{\hbar^2}{2m}|\nabla\Phi|^2 + V_{\text{ext}}(\mathbf{r})|\Phi|^2 + \frac{g}{2}|\Phi|^4\right] \tag{34}$$

and obtain the equation of motion variationally

$$i\hbar\frac{\partial\Phi}{\partial t} = \frac{\delta E}{\delta\Phi^*}.$$

Either way one obtains the Gross-Pitaevskii equation

$$i\hbar\frac{\partial\Phi}{\partial t} = \left(-\frac{\hbar^2}{2m}\nabla^2 + V_{\text{ext}}(\mathbf{r}) + g|\Phi(\mathbf{r},t)|^2\right)\Phi \tag{35}$$

which describes the time and space variation of the **order-parameter** $\Phi(\mathbf{r},t)$. Although this has the form of a "nonlinear Schrödinger equation" Φ is not a wavefunction in the normal sense, but a classical field. A steady-state solution can be obtained by assuming

$$\Phi(\mathbf{r},t) = e^{-i\mu t/\hbar}\phi(\mathbf{r}) \tag{36}$$

where μ is the chemical potential of the condensate, which then yields

$$\left[-\frac{\hbar^2}{2m}\nabla^2 + V_{\text{ext}}(\mathbf{r}) + g|\phi(\mathbf{r})|^2\right]\phi(\mathbf{r}) = \mu\phi(\mathbf{r}) \tag{37}$$

which is also known as the Gross-Pitaevskii equation.

It is clear from the method of derivation that the GP equation describes a "**mean-field**" theory, which should be a good approximation at low densities. such that $na^3 \ll 1$, and at low temperatures when the number of particles in the condensate is large.

I will illustrate the use of the GP equation with two examples. The first is rather simple [G1], a Bose gas confined to the half-space $x > 0$ *i.e.* with a planar boundary. The steady state GP equation

$$\frac{d^2\phi}{dx^2} - 8\pi a\phi^3 + 8\pi a n_0\phi = 0 \tag{38}$$

has the exact solution

$$\phi(x) = \sqrt{n_0} \tanh\left(\frac{x}{\sqrt{2}l}\right) \tag{39}$$

where $l = (8\pi a n_0)^{-\frac{1}{2}}$ is known as the **healing length**, which for typical parameters is of order 1 Å. This shows that the condensate density $|\phi|^2$ is depleted near the surface but becomes essentially constant over a short distance from the wall.

The Gross-Pitaevskii theory is used in much of the recent work on atomic BEC. Figure 3 shows results from a numerical solution of the GP equation for a spherical trap. In each case the dashed line is the ideal gas result. The parameter labelling the curves is $\zeta = Na/a_{h_0}$ with $a_{h_0} = \sqrt{\hbar/m\omega_0}$. For the atomic BEC experiments $a/a_{h_0} \approx 10^{-3}$ so $N \approx 10^3\zeta$. For repulsive interactions the results show a marked deviation from the Gaussian profile predicted for the ideal gas, with a distinct flattening in the central region. For attractive interactions (*e.g.* the Li experiments) the GP equation predicts that the condensate will become unstable if $N > N_{cr} = 0.575a_{h_0}/a$.

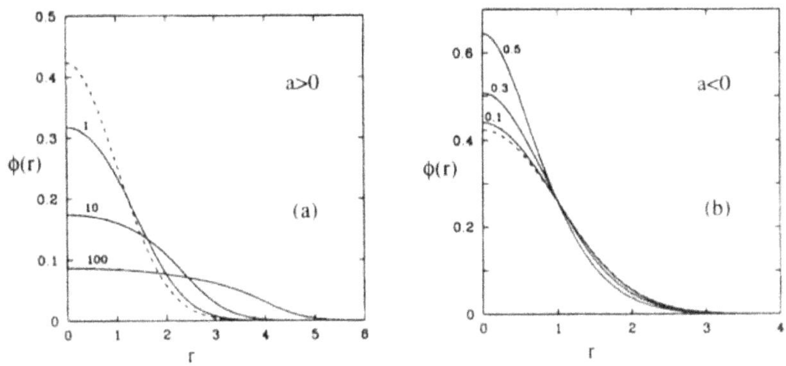

Figure 3. Condensate "wavefunction" $\phi(r)$ from numerical solution of the GP equation for a spherical harmonic trap, with distance in units of the oscillator length $\sqrt{\hbar/(m\omega_0)}$. The curves are labelled by values related to the number of particles N (see text). Case (a) is for positive scattering length (repulsive interactions), while case (b) is for attractive interactions (after Dalfovo et al. [G9]).

4 Strongly Interacting Bose Systems

The approaches discussed in the previous section will fail if the density of the system is too high or if the interactions are too strong. Mean-field theory does not correctly treat strong quantum and/or thermal fluctuations.

It becomes difficult to obtain reliable and systematic quantitative results. The most fundamental approach is via diagrammatic many-body theory expressed in terms of thermal Green's functions. The pioneering work is by Beliaev [26]. This leads to expressions which are formally exact, but approximations must be made to obtain specific analytic or numerical results. Ref. [G8] gives an account of this approach.

Renormalisation Group (RG) methods provide a way of describing the physics of strongly interacting systems near a critical point, and of obtaining numerically accurate results for universal properties. The idea was developed by Wilson [27] in the area of condensed matter systems/critical phenomena. It is not possible to describe this in any detail here (see *e.g.* [28]) but, for those unfamiliar with the method, the main steps are summarised as:

- parameterise the Hamiltonian H by a set of coupling constants $\{K_\alpha\}$ *e.g.* $T, \mu, g, \cdots$
- integrate out the short wavelength (large k) components to obtain a renormalised Hamiltonian $\tilde{H}$ with coupling constants $\{\tilde{K}_\alpha\}$ which obey a set of nonlinear RG relations

$$\tilde{K}_\alpha = f\{K_\beta\}$$

- identify the critical point with a **fixed point** $\tilde{K}_\alpha = K_\alpha$, and extract scaling fields, critical exponents, *etc.* by linearising the RG equations near the fixed point.

The philosophy is based on the idea that near a critical point the correlation length diverges and the small scale structure is unimportant for **universal** properties.

Fisher and Hohenberg [3] used an RG approach to investigate the superfluid transition in a dilute Bose gas in 2 dimensions. Although no BEC occurs the ideal Bose gas fixed point ($T = \mu = v = 0$) controls the low T behaviour. The critical temperature and superfluid density are shown to be accurately given by previous theories. More recently Bijlsma and Stoof [29] have used RG techniques to study the 3-dimensional atomic Bose gas. By numerically integrating the RG equations they obtain the critical temperature, the condensate and superfluid densities, and the $p - v$ phase diagram.

4.1 An Exactly Solvable System in 1-Dimension

The above approaches are still essentially perturbative, in the sense that the results are valid as long as some parameter remains small. It is therefore of fundamental importance to have a system where an exact solution can be obtained, not only for the insights this can give, but as a test case for approximate theories. Such a solvable model is provided by Lieb and Liniger [30], who were able to solve a 1-dimensional system of bosons with a repulsive δ-function potential, at $T = 0$,

$$H = -\sum_{i=1}^{N} \frac{\partial^2}{\partial x_i^2} + 2c \sum_{i<j} \delta(x_i - x_j) \tag{40}$$

where $\hbar = 2m = 1$ and c gives the strength of the potential.

The exact eigenfunctions can be written in the "Bethe ansatz" (BA) form

$$\psi(x_1, x_2, \cdots, x_N) = \sum_P a(P) \exp\left\{ i \sum_j k_{P_j} x_j \right\} \tag{41}$$

where $\{k_i\}$ are a set of N "quasi-momenta" and the sum is over all permutations of these. The quasi-momenta satisfy a set of coupled transcendental equations, the BA equations

$$k_j L = 2\pi I_j - \sum_{l=1}^{N} 2 \tan^{-1}\left(\frac{k_j - k_l}{c}\right) \tag{42}$$

where I_j are integers. The energy is then simply

$$E = \sum_{j=1}^{N} k_j^2. \tag{43}$$

In the thermodynamic limit $N, L \to \infty$, $N/L = n$ finite, it follows that

$$\frac{E_0}{L} = \int_{-K}^{K} \rho(k) k^2 dk \tag{44}$$

where $\rho(k)$, the density of roots, satisfies the integral equation

$$2c \int_{-K}^{K} \frac{\rho(p)}{c^2 + (p-k)^2} dp = 2\pi\rho(k) - 1 \tag{45}$$

with the momentum cutoff K defined implicitly by

$$\int_{-K}^{K} \rho(k) dk = n. \tag{46}$$

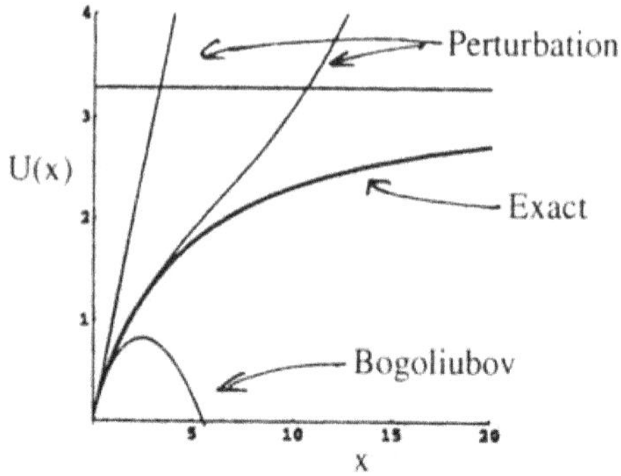

Figure 4. The function $u(x)$ in the result Eq.(47) for the ground state energy of exactly solvable 1-dimensional bose gas. The exact result provides a test of the accuracy of approximate methods (after Takahashi [32]).

These equations must be solved numerically.

It is usual to write the energy as

$$\frac{E_0}{N} = n^2 u(x) \tag{47}$$

where $u(x)$ is a monotonic increasing function of $x = c/n$, with the limiting values $u(0) = 0$ (free particles) and $u(\infty) = \pi^2/3$ (hard core bosons, [31]). Figure 4, taken from Takahashi [32], shows the exact result, compared with the Bogoliubov theory and perturbation calculations. The approximate calculations are accurate for small x, but fail completely for large x.

Yang and Yang [33] generalised this approach to finite temperatures and obtained an integral equation for the free energy. Although based on some unproved assumptions it appears to give the exact result.

4.2 Lattice Bose Systems

Progress can sometimes be made with intractable many-body problems by redefining them on a lattice of points. A model of this type, which is of much current interest, is the so-called Bose-Hubbard Model. This represents bosons with short-range (on-site) repulsive interactions moving in a periodic external potential. The Hamiltonian is written as

$$H = -t\sum_{\langle ij\rangle}(a_i^\dagger a_j + a_j^\dagger a_i) - \mu\sum_i n_i + V\sum_i n_i(n_i - 1) \tag{48}$$

where $a_i, a_i^\dagger$ are destruction and creation operators for bosons at site i $n_i = a_i^\dagger a_i$ is the site occupancy, the first term describes nearest neighbour hopping, and the on-site repulsion favours sites with $n_i = 0, 1$. The structure of the Hamiltonian is similar to the usual Hubbard model of interacting electrons - hence the somewhat quaint name. Such a model might describe aspects of liquid ^{4}He on a substrate, or absorbed into a porous medium such as Vycor. A related model has been proposed to describe granular superconductors or Josephson junction arrays.

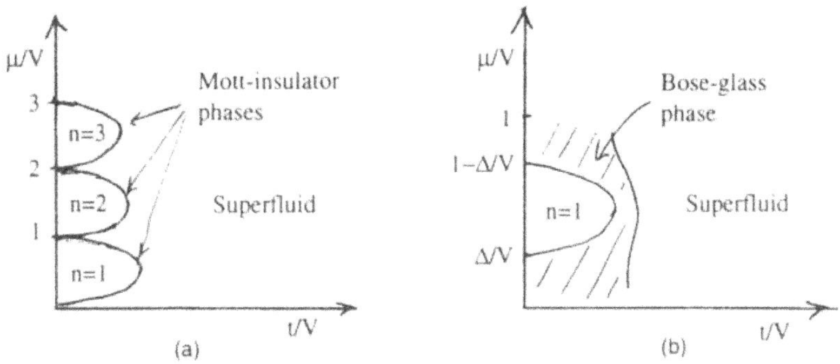

Figure 5. Schematic $T = 0$ phase diagrams for the Bose-Hubbard model (after Fisher *et al.* [34]). Case (a) is with no disorder; the Mott-insulator phases form lobes with $n = 1, 2, 3, \cdots$ bosons per site. Case (b) includes the effect of disorder; the Mott-insulator phase contracts and is separated from the superfluid phase by an intermediate "Bose-glass" phase.

The seminal paper is Fisher *et al.* [34] (FWGF), who introduced the model and identified important features of the phase diagram at $T = 0$. This is shown schematically in Fig.5a. For $t = 0$ (zero hopping) each site will contain n bosons, where $n - 1 < \mu/V < n$, in a "Mott insulator" phase. The insulating phase will persist for finite t in a series of lobes, which, for larger t will undergo a transition to a superfluid phase. FWGF also predict that, in the presence of disorder, a qualitative change will occur with the appearance of an intermediate "Bose glass" phase, which is still insulating because of localisation. This is shown schematically in Fig.5b.

A great deal of (largely) numerical work has been performed in recent years to test these predictions [35-37]. Batrouni and Scalettar [35] have used a quantum Monte Carlo method to study the 1-dimensional case, and find agreement with the FWGF predictions for both pure and disordered cases.

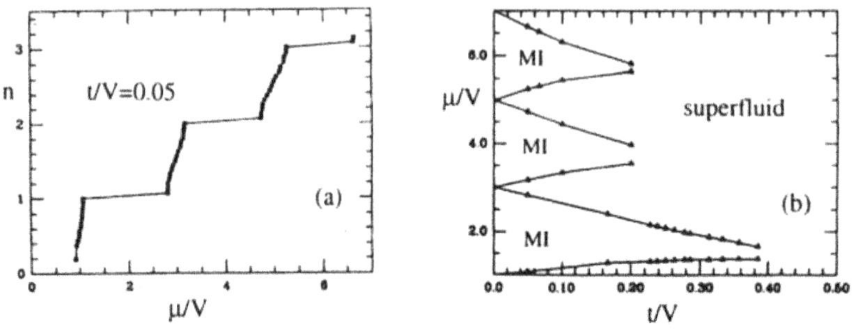

Figure 6. Numerical Quantum Monte Carlo results (Batrouni and Scalettar [35]) for the 1-dimensional Bose-Hubbard model. (a) Average number density versus chemical potential. The plateaux represent incompressible Mott-insulator phases. (b) The first three Mott lobes in the phase diagram.

Figure 6 shows their results for the phase diagram, showing the first three Mott lobes, and for the number density as a function of chemical potential μ. The latter clearly shows plateaus of zero compressibility $K = \partial n/\partial \mu$ within the Mott insulator phase. More recent work has considered also the 2-dimensional case.

5 Conclusion

I have attempted, in these three lectures, to cover some of the fundamental statistical mechanics of bose systems, including the phenomenon of Bose-Einstein condensation (BEC) in both uniform and non-uniform/finite systems. In the former case the phenomenon is most naturally treated in momentum space. Historically BEC was first considered in the ideal gas, but the full physical picture only arises when interactions are present. In finite systems, such as the trapped atomic gases, a description in real space is more natural. There is no "true" phase transition in these finite systems, but even for systems of as few as 1000 atoms, a sudden change to a condensed state occurs. I closed with some discussion of recent work on strongly interacting lattice models, which is closer to my own interests.

The field is currently very active on several fronts, with up to a dozen or more new papers weekly. It is an exciting time for both experimentalists and theorists.

Appendix A: Bose Integrals

The function $g_n(z)$ is defined by

$$g_n(z) = \frac{1}{\Gamma(n)} \int_0^\infty \frac{x^{n-1}}{z^{-1}e^x - 1} dx, \quad 0 \le z \le 1. \tag{A1}$$

and occurs for both integral and half-integral n. The special case

$$g_1(z) = -\ln(1-z) \tag{A2}$$

is expressible as a simple function. The other cases are not.

The $g_n(z)$ are monotonic increasing functions of z, as is clear from the Taylor series about $z = 0$

$$g_n(z) = \sum_{l=1}^{\infty} \frac{z^l}{l^n} = z + \frac{z^2}{2^n} + \frac{z^3}{3^n} + \cdots \tag{A3}$$

The values at $z = 1$ $(n > 1)$ are thus

$$g_n(1) = \sum_{l=1}^{\infty} \frac{1}{l^n} = \zeta(n) \tag{A4}$$

where $\zeta(n)$ is the Riemann zeta function:

Table 1. Values of $\zeta(n)$.

n	3/2	2	5/2	3	7/2	4
$\zeta(n)$	2.612	$\pi^2/6$	1.341	1.202	1.127	$\pi^4/90$

Expansions near $z = 1$ can be obtained in powers of $\alpha = -\ln z$. Examples are

$$g_{3/2}(z) = -3.54\alpha^{1/2} + 2.61 + 1.46\alpha - 0.104\alpha^2 + \cdots$$
$$g_{5/2}(z) = 2.36\alpha^{3/2} + 1.34 - 2.61\alpha - 0.730\alpha^2 + \cdots \tag{A5}$$

Further details can be found in Ref. [G1].

General References

[G1] R.K. Pathria, "Statistical Mechanics" (Pergamon, 1972).

[G2] K. Huang, "Statistical Mechanics" (Wiley, 1987).

[G3] K. Huang, in "Studies in Statistical Mechanics" Vol. 2, pp. 1-106, J. de Boer and G.E. Uhlenbeck, Eds. (North Holland 1964).

[G4] P. Nozières and D. Pines, "The Theory of Quantum Liquids" Vol. II (Addison-Wesley, 1990).

[G5] A. Griffin, "Excitations in a Bose-Condensed Liquid" (Cambridge, 1993).

[G6] A. Griffin, D.W. Snoke and S. Stringari (Eds.) "Bose-Einstein Condensation" (Cambridge, 1995).

[G7] A.S. Parkins and D.F. Walls, Phys. Reports **303**, 1-80 (1998).

[G8] H. Shi and A. Griffin, Phys. Reports **304**, 1-87 (1998).

[G9] F. Dalfovo, S. Giorgini, L.P. Pitaevskii and S. Stringari, Rev. Mod. Phys. **71**, 463-512 (1999).

References

[1] A. Einstein, Berline Ber. **1**, 3 (1925).

[2] P.C. Hohenberg, Phys. Rev. **158**, 383 (1967).

[3] D.S. Fisher and P.C. Hohenberg, Phys. Rev. B**37**, 4936 (1988).

[4] M.H. Anderson, *et al.*, Science **269**, 198 (1995).

[5] K.B. Davis *et al.*, Phys. Rev. Lett. **75**, 3969 (1995).

[6] C.C. Bradley *et al.*, Phys. Rev. Lett. **75**, 1687 (1995).

[7] V. Bagnato, D.E. Pritchard and D. Kleppner, Phys. Rev. A**35**, 4354 (1987).

[8] S. Grossman and M. Holthaus, Phys. Lett. A**208**, 188 (1995).

[9] V. Bagnato and D. Kleppner, Phys. Rev. A**44**, 7439 (1991).

[10] R.K. Pathria, Can. J. Phys. **61**, 228 (1983).

[11] W. Ketterle and N.J. van Druten, Phys. Rev. A**54**, 656 (1996).

[12] N.N. Bogoliubov, J. Phys. USSR **11**, 23 (1947).

[13] K.A. Brueckner and K. Sawada, Phys. Rev. **106**, 1117 (1957).

[14] T.D. Lee and C.N. Yang, Phys. Rev. **105**, 1119 (1957);

T.D. Lee, K. Huang and C.N. Yang, *ibid.* **106**, 1135 (1957).

[15] T.T. Wu, Phys. Rev. **115**, 1390 (1959).

[16] E. Braaten and A. Nieto, Eur. Phys. J. **B11**, 143 (1999).

[17] M. Holtzmann and W. Krauth, Phys. Rev. Lett. **83**, 2687 (1999).

[18] G. Baym *et al.*, Phys. Rev. Lett. **83**, 1703 (1999).

[19] K. Huang, Phys. Rev. Lett. **83**, 3770 (1999).

[20] O. Penrose and L. Onsager, Phys. Rev. **104**, 576 (1956).

[21] C.N. Yang, Rev. Mod. Phys. **34**, 694 (1962).

[22] P.W. Anderson, Rev. Mod. Phys. **38**, 298 (1966).

[23] P. Nozières and D. Saint James, J. Phys. (Paris) **43**, 1133 (1982).

[24] E.P. Gross, J. Math Phys. **4**, 195 (1963).

[25] L.P. Pitaevskii, Sov. Phys. JETP **13**, 451 (1961).

[26] S.T. Beliaev, Sov. Phys. JETP **7**, 289, 299 (1958).

[27] K.G. Wilson, Rev. Mod. Phys. **47**, 773 (1975).

[28] N. Goldenfeld, "Lectures on Phase Transitions and the Renormalisation Group" (Addison-Wesley, 1992).

[29] M. Bijlsma and H.T.C. Stoof, Phys. Rev. A**54**, 5085 (1996).

[30] E.H. Lieb and W. Liniger, Phys. Rev. **130**, 1605 (1963).

[31] M. Girardeau, J. Math. Phys. **1**, 516 (1960).

[32] M. Takahashi, "Thermodynamics of One-Dimensional Solvable Models" (Cambridge, 1999).

[33] C.N. Yang and C.P. Yang, J. Math. Phys. **10**, 1115 (1969).

[34] M.P.A. Fisher, P.B. Weichman, G. Grinstein and D.S. Fisher, Phys. Rev. B**40**, 546 (1989).

[35] G. Batrouni and R.T. Scalettar, Phys. Rev. B**46**, 9051 (1992).

[36] R.T. Scalettar, G.G. Batrouni, A.P. Kampf and G.T. Zimanyi, Phys. Rev. B**51**, 8467 (1995).

[37] T.D. Kühner and H. Monien, Phys. Rev. B**58**, R14741 (1998); N. Elstner and H. Monien, Phys. Rev. B**59**, 12184 (1999); M. Niemeyer, J.K. Freericks and H. Monien, Phys. Rev. B**60**, 2357 (1999).

CONDENSATE OSCILLATIONS, KINETIC EQUATIONS AND TWO-FLUID HYDRODYNAMICS IN A BOSE GAS

ALLAN GRIFFIN

Department of Physics, University of Toronto, Toronto, Ontario, Canada M5S 1A7
E-mail: griffin@physics.utoronto.ca
Web: http://www.physics.utoronto.ca/~griffin/

1 Introduction

Trapped Bose-condensed atomic gases [1,2] are remarkable because, in spite of the fact that these are very dilute systems, they exhibit robust coherent dynamic behaviour when perturbed. These quantum "wisps of matter" are a new phase of highly coherent matter. While binary collisions are very infrequent, the coherent mean field associated with the Bose condensate ensures that interactions play a crucial role in determining the collective response of these superfluid gases.

In our discussion of the theory of collective oscillations of atomic condensates, which is the main theme of these lectures, the macroscopic Bose wavefunction $\Phi(\mathbf{r},t)$ will play a central role. This wavefunction is the BEC order parameter. The initial attempts at defining this order parameter began with the pioneering work of Fritz London [3] in 1938, was further developed by Bogoliubov [4] in 1947 and finally formalized in the general quantum field theoretic formalism of Beliaev [5] in 1957. The first extension of these ideas to inhomogeneous Bose condensates was by Pitaevskii and, independently, by Gross in 1961, which led to the now famous Gross-Pitaevskii (GP) equation of motion for $\Phi(\mathbf{r},t)$. Most of this early work was limited to $T=0$ where, in a dilute Bose gas, one can assume all of the atoms are in the condensate.

In Section 2, I will first review the dynamics of a pure condensate at $T=0$, based on solving the linearized GP equation. This limit is especially appealing since one can ignore all the complications which arise from the presence of non-condensate atoms (the thermal cloud). We will discuss the normal mode solutions of the $T=0$ GP equation of motion using the "quantum hydrodynamic" formalism, which works in terms of the local condensate density $n_c(\mathbf{r},t)$ and superfluid velocity $\mathbf{v}_c(\mathbf{r},t)$. Within the Thomas-Fermi approximation, Stringari [6] has shown that the equations of motion for these two variables can be combined to give a wave equation for oscillations of the condensate.

In Section 3, we derive a generalized form of the GP equation for $\Phi(\mathbf{r}, t)$ which is valid at finite temperatures. [7] It involves terms which are coupled to the non-condensate component (the thermal cloud) and thus its solution in general depends on knowing the equations of motion for the dynamics of the non-condensate. In the present lectures, we will restrict ourselves to finite temperatures where the non-condensate can be described by a quantum kinetic equation for the single-particle distribution function $f(\mathbf{p}, \mathbf{r}, t)$. In this Boltzmann equation, the relatively high energy non-condensate atoms are simply free atoms moving in a self-consistent Hartree-Fock mean field.

A unique feature of a Bose-condensed gas is that the kinetic equation for $f(\mathbf{p}, \mathbf{r}, t)$ involves a collision integral (denoted by $C_{12}[f, \Phi]$) describing collisions between condensate and non-condensate atoms. The generalized GP equation for $\Phi(\mathbf{r}, t)$ also has a term which is related to the C_{12} collisions. This gives rise to damping of condensate fluctuations. In Section 3, we derive a finite T Stringari wave equation [8] with damping using the static Popov approximation. This means the thermal cloud is treated as always being in static thermal equilibrium, with $f(\mathbf{p}, \mathbf{r}, t) = f_0(\mathbf{p}, \mathbf{r})$ being given by the equilibrium Bose distribution.

In Section 4, we turn to a detailed treatment of the dynamics of the coupled condensate and non-condensate components, starting from the finite T generalized GP equation for $\Phi(\mathbf{r}, t)$ and our kinetic equation for $f(\mathbf{p}, \mathbf{r}, t)$. We derive a new set of two-fluid hydrodynamic equations, following the recent work by Zaremba, Nikuni and Griffin. [7] This derivation assumes the non-condensate is in *local* hydrodynamic equilibrium, induced by rapid collisions between the atoms in the thermal cloud. As a result, the non-condensate is completely described in terms of the non-condensate density $\tilde{n}(\mathbf{r}, t)$, velocity $\mathbf{v}_n(\mathbf{r}, t)$ and local pressure $\tilde{P}(\mathbf{r}, t)$. Both the condensate and non-condensate exhibit coupled coherent oscillations at the *same* frequency.

In Section 5, we review the famous two-fluid hydrodynamic equations first derived by Landau in 1941 and which form the basis of our understanding of the hydrodynamic behaviour of superfluid ^{4}He. [9,10] We discuss first sound and second sound and consider how they differ in superfluid ^{4}He and in a uniform Bose gas. We also point out the appearance of a new hydrodynamic zero frequency mode which appears in our generalized two-fluid equations. This mode is associated with the relaxation time for the condensate and non-condensate atoms to come into "diffusive" equilibrium with each other. Inclusion of damping due to thermal conductivity of the thermal cloud shows that this mode is the analogue of the well-known thermal diffusion mode in a classical gas.

We close this Introduction with some general references which may be useful to the reader. The 1998 Varenna Summer School lectures on BEC [2] has excellent articles on recent research on BEC in atomic gases, including very detailed reviews of the experimental work carried out by the JILA [11] and MIT [12] groups. For a general introduction to the theory of trapped Bose gases, we recommend the Reviews of Modern Physics [1] article by a leading theoretical group at the University of Trento. This authoritative review concentrates on thermodynamic properties and the collisionless dynamics. A recent long paper by Zaremba, Nikuni and the author [7] discusses the details of the derivation of the coupled hydrodynamic equations for the condensate and the thermal cloud. The present lectures may be viewed as an introduction to this paper, with more emphasis on the general structure of the derivation and the implications of this new set of superfluid equations.

2 Pure condensate dynamics (at $T = 0$)

This section will be largely a review of standard material [1] but will provide the starting point for generalizations in the following sections. As we have mentioned, at $T = 0$, one can assume that all atoms in a dilute Bose gas are described by the macroscopic wavefunction $\Phi(\mathbf{r}, t)$. This obeys the time-dependent Hartree equation of motion first written down 40 years ago by Pitaevskii and Gross,

$$i\hbar\frac{\partial\Phi}{\partial t}(\mathbf{r},t) = \left[-\frac{\hbar^2\nabla^2}{2m} + V_{ex}(\mathbf{r}) + V_H(\mathbf{r},t)\right]\Phi(\mathbf{r},t). \tag{1}$$

Here the trap harmonic potential is

$$V_{ex}(\mathbf{r}) = \frac{1}{2}m\omega_0^2 r^2 \qquad \text{(isotropic)} \tag{2}$$

and the self-consistent condensate Hartree potential is

$$V_H(\mathbf{r},t) = \int d\mathbf{r}' v(\mathbf{r}-\mathbf{r}')n_c(\mathbf{r},t) = g n_c(\mathbf{r},t). \tag{3}$$

As usual, since we are interested in extremely low energy atoms, we use the s-wave approximation and approximate the interatomic potential by the pseudopotential

$$\begin{aligned} v(\mathbf{r}-\mathbf{r}') &= \frac{4\pi a\hbar^2}{m}\delta(\mathbf{r}-\mathbf{r}') \\ &= g\delta(\mathbf{r}-\mathbf{r}'), \end{aligned} \tag{4}$$

where a is the correct s-wave scattering length. For further discussion of the effective interaction to use in the Gross-Pitaevskii (GP) equation, we refer to the lectures by Burnett in this volume. The condensate density is given by $n_c(\mathbf{r},t) = |\Phi(\mathbf{r},t)|^2$ and hence (1) reduces to a NLSE for $\Phi(\mathbf{r},t)$, namely

$$i\hbar\frac{\partial\Phi(\mathbf{r},t)}{\partial t} = \left[-\frac{\hbar^2\nabla_r^2}{2m} + V_{ex}(\mathbf{r}) + g|\Phi(\mathbf{r},t)|^2\right]\Phi(\mathbf{r},t). \qquad (5)$$

Since all atoms are in the identical quantum state, there is no exchange mean field in (1).

The $T=0$ GP equation in (5) has been the subject of literally hundreds of papers since the discovery of BEC in laser-cooled trapped atomic gases - and the equation appears in almost every chapter in this book. As discussed in recent reviews [1,2], it gives an excellent quantitative description of both the static and dynamic (linear and non-linear) behaviour in trapped Bose gases below about $T \lesssim 0.5T_{BEC}$. The accuracy can be of the order of a few percent, which is as much as one can expect since the non-condensate fraction of atoms at $T=0$ is also estimated to be about 1% or so. [13] The $T=0$ GP equation has been extended to deal with two-component Bose gases (involving two different atomic hyperfine states) and the effect of perturbations related to laser and rf fields (used to manipulate the Bose condensates) is easily incorporated. For further discussion and references, see the lectures by Ballagh in this book.

Our main purpose in this section is to use the GP equation (5) to discuss collective oscillations of a pure condensate. In Section 3, we will discuss the extension of this equation at finite T to deal with the effect of non-condensate atoms. We first briefly review the static equilibrium solution $\Phi_0(\mathbf{r})$ of the GP equation, given by

$$\left\langle\hat{\psi}(\mathbf{r},t)\right\rangle \equiv \Phi(\mathbf{r},t) = \Phi_0(\mathbf{r})e^{-i\mu t/\hbar}, \qquad (6)$$

where μ is the chemical potential of the condensate. The physics behind this can be seen from

$$\begin{aligned}\left\langle N-1|\hat{\psi}(\mathbf{r},t)|N\right\rangle &= e^{iE_{N-1}t/\hbar}\left\langle N-1|\hat{\psi}(\mathbf{r})|N\right\rangle e^{-iE_Nt/\hbar} \\ &= \left\langle N-1|\sqrt{N}|N-1\right\rangle e^{-i(E_N-E_{N-1})t/\hbar} \\ &= \sqrt{N}e^{-i\mu t/\hbar}. \end{aligned} \qquad (7)$$

Using (6) in (5) gives

$$\mu_{c0}\Phi_0(\mathbf{r}) = \left[-\frac{\hbar^2\nabla^2}{2m} + V_{ex}(\mathbf{r}) + g|\Phi_0(\mathbf{r})|^2\right]\Phi_0(\mathbf{r}). \qquad (8)$$

Assuming a vortex-free ground state, this GP equation for the static condensate wavefunction $\Phi_0(\mathbf{r}) = \sqrt{n_{c0}(\mathbf{r})}$ leads to the following expression for the equilibrium condensate chemical potential

$$\mu_{c0} = -\frac{\hbar^2\nabla^2\sqrt{n_{c0}(\mathbf{r})}}{2m\sqrt{n_{c0}(\mathbf{r})}} + V_{ex}(\mathbf{r}) + gn_{c0}(\mathbf{r}). \tag{9}$$

A standard approximation in solving (9) is to ignore the kinetic energy associated with the condensate amplitude $\sqrt{n_c}$, ie, neglect the $-\frac{\hbar^2\nabla^2}{2m}$ term. In this "Thomas-Fermi" (TF) approximation, the static GP equation for $\Phi_0(\mathbf{r})$ reduces to [1]

$$\left[V_{ex}(\mathbf{r}) + g|\Phi_0(\mathbf{r})|^2\right] = \mu_{c0}, \tag{10}$$

which is easily inverted to give the condensate density profile

$$\begin{aligned} n_{c0}(\mathbf{r}) &= \frac{1}{g}[\mu_{c0} - V_{ex}(\mathbf{r})] \\ &= \frac{1}{g}\left[\mu_{c0} - \frac{1}{2}m\omega_0^2 r^2\right]. \end{aligned} \tag{11}$$

Clearly in the TF approximation, the size of the condensate is R_{TF}, where

$$\mu_{c0} = \frac{1}{2}m\omega_0^2 R_{TF}^2. \tag{12}$$

One finds μ_{c0} from the condition $\int d\mathbf{r} n_{c0}(\mathbf{r}) = N = N_c$, which gives

$$\mu_{c0} = \frac{\hbar\omega_0}{2}\left[15\frac{N_c a}{a_{HO}}\right]^{2/5} \; ; \quad a_{HO} \equiv (\hbar/m\omega_0)^{1/2}. \tag{13}$$

We recall that the oscillator length a_{HO} is the size of the ground state Gaussian wavefunction of an atom in a parabolic potential. Combining (12) and (13) gives

$$\begin{aligned} R_{TF} &= a_{HO}\left(15\frac{N_c a}{a_{HO}}\right)^{1/5} \\ &\gg a_{HO}, \text{ if } \frac{N_c a}{a_{HO}} \gg 1. \end{aligned} \tag{14}$$

The TF approximation (11) for $n_{c0}(\mathbf{r})$ is very good for large N_c, except for a small region near the edge of condensate ($r \simeq R_{TF}$). In practice, for typical values of a_{HO} and a, one finds the TF approximation is very good for $N_c \gtrsim 10^4$ atoms. [1]

To discuss condensate fluctuations around the static equilibrium value of $\Phi_0(\mathbf{r})$, we first reformulate the time-dependent GP equation in terms of the condensate density and phase variables

$$\Phi(\mathbf{r},t) = \sqrt{n_c(\mathbf{r},t)}e^{i\theta(\mathbf{r},t)}. \tag{15}$$

Inserting this into the GP equation (5), gives

$$i\hbar\frac{\partial\sqrt{n_c}}{\partial t} - \sqrt{n_c}\frac{\hbar\partial\theta}{\partial t} = \mu_c(\mathbf{r},t)\sqrt{n_c} - i\hbar^2\frac{\sqrt{n_c}}{2m}\nabla^2\theta + \frac{\hbar^2\sqrt{n_c}}{2m}(\boldsymbol{\nabla}\theta)^2 - \frac{i\hbar^2}{m}(\boldsymbol{\nabla}\sqrt{n_c})\cdot\boldsymbol{\nabla}\theta, \tag{16}$$

where we have defined

$$\mu_c(\mathbf{r},t) \equiv -\frac{\hbar^2\nabla^2\sqrt{n_c}}{2m\sqrt{n_c}} + V_{ex}(\mathbf{r}) + gn_c(\mathbf{r},t). \tag{17}$$

Separating out the real and imaginary path gives two equations. The real part gives

$$\hbar\frac{\partial\theta}{\partial t} = -(\mu_c + \frac{1}{2}m\mathbf{v}_c^2), \tag{18}$$

where the condensate velocity field is defined by

$$m\mathbf{v}_c(\mathbf{r},t) \equiv \hbar\boldsymbol{\nabla}\theta(\mathbf{r},t). \tag{19}$$

The imaginary part of (16) gives

$$i\frac{\partial\sqrt{n_c}}{\partial t} = -\frac{\sqrt{n_c}}{2}\boldsymbol{\nabla}\cdot\mathbf{v}_c - \mathbf{v}_c\boldsymbol{\nabla}\sqrt{n_c}, \tag{20}$$

which can be rewritten in the form

$$\frac{\partial n_c}{\partial t} = -n_c\boldsymbol{\nabla}\cdot\mathbf{v}_c - \mathbf{v}_c\cdot\boldsymbol{\nabla}n_c = -\boldsymbol{\nabla}\cdot(n_c\mathbf{v}_c). \tag{21}$$

From now on, we shall set $\hbar = 1$ except in some final formulas.

In summary, we have shown that the time-dependent GP equation for the two-component order parameter $\Phi(\mathbf{r},t)$ is completely equivalent to the following two coupled equations for the condensate density $n_c(\mathbf{r},t)$ and velocity field $\mathbf{v}_c(\mathbf{r},t)$:

$$\begin{aligned}\frac{\partial n_c}{\partial t} &= -\boldsymbol{\nabla}\cdot(n_c\mathbf{v}_c)\\ m\left(\frac{\partial\mathbf{v}_c}{\partial t} + \frac{1}{2}\boldsymbol{\nabla}\mathbf{v}_c^2\right) &= -\boldsymbol{\nabla}\mu_c\,,\end{aligned} \tag{22}$$

where the position and time-dependent generalized condensate chemical potential μ_c is defined in (17). In this formulation, a complete description of the condensate dynamics is given in terms of two variables $n_c(\mathbf{r},t)$ and $\mathbf{v}_c(\mathbf{r},t)$, reminiscent of the hydrodynamic equations for a classical fluid. In the recent BEC literature, the equations in (22) are often referred to as the "hydrodynamic" theory. A better description would be to call it the "quantum hydrodynamic" theory, since it is equivalent to the mean-field GP equation. In later sections, we shall see that an exact GP equation taking into account the dynamics of the non-condensate leads to a generalized set of equations quite analogous to (22). The key equation (19) defining the superfluid velocity field of the condensate will turn out to be quite general. Needless to say, all aspects related to superfluidity of a trapped Bose gas (as compared to Bose condensation) are tied to the fact that the condensate exhibits motion related to the gradient of a phase (which means that, ignoring vortices, the condensate motion is irrotational, $\nabla \times \mathbf{v}_c(\mathbf{r},t) = 0$).

Stringari [6] first pointed out that within the *dynamic* TF approximation, one could combine the two equations in (22) into a single condensate wave equation. This approach will be the basis of our development in these lectures. Neglecting the kinetic energy term $-\nabla^2\sqrt{n_c}$ in (17) as small compared to the condensate interaction energy gn_c, we linearize the resulting equations around the equilibrium values

$$\begin{aligned} n_c &= n_{c0} + \delta n_c \\ \mathbf{v}_c &= \mathbf{v}_{c0} + \delta \mathbf{v}_c \,. \end{aligned} \tag{23}$$

We obtain

$$\begin{aligned} \frac{\partial \delta n_c}{\partial t} &= -\nabla \cdot (n_{c0}\delta \mathbf{v}_c) - \nabla \cdot (\mathbf{v}_{c0}\delta n_c) \\ \frac{\partial \delta \mathbf{v}_c}{\partial t} &= -\nabla \left(\mu'_{c0} + g\delta n_c + m\mathbf{v}_{c0}\cdot \delta \mathbf{v}_c\right) , \end{aligned} \tag{24}$$

where $\mu'_{c0} \equiv \mu_{c0} + \frac{1}{2}m\mathbf{v}_{c0}^2$ is independent of position. Assuming that there is no vortex in the solution of the static GP equation (ie, $\mathbf{v}_{c0} = 0$), (24) reduces to the following coupled linearized equations for δn_c and $\delta \mathbf{v}_c$:

$$\begin{aligned} \frac{\partial \delta n_c}{\partial t} &= -\nabla \cdot (n_{c0}(\mathbf{r})\delta \mathbf{v}_c) \\ \frac{\partial \delta \mathbf{v}_c}{\partial t} &= -\frac{g}{m}\nabla \delta n_c . \end{aligned} \tag{25}$$

These can be combined to give the well known $T = 0$ Stringari wave equa-

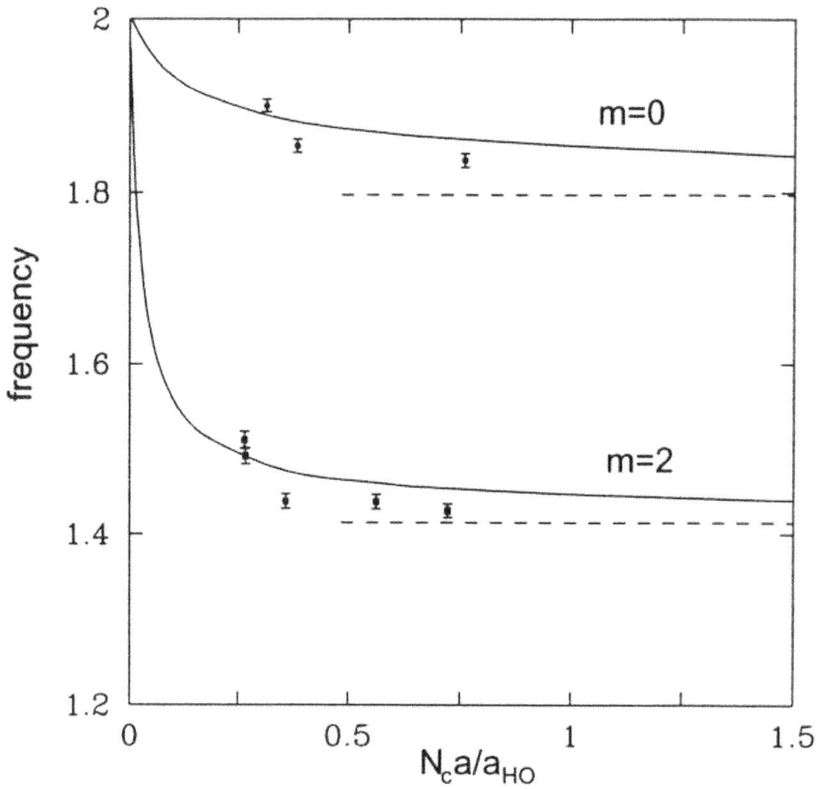

Figure 1. Plot of the lowest condensate mode frequencies (monopole and quadrupole) as a function of N_c. The full curves are the solutions of the coupled Bogoliubov equations[14] and the dashed lines are the Stringari TF approximation Ref. [6]. The data points are for a JILA trap with $\lambda = \sqrt{8}$ [11]. For these trap parameters, N_c varies from approximately 10^3 to 10^4 atoms. For further details, see Fig.14 of Ref. [1] .

tion [6,1]

$$\frac{\partial^2 \delta n_c}{\partial t^2} = \frac{g}{m} \nabla \cdot [n_{c0}(\mathbf{r}) \nabla \delta n_c] . \tag{26}$$

Since the derivation used the TF approximation (where $n_{c0}(\mathbf{r})$ is given by (11) for $r \leq R_{TF}$), one can rewrite (26) in the equivalent form (for an isotropic trap potential)

$$\frac{\partial^2 \delta n_c}{\partial t^2} = \frac{\mu_{c0}}{m} \nabla \cdot \left\{ \left[[1 - \frac{r^2}{R_{TF}^2} \right] \nabla \delta n_c \right\}, \quad r \leq R_{TF}, \tag{27}$$

where μ_{c0} is given by (13). It turns out that the normal mode solutions $\delta n_c(\mathbf{r}, t) = \delta n_\omega(\mathbf{r}) e^{-i\omega t}$ of (27) have frequencies which are independent of the interaction strength g or the value of N_c. This is a feature of the underlying TF approximation, which typically starts to breakdown (as noted earlier) when $N_c \lesssim 10^4$ atoms. This is shown by explicit numerical solutions [13,14] of the coupled Bogoliubov equations of motion which describe the normal mode solutions of the linearized GP equation when we take the kinetic energy of the condensate amplitude $\sqrt{n_c}$ fully into account. For values of $N_c \lesssim 10^4$, the normal mode frequencies depend significantly on the magnitude of N_c, as shown in Fig. 1.

We also note that (26) can be equally well rewritten in terms of the superfluid velocity $\delta \mathbf{v}_c$ or, equivalently, the phase fluctuations $\delta\theta$. This emphasizes that the measured condensate density fluctuations [11,12] are directly related to the existence of phase fluctuations. The existence of collective modes of a pure condensate may thus be viewed already as "evidence" of superfluidity, the latter being always a consequence of the phase coherence of the macroscopic wavefunction given in (15) which gives rise to the irrotational velocity in (19).

We conclude this section with several examples [1,6,15] of condensate normal modes based on solving the $T = 0$ Stringari equation (26). A wonderful aspect about the collective oscillations of a condensate in a trapped gas is you can "see" them. As Ketterle has remarked, these condensates are *robust* - one can kick them, shake them and these "wisps" of Bose-condensed matter keep their integrity. [11,12] The uniform Bose-condensed gas is especially simple, since $\delta n_c = \delta n_{k\omega} e^{i(\mathbf{k}\cdot\mathbf{r} - \omega t)}$. This gives

$$-\omega^2 \delta n_{k\omega} = \frac{g n_{c0}}{m} (-k^2) \delta n_{k\omega}, \tag{28}$$

or $\omega^2 = c_0^2 k^2$, with $c_0 \equiv \sqrt{g n_{c0}/m}$. This recovers the well-known Bogoliubov phonon oscillations [4] of a uniform Bose condensate. The neglect of the kinetic energy in our TF approximation precludes us from obtaining the particle-like behaviour at large values of the wavevector k. [15]

The Kohn (or sloshing) mode corresponds to the oscillation of the centre-of-mass of the static condensate profile with the trap frequency ω_0. This mode is described by

$$n_c(\mathbf{r}, t) = n_{c0}(\mathbf{r} - \boldsymbol{\eta}(t)), \tag{29}$$

where $\frac{d\boldsymbol{\eta}(t)}{dt} \equiv \mathbf{v}_c(t)$ and

$$\frac{d^2\boldsymbol{\eta}(t)}{dt^2} = -\omega_0^2\boldsymbol{\eta}(t). \tag{30}$$

The proof is simple. Linearizing (29), we find

$$n_c(\mathbf{r},t) = n_{c0}(\mathbf{r}) - \boldsymbol{\eta}\cdot\boldsymbol{\nabla}n_{c0}(\mathbf{r}), \tag{31}$$

which gives the explicit form for the condensate fluctuation,

$$\delta n_c(\mathbf{r},t) = \frac{1}{g}m\omega_0^2\mathbf{r}\cdot\boldsymbol{\eta}(t). \tag{32}$$

Inserting this into the Stringari equation (26), we obtain

$$\begin{aligned} -\boldsymbol{\nabla}n_{c0}\cdot\frac{d^2\boldsymbol{\eta}(t)}{dt^2} &= \omega_0^2\boldsymbol{\nabla}\cdot[n_{c0}\boldsymbol{\nabla}(\boldsymbol{\eta}(t)\cdot\mathbf{r})] \\ &= \omega_0^2\boldsymbol{\nabla}\cdot[n_{c0}\boldsymbol{\eta}] \\ &= \omega_0^2\boldsymbol{\eta}\cdot\boldsymbol{\nabla}n_{c0}. \end{aligned} \tag{33}$$

This confirms that the centre-of-mass position $\boldsymbol{\eta}(t)$ of the static condensate distribution satisfies the SHO equation (30) with frequency ω_0.

The breathing (or monopole) condensate normal mode corresponds to a velocity fluctuation of the form

$$\delta\mathbf{v}_c(\mathbf{r},t) = A\mathbf{r}\,e^{-i\omega t}\ (r \le R_{TF}). \tag{34}$$

Using $\boldsymbol{\nabla}\cdot\mathbf{r} = 3$ and $\boldsymbol{\nabla}\cdot(r^2\mathbf{r}) = 5r^2$, it is easy to verify from the continuity equation in (22) that the associated density fluctuation is

$$\begin{aligned} -i\omega\delta n_c &= -\boldsymbol{\nabla}\cdot[n_{c0}(\mathbf{r})\delta\mathbf{v}_c] \\ &= -\frac{\mu_{c0}}{g}A\left[3 - 5\frac{r^2}{R_{TF}^2}\right]e^{-i\omega t}, \end{aligned} \tag{35}$$

or

$$\delta n_\omega(\mathbf{r}) \equiv B\left(1 - \frac{5}{3}\frac{r^2}{R_{TF}^2}\right), \tag{36}$$

where μ_{c0} is given by (12). Inserting this into the Stringari wave equation (27) gives

$$\begin{aligned} -\omega^2\delta n_\omega(\mathbf{r}) &= \frac{\mu_{c0}}{m}\boldsymbol{\nabla}\cdot\left[\left(1 - \frac{r^2}{R_{TF}^2}\right)B\frac{5}{3}\frac{2\mathbf{r}}{R_{TF}^2}\right] \\ &= -5\omega_0^2\delta n_\omega(\mathbf{r}). \end{aligned} \tag{37}$$

Thus the breathing mode for an isotropic parabolic trap has a frequency $\omega = \sqrt{5}\omega_0$. As noted earlier, this mode frequency is not explicitly dependent on the interaction strength g. However the underlying theory is very dependent on mean-field effects. Moreover, we recall that a *non-interacting* trapped Bose gas has a breathing mode with a frequency $\omega = 2\omega_0$ at all temperatures. [1]

As a final example, we consider the so-called "surface" modes of a $T = 0$ condensate, as described by [6]

$$\delta \mathbf{v}_c(\mathbf{r}, t) = A\nabla \left[r^l Y_{lm}(\theta, \phi)\right] e^{-i\omega t}. \tag{38}$$

This mode corresponds to phase fluctuations of the form $\delta\theta_\omega = mA\, r^l Y_{lm}(\theta, \phi)$. One may easily verify that $\nabla \cdot \delta\mathbf{v}_c = 0$ and hence from (25) we obtain

$$\frac{\partial^2 \delta \mathbf{v}_c}{\partial t^2} = -\omega_0^2 \nabla(\delta \mathbf{v}_c \cdot \mathbf{r}). \tag{39}$$

This is equivalent to

$$\frac{\partial^2 \delta\theta}{\partial t^2} = -\omega_0^2 (\nabla \delta\theta) \cdot \mathbf{r} \tag{40}$$

or

$$-\omega^2 \delta\theta_\omega(\mathbf{r}) = -\omega_0^2 l \delta\theta_\omega(\mathbf{r}). \tag{41}$$

Thus the surface oscillations of the condensate phase have a frequency given by $\omega = \sqrt{l}\omega_0 (l = 1, 2, 3, \ldots)$.

One great advantage of the quantum hydrodynamic formalism (within the TF approximation) is that it is easy to also treat the $T = 0$ normal modes of an anisotropic parabolic well. We refer to the literature for further discussion. [1,6,15] We also note that it is straightforward to derive a Stringari wave equation for the oscillations of a vortex state (this has been done by Svidzinsky and Fetter [16]).

3 Generalized GP equation at finite temperatures

In this section, we generalize the $T = 0$ GP equation of Section 2 to finite T, so that it includes the effect of the coupling of the condensate to the non-condensate degrees of freedom. We concentrate in this Section on how the $T = 0$ condensate modes are renormalized and damped by the thermal cloud. We defer discussion of the appearance of new collective modes mainly associated with the thermal cloud to sections 4 and 5.

Up to the beginning of 2000, the study of the dynamics of a trapped Bose gas at finite temperatures has been largely ignored by experimentalists but actively studied by many theorists (especially in the last year or so). One of the reasons for the lack of finite temperature data is that there are so many interesting phenomena to study at $T \simeq 0$! However, another reason seems to be the implicit belief that the presence of the non-condensate just complicates the behaviour of a pure $T = 0$ condensate - but is not the source of any interesting new physics. In this and subsequent sections, I hope to argue that this "belief" is quite wrong. The coupling of the condensate and non-condensate degrees of freedom at finite T leads to a new two-component system in which *both* components can exhibit coherent behaviour, analogous to the well-known superfluid macroscopic phenomena appearing in liquid ^{4}He.

We shall see that, as expected, the finite temperature GP equation of motion for $\Phi(\mathbf{r}, t)$ is not closed. Its general solution involves knowing the equations of motion for the non-condensate. We will work within an approximation where the non-condensate atoms can be described by a quantum Boltzmann transport equation for the single-particle distribution function $f(\mathbf{p}, \mathbf{r}, t)$, with

$$\tilde{n}(\mathbf{r}, t) = \int \frac{d\mathbf{p}}{(2\pi)^3} f(\mathbf{p}, \mathbf{r}, t). \tag{42}$$

The details of this kinetic equation will be developed in Section 4. In this section, we concentrate on incorporating the effect of a *static* thermal cloud into the finite T dynamics of the condensate.

The theory of interacting Bose-condensed fluids is most usefully discussed using quantum field operators. This procedure was formalized by Beliaev [5] in 1957 and developed by Bogoliubov [17], Gavoret and Nozières [18], Martin and Hohenberg [19], and others in the early 1960's. We recall:

$$\begin{aligned} \hat{\psi}^\dagger(\mathbf{r}) &= \text{creates atom at } \mathbf{r} \\ \hat{\psi}(\mathbf{r}) &= \text{destroys atom at } \mathbf{r}. \end{aligned} \tag{43}$$

These quantum field operators satisfy the usual Bose commutation relations, such as

$$\left[\hat{\psi}(\mathbf{r}), \hat{\psi}^\dagger(\mathbf{r}')\right] = \delta(\mathbf{r} - \mathbf{r}'). \tag{44}$$

All observables can be written in terms of these quantum field operators, such as the density $\hat{n}(\mathbf{r}) = \hat{\psi}^\dagger(\mathbf{r})\hat{\psi}(\mathbf{r})$ and interaction energy

$$\begin{aligned} \hat{V} &= \frac{1}{2}\int d\mathbf{r} \int d\mathbf{r}' \hat{\psi}^\dagger(\mathbf{r}')\hat{\psi}^\dagger(\mathbf{r}) v(\mathbf{r} - \mathbf{r}')\hat{\psi}(\mathbf{r}')\hat{\psi}(\mathbf{r}) \\ &= \frac{1}{2} g \int d\mathbf{r} \hat{\psi}^\dagger(\mathbf{r})\hat{\psi}^\dagger(\mathbf{r})\hat{\psi}(\mathbf{r})\hat{\psi}(\mathbf{r}). \end{aligned} \tag{45}$$

The crucial idea due to Bogoliubov [4,17] and later generalized by Beliaev [5] is to separate out the condensate component of the field operators,

$$\hat{\psi}(\mathbf{r}) = \left\langle \hat{\psi}(\mathbf{r}) \right\rangle + \tilde{\psi}(\mathbf{r}), \tag{46}$$

where

$$\left\langle \hat{\psi}(\mathbf{r}) \right\rangle \equiv \Phi(\mathbf{r}) = \text{Bose macroscopic wavefunction.}$$

This quantity plays the role of the "order parameter" for the Bose superfluid phase transition:

$$\begin{aligned} \Phi(\mathbf{r}) &= 0 \quad T > T_c \\ &\neq 0 \quad T < T_c. \end{aligned}$$

We note that $\Phi(\mathbf{r}) \equiv \sqrt{n_c}e^{i\theta}$ is a 2-component order parameter. Clearly, $\Phi(\mathbf{r})$ is not simply related to the many-particle wavefunctions $\Psi(\mathbf{r}_1, \mathbf{r}_2, \ldots \mathbf{r}_N)$. The thermal average in $\left\langle \hat{\psi}(\mathbf{r}) \right\rangle$ involves a small symmetry-breaking perturbation to allow $\Phi(\mathbf{r})$ to be finite,

$$\hat{H}_{SB} = \lim_{\eta \to 0} \int d\mathbf{r} \left[\eta(\mathbf{r}) \hat{\psi}^\dagger(\mathbf{r}) + \eta^*(\mathbf{r}) \hat{\psi}(\mathbf{r}) \right]. \tag{47}$$

The philosophy behind the concept of symmetry-breaking was extensively discussed by Bogoliubov [17], for a variety of condensed matter systems, in an article which is still highly recommended.

It is useful to make a few comments on the physics behind $\Phi(\mathbf{r}, t)$. $\Phi(\mathbf{r}, t)$ is a *coherent* state, with a "clamped" value of phase - rather than a Fock-state of fixed N, with no well-defined phase. $\Phi(\mathbf{r}, t)$ acts like a *classical* field, since quantum fluctuations are negligible when N_c is large. Probably P.W. Anderson [20] deserves the greatest credit for understanding (in the period 1958-1963) the new physics behind working with a broken-symmetry state $\Phi(\mathbf{r}, t)$, both in BCS superconductors and in superfluid ^{4}He. It captures the physics of the new phase of matter (such as the occurence of the Josephson effect) and the associated superfluidity. The symmetry-breaking perturbation (47) allows the system to internally set up off-diagonal symmetry-breaking fields, which persist even when the external symmetry-breaking perturbation in (47) is set to zero at the end ($\eta \to 0$). The same sort of physics is the basis of the BCS theory of superconductors.

One can formulate the GP and Bogoliubov approximations directly in terms of a variational many-particle wavefunction (see the lectures by Leggett in this volume). However, such formulations are limited to simple mean-field approximations. The explicit introduction of the broken-symmetry order parameter $\Phi(\mathbf{r},t)$ gives a systematic way [5,17,19,21] of isolating the role of the Bose condensate in a general treatment of an interacting Bose-condensed fluid. As we shall see, the resulting formalism allows one to deal with questions related to damping as well as superfluidity in both the collisionless and hydrodynamic regions.

The exact Heisenberg equation of motion for the field operator is

$$i\frac{\partial\hat{\psi}(\mathbf{r},t)}{\partial t} = \left[-\frac{\nabla^2}{2m} + V_{ex}(\mathbf{r}) + \delta U(\mathbf{r},t)\right]\hat{\psi}(\mathbf{r},t) + \eta(\mathbf{r}) + g\hat{\psi}^\dagger(\mathbf{r},t)\hat{\psi}(\mathbf{r},t)\hat{\psi}(\mathbf{r},t), \tag{48}$$

where $\delta U(\mathbf{r},t)$ is a small time-dependent driving potential. This gives an *exact* equation of motion for $\Phi(\mathbf{r},t) \equiv \left\langle\hat{\psi}(\mathbf{r},t)\right\rangle$,

$$i\frac{\partial\Phi(\mathbf{r},t)}{\partial t} = \left[-\frac{\nabla^2}{2m} + V_{ex}(\mathbf{r}) + \delta U(\mathbf{r},t)\right]\Phi(\mathbf{r},t) + \eta(\mathbf{r}) + g\left\langle\hat{\psi}^\dagger(\mathbf{r},t)\hat{\psi}(\mathbf{r},t)\hat{\psi}(\mathbf{r},t)\right\rangle, \tag{49}$$

with [(using the decomposition (46)]

$$\hat{\psi}^\dagger\hat{\psi}\hat{\psi} = |\Phi|^2\Phi + 2|\Phi|^2\tilde{\psi} + \Phi^2\tilde{\psi}^\dagger + \Phi^*\tilde{\psi}\tilde{\psi} + 2\Phi\tilde{\psi}^\dagger\tilde{\psi} + \tilde{\psi}^\dagger\tilde{\psi}\tilde{\psi}. \tag{50}$$

Taking the symmetry-breaking average of (50), one finds

$$\left\langle\hat{\psi}^\dagger\hat{\psi}\hat{\psi}\right\rangle = n_c\Phi + \tilde{m}\Phi^* + 2\tilde{n}\Phi + \left\langle\tilde{\psi}^\dagger\tilde{\psi}\tilde{\psi}\right\rangle, \tag{51}$$

where

$$\begin{aligned} n_c(\mathbf{r},t) &\equiv |\Phi(\mathbf{r},t)|^2 = \text{condensate density} \\ \tilde{n}(\mathbf{r},t) &\equiv \left\langle\tilde{\psi}^\dagger(\mathbf{r},t)\tilde{\psi}(\mathbf{r},t)\right\rangle = \text{non-condensate density} \\ \tilde{m}(\mathbf{r},t) &\equiv \left\langle\tilde{\psi}(\mathbf{r},t)\tilde{\psi}(\mathbf{r},t)\right\rangle = \text{off-diagonal (anomalous) density}. \end{aligned} \tag{52}$$

Using (51) in (49), our "exact" equation of motion for $\Phi(\mathbf{r},t)$ is [7]

$$i\frac{\partial\Phi}{\partial t} = \left[-\frac{\nabla^2}{2m} + V_{ex} + gn_c(\mathbf{r},t) + 2g\tilde{n}(\mathbf{r},t)\right]\Phi + g\tilde{m}(\mathbf{r},t)\Phi^* + g\left\langle\tilde{\psi}^\dagger(\mathbf{r},t)\tilde{\psi}(\mathbf{r},t)\tilde{\psi}(\mathbf{r},t)\right\rangle. \tag{53}$$

We now consider various approximations to the generalized GP equation in (53):

(a) The Hartree-Fock-Bogoliubov (HFB) approximation for Φ corresponds to neglecting the three-field correlation function $\left\langle\tilde{\psi}^\dagger\tilde{\psi}\tilde{\psi}\right\rangle$ but keeping the n_c, $\tilde{n}$ and $\tilde{m}$ fluctuations. This HFB has been exhaustively treated in many recent papers, in conjunction with the separate equations of motion for $\tilde{n}$ and $\tilde{m}$. [22,23,24] This HFB approximation can be used to generate the same normal mode spectrum which the Beliaev second-order self-energy approximation gives at finite T. [25]

(b) The *dynamic* Popov approximation corresponds to ignoring both $\left\langle\tilde{\psi}^\dagger\tilde{\psi}\tilde{\psi}\right\rangle$ and $\tilde{m} = \left\langle\tilde{\psi}\tilde{\psi}\right\rangle$ in (53). Theories of this kind involve coupled equations for Φ and $\tilde{n}$. [26]

(c) The *static* Popov approximation [22,13,1] involves even a further simplification, namely it ignores fluctuations in the density $\tilde{n}(\mathbf{r},t)$ of the thermal cloud,

$$\tilde{n}(\mathbf{r},t) \simeq n_{c0}(\mathbf{r}). \tag{54}$$

Using (54) in (53) corresponds to treating the dynamics of the condensate moving in the *static* mean field of the non-condensate thermal cloud, ie,

$$i\hbar\frac{\partial\Phi}{\partial t} = \left[-\frac{\nabla^2}{2m} + V_{ex}(\mathbf{r}) + 2g\tilde{n}_0(\mathbf{r}) + gn_c(\mathbf{r},t)\right]\Phi(\mathbf{r},t). \tag{55}$$

There is a considerable literature based on this static Popov-approximation. [1] In a nutshell, in this section, we will discuss an extension of (55) which involves including the damping associated with the $\langle\tilde{\psi}^\dagger\tilde{\psi}\tilde{\psi}\rangle$ term, but again treating the thermal cloud statically.

As in Section 2, we first rewrite (53) in terms of the condensate amplitude and phase. Inserting (15) into (53) and following the same procedure which led to the $T = 0$ equations in (22), we obtain:

$$\frac{\partial n_c}{\partial t} + \nabla\cdot n_c\mathbf{v}_c = 2g\mathrm{Im}\left[\Phi^{*2}\tilde{m} + \Phi^*\langle\tilde{\psi}^\dagger\tilde{\psi}\tilde{\psi}\rangle\right] \tag{56}$$

$$\frac{\partial\theta}{\partial t} = -\left(\mu_c + \frac{(\nabla\theta)^2}{2m}\right), \tag{57}$$

where now (compare with (17))

$$\mu_c(\mathbf{r},t) = -\frac{\nabla^2\sqrt{n_c}}{2m\sqrt{n_c}} + V_{ex} + gn_c + 2g\tilde{n} + \frac{g}{n_c} Re\left[\Phi^{*2}\tilde{m} + \Phi^*\left\langle\tilde{\psi}^\dagger\tilde{\psi}\tilde{\psi}\right\rangle\right]. \tag{58}$$

Formally, (53) and hence (56) - (58) are *exact* results. In these lectures, we will limit ourselves to finite temperatures where the dominant thermal excitations are well approximated by high energy non-condensate atoms moving in a self-consistent dynamic HF mean-field

$$\begin{aligned}\tilde{\varepsilon}_p(\mathbf{r},t) &= \frac{p^2}{2m} + V_{ex}(\mathbf{r}) + 2g\left[n_c(\mathbf{r},t) + \tilde{n}(\mathbf{r},t)\right] \\ &\equiv \frac{p^2}{2m} + U(\mathbf{r},t).\end{aligned} \tag{59}$$

In this region, we shall neglect the anomalous pair correlations among the thermal atoms (ie, $\tilde{m} = 0$). One can show that both $\tilde{m}$ and $\langle\tilde{\psi}^\dagger\tilde{\psi}\tilde{\psi}\rangle$ in (56) and (58) are of order g. Thus one sees the correction terms in (56) and (58) involving these functions are of order g^2. In fact, calculation shows that to order g, $\langle\tilde{\psi}^\dagger\tilde{\psi}\tilde{\psi}\rangle$ is imaginary. Our procedure is to ignore the $O(g^2)$ terms in (56) and (58) except those which are imaginary and hence are a source of damping of condensate motion.

The end result of this approach is that we are left with

$$\begin{aligned}\frac{\partial n_c}{\partial t} + \nabla\cdot n_c\mathbf{v}_c &= -\Gamma_{12}[f,\Phi] \\ m\left(\frac{\partial \mathbf{v}_c}{\partial t} + \frac{1}{2}\nabla\mathbf{v}_c^2\right) &= -\nabla\mu_c\,,\end{aligned} \tag{60}$$

with

$$\mu_c(\mathbf{r},t) = -\frac{\nabla^2\sqrt{n_c}}{2m\sqrt{n_c}} + V_{ex}(\mathbf{r}) + gn_c(\mathbf{r},t) + 2g\tilde{n}(\mathbf{r},t) \tag{61}$$

and we have defined the new function

$$\Gamma_{12}[f,\Phi] \equiv -2gIm\left[\Phi^*\langle\tilde{\psi}^\dagger\tilde{\psi}\tilde{\psi}\rangle\right]. \tag{62}$$

We see that Γ_{12} plays the role of a source term in the condensate continuity equation in (60). It depends on both $\Phi(\mathbf{r},t)$ and the single-particle distribution function $f(\mathbf{p},\mathbf{r},t)$. It is useful to note that (60) and (61) are equivalent to the approximate GP equation (see also Refs. [28,29])

$$i\hbar\frac{\partial\Phi}{\partial t} = \left[-\frac{\hbar^2\nabla^2}{2m} + V_{ex}(\mathbf{r}) + gn_c(\mathbf{r},t) + 2g\tilde{n}(\mathbf{r},t) - i\hbar R(\mathbf{r},t)\right]\Phi\,, \tag{63}$$

where the dissipative function is

$$R(\mathbf{r},t) \equiv \frac{\Gamma_{12}[f,\Phi]}{2n_c(\mathbf{r},t)} \sim 0(g^2). \tag{64}$$

To proceed, we need to calculate the field correlation $\langle\tilde{\psi}^\dagger\tilde{\psi}\tilde{\psi}\rangle$ which determines Γ_{12} in (62). This has been done in Appendix A of Ref. [7], working to lowest order in g at finite T where the single-particle spectrum in (59) is adequate. The result is

$$\begin{aligned}\langle\tilde{\psi}^\dagger\tilde{\psi}\tilde{\psi}\rangle = -ig\frac{\Phi(\mathbf{r},t)}{(2\pi)^5}&\int d\mathbf{p}_1\int d\mathbf{p}_2\int d\mathbf{p}_3\\ &\times\delta(\mathbf{p}_c+\mathbf{p}_1-\mathbf{p}_2-\mathbf{p}_3)\delta(\epsilon_c+\tilde{\epsilon}_1-\tilde{\epsilon}_2-\tilde{\epsilon}_3)\\ &\times[f_1(1+f_2)(1+f_3)-(1+f_1)f_2f_3],\end{aligned} \tag{65}$$

where $f_1 \equiv f(\mathbf{p},\mathbf{r},t), \epsilon_c \equiv \mu_c(\mathbf{r},t)+\frac{1}{2}m\mathbf{v}_c^2(\mathbf{r},t)$ is the local condensate atom energy and $\mathbf{p}_c \equiv m\mathbf{v}_c$ is the condensate atom momentum. In Section 4, we shall see that (65) is closely related to the collision integral $C_{12}[f,\Phi]$ which enters the Boltzmann equation for $f(\mathbf{p},\mathbf{r},t)$ and describes collisions with the non-condensate atoms in which one condensate atom is involved. We note that to leading order, the expression in (65) is imaginary. Using this in (62), we find

$$\begin{aligned}\Gamma_{12}(\mathbf{r},t) = 2g^2\frac{n_c(\mathbf{r},t)}{(2\pi)^5}&\int d\mathbf{p}_1\int d\mathbf{p}_2\int d\mathbf{p}_3\delta(\mathbf{p}_c+\mathbf{p}_1-\mathbf{p}_2-\mathbf{p}_3)\\ &\times\delta(\epsilon_c+\tilde{\epsilon}_1-\tilde{\epsilon}_2-\tilde{\epsilon}_3)\\ &\times[f_1(1+f_2)(1+f_3)-(1+f_1)f_2f_3].\end{aligned} \tag{66}$$

As a first application of our new equations in (60), we will ignore the dynamics of the non-condensate cloud. That is, we will assume that the condensate interacts with a *static* thermal cloud in thermal equilibrium

$$f(\mathbf{p},\mathbf{r},t) \simeq f^0(\mathbf{p},\mathbf{r}) = \frac{1}{e^{\beta[p^2/2m+U_0(\mathbf{r})-\tilde{\mu}_0]}-1}, \tag{67}$$

where $\tilde{\mu}_0$ is the equilibrium chemical potential of the non-condensate atoms. In Section 4, we will discuss the C_{22} collisions which produce this static equilibrium Bose distribution. The value of $\tilde{\mu}_0$ is known since it must equal the condensate equilibrium chemical potential μ_{c0} given by the static solution of the generalized GP equation. Within the TF approximation, (61) reduces to

$$\mu_{c0} = V_{\rm ex}(\mathbf{r})+gn_{c0}(\mathbf{r})+2g\tilde{n}_0(\mathbf{r}) \tag{68}$$

and hence we see that (59) reduces to

$$\begin{aligned}\tilde{\epsilon}_p(\mathbf{r}) - \tilde{\mu}_0 &= \frac{p^2}{2m} + U_0(\mathbf{r}) - \tilde{\mu}_0 \\ &= \frac{p^2}{2m} + gn_{c0}(\mathbf{r}) \ . \end{aligned} \tag{69}$$

This result is consistent with the correct Bogoliubov "excitation energy." [1,15] The static non-condensate density in (68) is given by

$$\tilde{n}_0(\mathbf{r}) = \frac{1}{\Lambda^3} g_{3/2}\left(z_0 = e^{-\beta g n_{c0}(\mathbf{r})}\right) . \tag{70}$$

We can now calculate $\Gamma_{12}(\mathbf{r},t)$ using the equilibrium Bose distribution (67) for the thermal cloud. One finds [8] (for a related calculation, see Ref. [7]).

$$\Gamma^0_{12}(\mathbf{r},t) = \frac{n_c(\mathbf{r},t)}{\tau_{12}(\mathbf{r},t)}\left[e^{-\beta(\tilde{\mu}_0 - \epsilon_c(\mathbf{r},t) - \frac{1}{2}mv_c^2)} - 1\right], \tag{71}$$

where we have defined a C_{12} collision time

$$\begin{aligned}\frac{1}{\tau_{12}(\mathbf{r},t)} \equiv \frac{2g^2}{(2\pi)^5}\int d\mathbf{p}_1 \int d\mathbf{p}_2 \int d\mathbf{p}_3 \delta(\mathbf{p}_c + \mathbf{p}_1 - \mathbf{p}_2 - \mathbf{p}_3) \\ \times\ \delta(\varepsilon_c + \tilde{\varepsilon}_{p_1} - \tilde{\varepsilon}_{p_2} - \tilde{\varepsilon}_{p_3})(1 + f_1^0) f_2^0 f_3^0 \ . \end{aligned} \tag{72}$$

$\Gamma_{12}[f_0, \Phi]$ in (71) still depends on $n_c(\mathbf{r},t)$ and $\mathbf{v}_c(\mathbf{r},t)$ of the condensate through μ_c, $\mathbf{p}_c$ and ϵ_c. We note that when the condensate is in static equilibrium, $\mu_c(\mathbf{r},t) \to \mu_{c0}$ and $\mathbf{v}_{c0} = 0$. In this case, $\mu_{c0} = \tilde{\mu}_0$ and the expression in the square bracket in (71) is seen to vanish. Thus $\Gamma_{12}(f_0, \Phi_0) = 0$, as it should when both components are in static thermal equilibrium.

Summarizing, at this point we have a closed set of equations which can be used to describe the dynamics of the condensate in a trapped Bose gas which include the interactions with a *static* equilibrium thermal cloud. These equations are

$$\frac{\partial n_c}{\partial t} + \nabla \cdot n_c \mathbf{v}_c = -\Gamma_{12}[f^0, \Phi] \tag{73}$$

$$m\left(\frac{\partial}{\partial t} + \mathbf{v}_c \cdot \boldsymbol{\nabla}\right)\mathbf{v}_c = -\boldsymbol{\nabla}\mu_c \, , \tag{74}$$

where $\Gamma_{12}[f^0, \Phi]$ is given explicitly by the expression in (71) and

$$\mu_c(\mathbf{r},t) = -\frac{\nabla^2\sqrt{n_c}}{2m\sqrt{n_c}} + V_{ex}(\mathbf{r}) + 2g\tilde{n}_0(\mathbf{r}) + gn_c(\mathbf{r},t). \tag{75}$$

From (73) and (74), we can obtain linearized equations of motion for the condensate fluctuations δn_c and $\delta \mathbf{v}_c$. We use the fact that, to lowest order in the fluctuations from static equilibrium, (71) reduces to

$$\delta\Gamma_{12}^0 = \frac{\beta n_{c0}(\mathbf{r})}{\tau_{12}^0(\mathbf{r})}\delta\mu_c(\mathbf{r},t), \tag{76}$$

where the "equilibrium" C_{12} collision rate [using (69) in (72)] is defined by

$$\frac{1}{\tau_{12}^0(\mathbf{r})} \equiv \frac{2g^2}{(2\pi)^5}\int d\mathbf{p}_1 \int d\mathbf{p}_2 \int d\mathbf{p}_3 \delta(\mathbf{p}_1 - \mathbf{p}_2 - \mathbf{p}_3)$$
$$\times\, \delta\Big(\frac{p_1^2 - p_2^2 - p_3^2}{2m} - gn_{c0}\Big)(1 + f_1^0)f_2^0 f_3^0\,. \tag{77}$$

In the static TF approximation, we recall that the equilibrium distribution is given by $f_i^0 = [e^{\beta(p_i^2/2m + gn_{c0})} - 1]^{-1}$.

In the present discussion, we further restrict ourselves to the dynamic Thomas-Fermi limit valid for large N_c,

$$\frac{\partial \delta n_c}{\partial t} + \nabla\cdot(n_{c0}\delta\mathbf{v}_c) = -\frac{1}{\tau'}\delta n_c \tag{78}$$

$$m\frac{\partial \delta\mathbf{v}_c}{\partial t} = -g\nabla\delta n_c\,. \tag{79}$$

The collision time $\tau'(\mathbf{r})$ describes collisions between the condensate and non-condensate atoms when the condensate is perturbed away from equilibrium,

$$\frac{1}{\tau'(\mathbf{r})} = \frac{gn_{c0}(\mathbf{r})}{k_B T}\frac{1}{\tau_{12}^0(\mathbf{r})}\,. \tag{80}$$

The new term on the right-hand side of (78) causes damping of the condensate fluctuations due to the lack of collisional detailed balance between the condensate and the static thermal cloud. We note that this collision time $\tau'(\mathbf{r})$ is only a function of the position $\mathbf{r}$ through its dependence on the static condensate density $n_{c0}(\mathbf{r})$.

We can easily combine (78) and (79) to obtain what we shall refer to as the finite T Stringari wave equation [8]

$$\frac{\partial^2 \delta n_c}{\partial t^2} - \frac{g}{m}\nabla\cdot(n_{c0}\nabla\delta n_c) = -\frac{1}{\tau'}\frac{\partial \delta n_c}{\partial t}\,. \tag{81}$$

If we neglect the right-hand side, we obtain the undamped finite T Stringari normal modes $\delta n_c(\mathbf{r},t) = \delta n_i(\mathbf{r})e^{-i\omega_i t}$ given by the solution of

$$-\frac{g}{m}\nabla\cdot[n_{c0}(\mathbf{r})\nabla\delta n_i(\mathbf{r})] = \omega_i^2\delta n_i(\mathbf{r}). \tag{82}$$

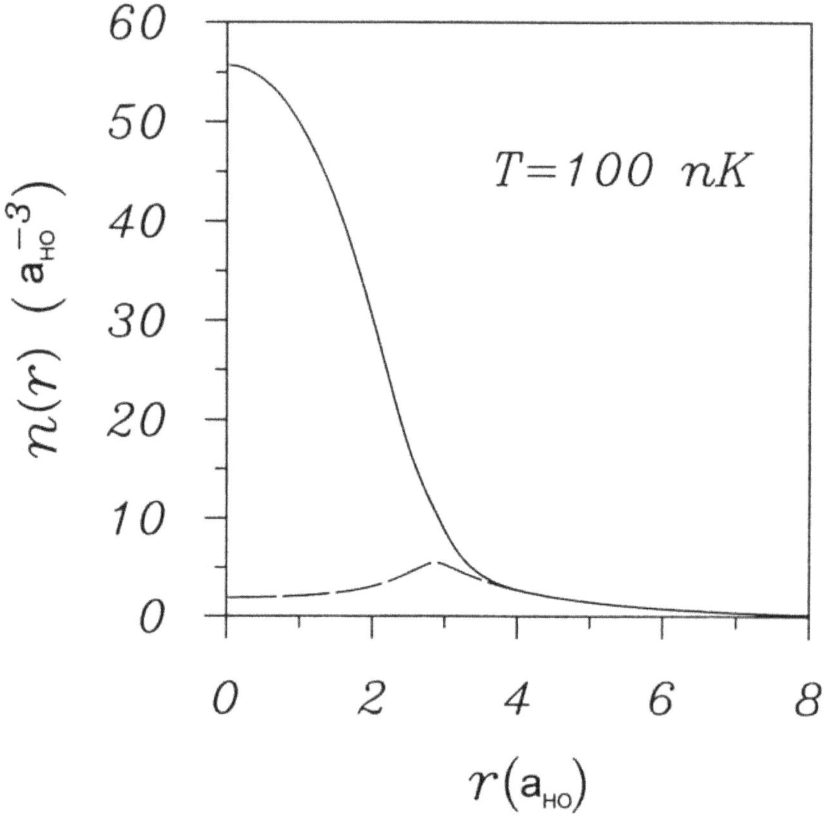

Figure 2. Density of atoms as a function of position in an isotropic parabolic trap containing 5000 ^{87}Rb atoms at T = 100 nK, with T_{BEC} = 149 nK. At this temperature, the total number of atoms in the condensate and non-condensate are equal. The solid line is the total density and the dashed line is the non-condensate density. The TF approximation is not used. The parameter a_{HO} is the oscillator length. See Ref. [7].

As has been noted by several authors in recent papers [1,24], $n_{c0}(\mathbf{r})$ at finite T can be well approximated by the TF condensate profile at $T = 0$ but with the number of atoms in the condensate $N_c(T)$ now being a function of temperature. This is because the static mean field of the non-condensate plays a minor role. In the regions where the condensate density is finite, we effectively always have $n_{c0}(\mathbf{r}) \gg \tilde{n}_0(\mathbf{r})$, as illustrated in Fig. 2. With this approximation for $n_{c0}(\mathbf{r})$, the solutions of the finite T Stringari equation (81) will be identical to those at $T = 0$ (see Section 2), since the $T = 0$ Stringari frequencies do not depend on the magnitude of N_c. Of course, as shown by calculations solving the coupled Bogoliubov equations [1,14], when $N_c \lesssim 10^4$ the TF approximation breaks down. Thus the condensate collective mode frequencies will always become temperature dependent close to $T_{\rm BEC}$, where the TF approximation is no longer valid. [13]

We can use the undamped Stringari modes given by (82) as a basis set to solve (81) and find the damping of these modes. Writing $\delta n_c(\mathbf{r}) = \sum_i c_i \delta n_i(\mathbf{r})$, and using the orthonormality condition $\int d\mathbf{r} \delta n_i(\mathbf{r}) \delta n_j(\mathbf{r}) = \delta_{ij}$, one obtains the following algebraic equations for the coefficients c_i

$$\omega^2 c_i = \omega_i^2 c_i - i\omega \sum_j \gamma_{ij} c_j \,, \tag{83}$$

where

$$\gamma_{ij} \equiv \int d\mathbf{r} \delta n_i(\mathbf{r}) \delta n_j(\mathbf{r}) / \tau'(\mathbf{r}) \,. \tag{84}$$

Assuming the damping is small (our present discussion is for the collisionless region), (83) is easily solved using perturbation theory by setting $\gamma_{ij} = 0$ for $j \neq i$. This gives the damped Stringari frequency (to lowest order) $\Omega_i = \omega_i - i\Gamma_i$, with

$$\Gamma_i \equiv \frac{\gamma_{ii}}{2} = \frac{1}{2} \int d\mathbf{r} \frac{\delta n_i(\mathbf{r})^2}{\tau'(\mathbf{r})} \,. \tag{85}$$

This result for Γ_i is reasonable, namely it involves a spatial average over $1/\tau'(\mathbf{r})$ weighted with respect to the undamped density fluctuations of the Stringari wave equation (81). Calculation shows that the effect of coupling to other modes ($\gamma_{ij} \neq 0$) is extremely small.

Before discussing the trapped gas, it is useful to first apply our theory to a homogeneous gas, where τ' is independent of position. In this case, (85) simply reduces to $\Gamma_i = 1/2\tau'$. Although our model applies only to the collisionless region, it is useful to compare the inter-component collision time for a uniform gas in the collisionless and hydrodynamic regimes. In Section 5, we show that the inter-component collision time τ_μ in the hydrodynamic region [7] is given by $\tau_\mu = \sigma_H \tau'$, where the temperature-dependent factor σ_H depends on various thermodynamic functions [σ_H is given explicitly by (165) in Section 4]. In Fig. 3, we compare $1/\tau'$ and $1/\tau_\mu$ as functions of T. We see that σ_H dramatically alters the inter-component relaxation rate $1/\tau_\mu$ appropriate to the hydrodynamic regime, as compared to $1/\tau'$ involved in the

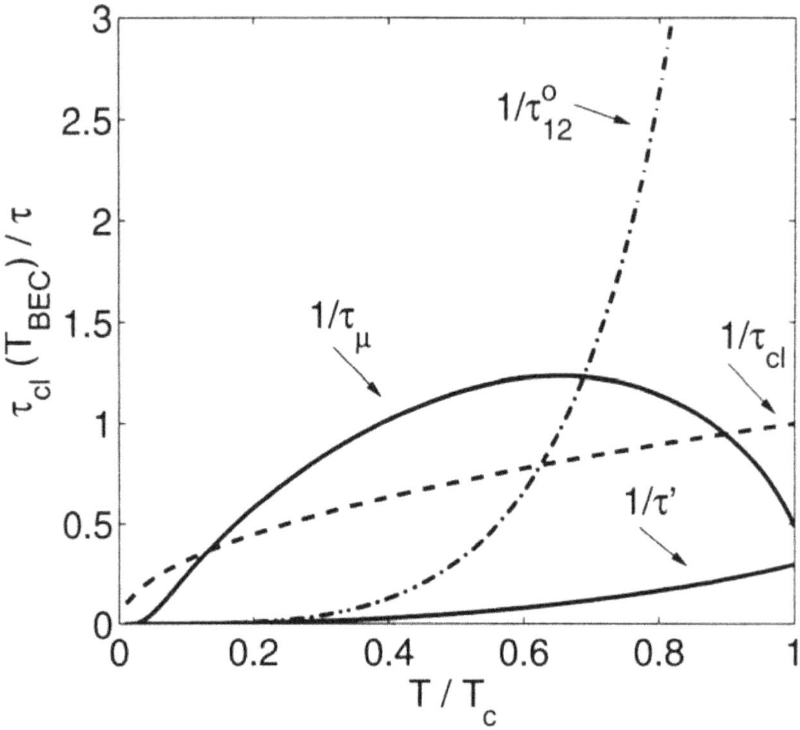

Figure 3. Collision rates in a homogeneous Bose-condensed gas, normalized to the collision time (95) in a classical gas at the BEC transition temperature $\tau_{cl}^{-1}(T_{BEC})$. We have taken $gn = 0.1k_BT_{BEC}$, where n is the total density. See also Ref. [8].

collisionless regime. For completeness, in Fig. 3 we also plot the often-used classical collision time as well as τ_{12}^0 defined in (77).

We briefly discuss some numerical calculations [8] of the inter-component damping using the above formalism. We consider ^{87}Rb atoms in a spherically symmetric trap with frequency $\nu_0 = 10$ Hz and $N = 2 \times 10^6$. In the collisionless limit, we require $\omega_i\tau_{cl} \gg 1$. We obtain an upper limit on $1/\tau_{cl}$ by taking the density in the center of the trap $n(r = 0)$, which gives $1/\tau_{cl} = 8a^2N\omega_0^3m/(\pi k_BT)$. For the parameters we use, $\omega_{10}\tau_{cl} \approx 19$ (compared to $\omega_{02}\tau_{cl} \approx 20$ for the JILA data [30] and $\omega_{02}\tau_{cl} \approx 2$ for the MIT data [31] on collective oscillations at finite T).

In Fig. 4 we plot the damping rate Γ_{10} for the breathing mode ($n = 1$,

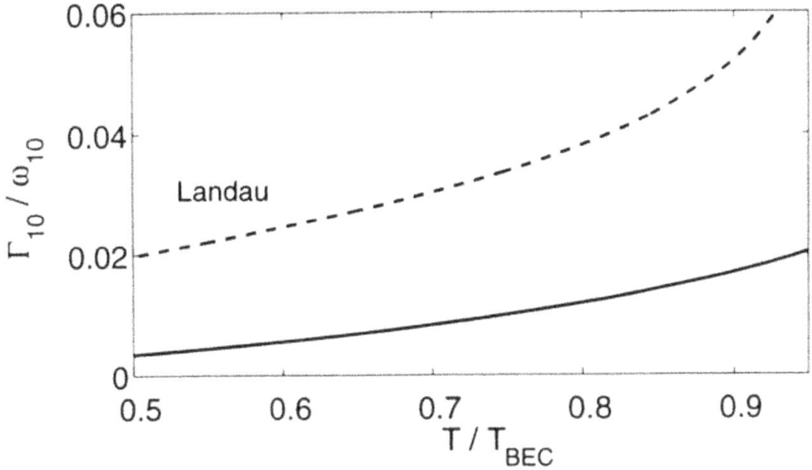

Figure 4. Breathing mode damping rates vs. temperature. In these plots, the damping rates are normalized to the mode frequencies and we only plot up to $T = 0.95T_{\rm BEC}$, above which the Thomas-Fermi approximation will start to break down. We show the inter-component damping rate given by (80) and (85) for the breathing mode ($n = 1, l = 0$) of a spherically symmetric trap, as well as the Landau damping. [45] For further details, see Ref. [8].

$l = 0$) as a function of temperature up to $T = 0.95T_{\rm BEC}$, where $N_c \approx 7 \times 10^4$. At higher temperatures, the Thomas-Fermi approximation will start to break down and the mode frequencies become temperature dependent.

The damping of condensate modes we have discussed in this section is due to the fact that the condensate is out of diffusive equilibrium with the thermal cloud. At finite T, the collective oscillations of the coupled condensate and non-condensate can be generally split into two classes. One mode mainly involves (out-of-phase) motion of the condensate and is the finite T generalization of the analogous $T = 0$ condensate mode. [6] For the same symmetry, there is another mode which mainly involves the (in-phase) motion of the non-condensate thermal cloud. This mode is naturally viewed as a generalization of the $T > T_{BEC}$ thermal cloud oscillation to temperatures below T_{BEC}. These two types of modes (for a given symmetry) have been obtained [7] in the collision-dominated hydrodynamic region, when the dynamics of the thermal cloud is fully allowed for. Treating (to first approximation) the thermal cloud (the non-condensate component) as always in *static* equilibrium in this section means that our theory is only applicable to condensate oscillations which involve motions which are out-of-phase with the thermal cloud.

As we discuss in Section 4, the possibility that the condensate may be out of diffusive equilibrium with the non-condensate is a general feature of the dynamics of trapped Bose-condensed gases. In the hydrodynamic two-fluid region, it leads to a new relaxational mode. [32] This effect is missed in the well-known form of the two-fluid hydrodynamics developed by Landau [9,10], where one (implicitly) assumes that the normal and superfluid components are always in local diffusive equilibrium. We will return to this question in Sections 4 and 5.

Of course, in addition to the inter-component relaxation discussed in this section, one *also* has Landau and Beliaev damping of condensate oscillations. [25,24,33] These mechanisms are treated in more detail by Burnett in this book. Landau damping is briefly discussed in Section 4.

Clearly one can extend the formalism of this section to any problem based on the $T = 0$ GP equation. In particular, it can be used to discuss this kind of inter-component damping at finite T of oscillations in two-component Bose gases, vortex dynamics [15,16], and Josephson oscillations between two traps.

4 Dynamics of the coupled condensate and non-condensate at finite temperatures

In contrast to Section 3, we now grapple with the dynamics of the thermal cloud. As we noted in Section 3, in dealing with this very complicated problem, we will be quite modest and treat the non-condensate using the simplest microscopic model which captures the important physics. We limit ourselves to finite T, where the non-condensate atoms can be described by the particle-like HF spectrum given by (59). For trapped Bose gases, this spectrum is probably adequate down to quite low temperatures, for reasons discussed in an important paper by the Trento group. [34] To extend our present analysis to very low temperatures is much more involved since then the excitations of the thermal cloud take on a collective aspect (ie, a Bogoliubov-type quasiparticle spectrum must be used). [23]

The single-particle spectrum (59) is appropriate in the semi-classical limit. where we can use the single-particle distribution function $f(\mathbf{p},\mathbf{r},t)$ given by the solution of a kinetic equation. This procedure generalizes the approach of Boltzmann (1880's) for describing binary collisions in a classical gas. Such a quantum Boltzmann equation for a trapped Bose-condensed gas at finite temperatures has been derived and extensively discussed by Zaremba, Nikuni and the author. [7] The conditions of validity are

$$k_BT \gg gn_c \,, \; k_BT \gg \hbar\omega_0, \tag{86}$$

where $\hbar\omega_0$ is the spacing of the energy levels of the harmonic trap. More general but less explicit discussions of kinetic equations are given in Refs. [28,29]. Related work of Gardiner and coworkers [35] is discussed in the lectures by Ballagh in this book. The quantum kinetic equation we use is given by

$$\begin{aligned} \frac{\partial f(\mathbf{p},\mathbf{r},t)}{\partial t} &+ \frac{\mathbf{p}}{m}\cdot\nabla_r f(\mathbf{p},\mathbf{r},t) - \nabla_r U(\mathbf{r},t)\cdot\nabla_p f(\mathbf{p},\mathbf{r},t) \\ &= C_{22}[f] + C_{12}[f]. \end{aligned} \tag{87}$$

The right hand side describes how binary collisions effect the value of the single-particle distribution function $f(\mathbf{p},\mathbf{r},t)$. The effective time-dependent HF potential $U(\mathbf{r},t)$ is defined in (59). The effect of collisions between *excited* atoms in the non-condensate is described by:

$$\begin{aligned} C_{22}[f] = \frac{2g^2}{(2\pi)^5}&\int d\mathbf{p}_2\int d\mathbf{p}_3\int d\mathbf{p}_4\delta(\mathbf{p}+\mathbf{p}_2-\mathbf{p}_3-\mathbf{p}_4) \\ &\times\delta\left(\tilde{\varepsilon}_p+\tilde{\varepsilon}_{p_2}-\tilde{\varepsilon}_{p_3}-\tilde{\varepsilon}_{p_4}\right) \\ &\times\left[(1+f)(1+f_2)f_3f_4 - ff_2(1+f_3)(1+f_4)\right]. \end{aligned} \tag{88}$$

We recall that creating a Boson gives a factor $(1+f)$ and destroying a Boson gives f. In the classical high temperature limit, $f \ll 1$ and the collision integral C_{22} in (88) becomes much simpler.

In addition to C_{22} collisions, we also have collisions which involve *one* condensate atom:

$$\begin{aligned} C_{12}[f] = \frac{2g^2}{(2\pi)^2}&\int d\mathbf{p}_1\int d\mathbf{p}_2\int d\mathbf{p}_3\delta(m\mathbf{v}_c+\mathbf{p}_1-\mathbf{p}_2-\mathbf{p}_3) \\ &\times\delta\left(\varepsilon_c+\tilde{\varepsilon}_{p1}-\tilde{\varepsilon}_{p2}-\tilde{\varepsilon}_{p3}\right)\left[\delta\left(\mathbf{p}-\mathbf{p}_1\right)-\delta(\mathbf{p}-\mathbf{p}_2)-\delta(\mathbf{p}-\mathbf{p}_3)\right] \\ &\times\left[n_c(1+f_1)f_2f_3 - n_cf_1(1+f_2)(1+f_3)\right]. \end{aligned} \tag{89}$$

For convenience, we recall that the condensate atom has

$$\begin{aligned} &\text{energy: } \varepsilon_c = \mu_c + \frac{1}{2}mv_c^2 \;;\; \mu_c = V_{ex} + gn_c + 2g\tilde{n} \\ &\text{momentum: } \mathbf{p}_c = m\mathbf{v}_c \end{aligned} \tag{90}$$

We note the *key* difference between C_{12} and C_{22} collisions:

- C_{22} and C_{12} both conserve energy and momentum in collisions.
- C_{12} does *not* (but C_{22} does) conserve the number of condensate atoms. C_{12} describes how atoms are "kicked" in and out of the condensate.

It turns out in the generalized GP equation given in (63), the damping term $-iR(\mathbf{r},t)$ is closely related to $C_{12}[f]$. This makes sense, since the C_{12} collisions modify the condensate described by $\Phi(\mathbf{r},t)$. One easily may verify that [comparing (66) and (89)]

$$\Gamma_{12}[f,\Phi] = \int \frac{d\mathbf{p}}{(2\pi)^3} C_{12}[f(\mathbf{p},\mathbf{r},t)]. \tag{91}$$

More precisely, the three-field correlation function in the exact equation of motions [see (56) and (58)] is related to $C_{12}[f]$ by

$$\int \frac{d\mathbf{p}}{(2\pi)^3} C_{12}[f] = 2g\sqrt{n_c}\; Im\left\langle \tilde{\psi}^\dagger \tilde{\psi} \tilde{\psi} \right\rangle. \tag{92}$$

We have to solve the equation of motions for $f(\mathbf{p},\mathbf{r},t)$ and $\Phi(\mathbf{r},t)$ treating $C_{12}[f]$ very carefully. There will be an exchange of atoms between the $\tilde{n}(\mathbf{r},t)$ and $n_c(\mathbf{r},t)$ components through the C_{12} collisions. We can use these coupled equations for a variety of problems. Recently these coupled equations have been used in two problems:

(a) Derivation of a generalized set of two-fluid hydrodynamic equations. [7]

(b) Discussion of the rate of growth of a condensate [36] due to a sudden quench in which the high energy spectrum of the thermal cloud distribution is suddenly removed. This work has many points of contact with the formalism reviewed by Ballagh in this book for condensate growth. [35]

In this section, we will linearize these equations and consider the collective oscillations of the combined system composed of condensate and non-condensate. [35] It is useful to introduce *two* regimes to describe collective modes in interacting systems [21,37]:

I. Collisionless (produced by *mean fields*)

$$\omega\tau_R \gg 1 \quad \text{or} \quad T \ll \tau_R \quad \left(\omega \equiv \frac{2\pi}{T}\right) \tag{93}$$

II. Hydrodynamic (produced by *collisions*)

$$\omega\tau_R \ll 1 \quad \text{or} \quad T \gg \tau_R, \tag{94}$$

where τ_R is some appropriate relaxation time. What should we use for τ_R? For a uniform classical gas, this is the collision time [38,32]

$$\frac{1}{\tau_c} = \sqrt{2}\tilde{n}\sigma\bar{v} \tag{95}$$

where

$$\sigma = 8\pi a^2 \text{ (for Bose atoms); } a = s\text{-wave scattering length.}$$

$$\bar{v} \simeq \text{ average velocity of atoms } \sim \sqrt{\frac{k_B T}{m}}.$$

$$\tilde{n} = \text{ density of excited atoms.}$$

Even for a Bose-condensed gas, taking $\tau_R \sim \tau_c$ is a reasonable first estimate. To get into the interesting hydrodynamic region ($\omega\tau_R \ll 1$), we need a small value of τ_R, ie, a large density $\tilde{n}$ or a large collision cross-section σ (perhaps by working near a Feshbach resonance). We note that current BEC experiments deal with $N \simeq 10^6$ atoms or larger and thus the kind of density of thermal atoms needed to be in the hydrodynamic region now seem achievable.

Let us look at the kinetic equation (87), writing it in the schematic form:

$$\hat{\mathcal{L}} f = C_{22}[f] + C_{12}[f]. \tag{96}$$

In the collisionless region, we need only solve $\hat{\mathcal{L}} f = 0$. For comparison with the collision-dominated hydrodynamic domain, we briefly discuss some features of this collisionless domain, limiting ourselves to a *uniform* gas for simplicity. This enables us to understand how Landau and Beliaev damping arise from the collisionless Boltzmann equation, in contrast to the inter-component collisional damping (discussed in Section 3) which has its origin in the $C_{12}[f, \Phi]$ collision integral.

We consider linear response theory [37,39] for the deviations of $f(\mathbf{p}, \mathbf{r}, t)$ from thermal equilibrium

$$f(\mathbf{p}, \mathbf{r}, t) = f_0(\mathbf{p}, \mathbf{r}) + \delta f(\mathbf{p}, \mathbf{r}, t). \tag{97}$$

For a uniform normal Bose gas to first order in δf, the collisionless kinetic equation reduces to

$$\frac{\partial \delta f}{\partial t} + \frac{\mathbf{p}}{m} \cdot \nabla_r \delta f - \nabla_r [2g\delta n(\mathbf{r}, t)] + \delta U(\mathbf{r}, t)] \cdot \nabla_p f_0(\mathbf{p}) = 0. \tag{98}$$

Assuming $\delta U(\mathbf{r}, t) = \delta U_{k\omega} e^{i(\mathbf{k}\cdot\mathbf{r} - \omega t)}$, then we have $\delta f(\mathbf{p}, \mathbf{r}, t) = \delta f_{k\omega}(\mathbf{p}) e^{i(\mathbf{k}\cdot\mathbf{r} - \omega t)}$, where $\delta f_{k\omega}(\mathbf{p})$ satisfies

$$-i\omega \delta f(\mathbf{p}) + i\frac{\mathbf{p}\cdot\mathbf{k}}{m}\delta f(\mathbf{p}) - [2g\delta n_{k\omega} + \delta U_{k\omega}] i\mathbf{k} \cdot \nabla_p f_0(\mathbf{p}) = 0 \,, \tag{99}$$

where

$$f_0(p) = \frac{1}{e^{\beta(\frac{p^2}{2m} + 2gn_0 - \mu_0)} - 1} \equiv f_0(\epsilon_p) \tag{100}$$

and

$$\delta n_{k\omega} \equiv \int \frac{d\mathbf{p}}{(2\pi)^3} \delta f_{k\omega}(\mathbf{p}). \tag{101}$$

Solving (99), one finds

$$\delta n_{k\omega} = -\int \frac{d\mathbf{p}}{(2\pi)^3} \frac{\mathbf{p}\cdot\mathbf{k}}{m} \frac{\partial f_0(\epsilon_p)}{\partial \epsilon_p} \frac{[2g\delta n_{k\omega} + \delta U_{k\omega}]}{\omega - \frac{\mathbf{p}\cdot\mathbf{k}}{m}} \tag{102}$$

or

$$\delta n_{k\omega} = \frac{\chi^0_{nn}(\mathbf{k},\omega)}{1 - 2g\chi^0_{nn}(k,\omega)} \delta U_{k\omega} \equiv \chi_{nn}(\mathbf{k},\omega)\delta U_{k\omega}. \tag{103}$$

Here

$$\chi^0_{nn}(\mathbf{k},\omega) = -\int \frac{d\mathbf{p}}{(2\pi)^3} \frac{\mathbf{p}\cdot\mathbf{k}}{m} \frac{\partial f_0(\epsilon_p)}{\partial \epsilon_p} \frac{1}{\omega - \frac{\mathbf{p}\cdot\mathbf{k}}{m}} \tag{104}$$

is the $k \to 0$ limit of the density response function of a non-interacting uniform Bose gas.

This shows how one can use the collisionless kinetic equation to find the density response function $\chi_{nn}(\mathbf{k},\omega)$ of a uniform Bose gas. The density fluctuations are given by the poles of $\chi_{nn}(\mathbf{k},\omega)$ in (103). In this RPA with exchange, the solutions of

$$1 - 2g\chi^0_{nn}(\mathbf{k},\omega) = 0 \tag{105}$$

are called "zero sound" modes (following Landau's terminology in the analogous calculation for an interacting Fermi gas [37]). It is clear that the solutions $\omega(k)$ of (105) will be Landau damped, with a width related to

$$\begin{aligned} &- 2g\mathrm{Im}\chi^0_{nn}(\mathbf{k},\omega + i0^+) \\ &= 2g\pi \int \frac{d\mathbf{p}}{(2\pi)^3} [f_0(\epsilon_p) - f_0(\epsilon_{p+k})]\delta[\omega(k) - (\epsilon_{p+k} - \epsilon_p)]. \end{aligned} \tag{106}$$

Clearly this damping will come from the zero sound mode absorbing a thermal excitation of energy ϵ_p to create a thermal excitation ϵ_{p+k},

$$\hbar\omega(k) = \epsilon_{p+k} - \epsilon_p \ . \tag{107}$$

We might also remark that if this linear response calculation is extended to deal with a Bose-condensed gas, one finds a new damping mechanism because now $\chi^0_{nn}(\mathbf{k},\omega)$ describes non-interacting Bogoliubov excitations E_p. In addition to the contributions of the kind given by (107), one finds $\chi^0_{nn}(\mathbf{k},\omega)$ now has additional poles at

$$\hbar\omega(k) = E_{p+k} + E_p. \tag{108}$$

These give rise to "Beliaev" damping, where the zero sound mode can decay into two Bogoliubov excitations. In the limit of $T \to 0$, Beliaev damping dominates over Landau damping. It was first calculated at $T = 0$ by Beliaev [40] in 1957 in his classic study of a weakly interacting Bose gas.

Several recent papers have discussed Landau and Beliaev damping of collective modes in a trapped Bose gas. [41,33,24,42,43] This problem has been nicely formulated directly in terms of response functions by Minguzzi and Tosi [26] in conjunction with mean fields produced by the condensate and non-condensate density fluctuations. Giorgini [24] has given a physically appealing discussion working directly with the coupled mean field equations of motion for $\Phi(\mathbf{r},t), \tilde{n}(\mathbf{r},t)$ and $\tilde{m}(\mathbf{r},t)$. The approach of the Oxford group is reviewed in Burnett's lectures in this book.

In this connection, a simple treatment of Landau damping of condensate collective modes can be given starting from our generalized GP equation in (63). In Section 3, we ignored the fluctuations in the HF term $2g\tilde{n}(\mathbf{r},t)$ and concentrated on the damping associated with the $-iR(\mathbf{r},t)$ term in (63). A simple way of including the $\delta\tilde{n}(\mathbf{r},t)$ fluctuations induced by the mean field associated with the condensate fluctuations is to use (see Refs. [26,8])

$$\delta\tilde{n}(\mathbf{r},\omega) = \int d\mathbf{r}' \chi^0_{\tilde{n}\tilde{n}}(\mathbf{r},\mathbf{r}',\omega) 2g\delta n_c(\mathbf{r}',\omega). \tag{109}$$

Here $\chi^0_{\tilde{n}\tilde{n}}$ is the density response function for a non-interacting gas of atoms with the equilibrium HF spectrum given by (69) and the chemical potential $\tilde{\mu}_0$ is equal to μ_{c0} in (68). For a uniform Bose gas, the condensate collective modes given by (63) are found to be

$$\omega^2 = c_0^2 k^2 [1 + 4g\chi^0_{\tilde{n}\tilde{n}}(\mathbf{k},\omega)] - \frac{i\omega}{\tau'} . \tag{110}$$

Here $c_0 = \sqrt{g n_{c0}/m}$ is the Bogoliubov phonon velocity [see (28)] and τ' is defined in (80). In the long wavelength limit, one finds [8,43,44]

$$\lim_{k\to 0} -\mathrm{Im}\,\chi^0_{\tilde{n}\tilde{n}}(\mathbf{k},\omega = c_0 k) = \frac{m k_B T}{2\pi c_0}\frac{1}{3}, \tag{111}$$

where the $1/3$ factor arises from the exchange term in the single-particle self-energy. Thus one obtains

$$\omega = c_0 k - i(\Gamma_L + \frac{1}{2\tau'}), \tag{112}$$

where the Landau damping is given by (RPA with exchange)

$$\Gamma_L = \frac{4}{3} a\, k_B T k. \tag{113}$$

The exact result for Γ_L obtained from a careful evaluation of the Beliaev self-energies [25,33,45] at finite T is the same as (113), apart from a slightly smaller numerical coefficient (4/3 is replaced by $3\pi/8$). This difference is due to our neglect of the fluctuations associated with the anomalous pair density $\tilde{m}(\mathbf{r},t)$. Such terms [24] are in the exact equation for $\Phi(\mathbf{r},t)$ given in (53) but were neglected in deriving (63).

To conclude this discussion of different kinds of damping in the collisionless limit, we can also include the effect of the collision terms in (87) as a perturbative correction to the solution determined by $\hat{\mathcal{L}}f = 0$. For a Bose gas *above* T_{BEC}, one might use the simple "relaxation time" approximation [46]

$$C_{22}[f] \simeq -(f - f_0)/\tau_{22} = -\delta f/\tau_{22} \ . \tag{114}$$

The inclusion of such a term in (98) simply involves the change

$$\frac{\partial \delta f}{\partial t} \rightarrow \frac{\partial \delta f}{\partial t} + \frac{\delta f}{\tau_{22}}, \tag{115}$$

and hence one finds (103) again, with the replacement $\omega \rightarrow \omega + i/\tau_{22}$ in the denominator of (104). The collision time τ_{22} is the analogue of the classical expression given in (95). One sees that including $C_{22}[f]$ in the simple relaxation time approximation (114) only slightly alters the Landau damping given by (106). This is quite different from the contribution arising from $C_{12}[f]$ in so far as this enters directly into the generalized GP equation (see Section 3).

We now leave the discussion of the collisionless domain and turn to the collision-dominated hydrodynamic region. In this case, the collisions between the non-condensate atoms are assumed to be sufficiently rapid that the form of $f(\mathbf{p},\mathbf{r},t)$ is largely determined by the requirement that

$$C_{22}[f,\Phi] = 0. \tag{116}$$

That is, the collisions are so strong they force the system to be in *local* equilibrium. The unique solution (denoted by $\tilde{f}$) of the equation (116) is well-known to be given by

$$\tilde{f}(\mathbf{p},\mathbf{r},t) = \frac{1}{e^{\beta[\frac{(\mathbf{p}-m\mathbf{v}_n)^2}{2m}+U(\mathbf{r},t)-\tilde{\mu}(\mathbf{r},t)]} - 1}, \tag{117}$$

where $\mathbf{v}_n$ is the average local velocity and $\tilde{\mu}$ is the local chemical potential of the thermal atoms. This local equilibrium Bose distribution involves the local variables $\beta, \mathbf{v}_n, \tilde{\mu}$ and U, all of which depend on $(\mathbf{r},t)$. Why must $\tilde{f}$ have the form in (117)? To satisfy $C_{22}[\tilde{f}_1] = 0$, we must have [see (88)]

$$(1+\tilde{f}_1)(1+\tilde{f}_2)\tilde{f}_3\tilde{f}_4 - \tilde{f}_1\tilde{f}_2(1+\tilde{f}_3)(1+\tilde{f}_4) = 0, \tag{118}$$

and this requires that f be given by the Bose distribution (117). Here we have used the fact that a Bose distribution satisfies

$$f(x) \equiv \frac{1}{e^x - 1} = -[f(-x) + 1] \tag{119}$$

and energy and momentum conservation

$$\left.\begin{array}{c} \mathbf{p}_1 + \mathbf{p}_2 = \mathbf{p}_3 + \mathbf{p}_4 \\ \\ \tilde{\varepsilon}_{p_1} + \tilde{\varepsilon}_{p_2} = \tilde{\varepsilon}_{p_3} + \tilde{\varepsilon}_{p_4} \end{array}\right\}. \tag{120}$$

However, while the fact that $\tilde{f}$ is given by the local equilibrium Bose distribution in (117) ensures that $C_{22}[\tilde{f}] = 0$, one finds [7,32] that $C_{12}[\tilde{f}] \neq 0$. More precisely, the term in (89) reduces to

$$\begin{aligned} &\left[(1+\tilde{f}_1)\tilde{f}_2\tilde{f}_3 - \tilde{f}_1(1+\tilde{f}_2)(1+\tilde{f}_3)\right] \\ &\propto \left[e^{-\beta[\tilde{\mu}-\mu_c-\frac{1}{2}m(\mathbf{v}_n-\mathbf{v}_c)^2]} - 1\right](1+\tilde{f}_1)\tilde{f}_2\tilde{f}_3, \end{aligned} \tag{121}$$

using energy and momentum conservation. The expression in the square bracket in (121) only vanishes if the condensate and non-condensate are in *diffusive* equilibrium, which requires that the chemical potentials must be equal

$$\tilde{\mu} = \mu_c + \frac{1}{2}m(\mathbf{v}_n - \mathbf{v}_c)^2. \tag{122}$$

When we *perturb* the system, this may not be true, ie, the two components may be out of diffusive equilibrium.

We can now derive hydrodynamic equations for the non-condensate by taking moments of the Boltzmann equation, the standard procedure used in classical gases. [38] The first moment gives a continuity equation with a source term:

$$\int d\mathbf{p}\left\{\mathcal{L}\tilde{f} = C_{12}[\tilde{f}]\right\} \rightarrow \frac{\partial \tilde{n}}{\partial t} = -\nabla\cdot(\tilde{n}\mathbf{v}_n) + \Gamma_{12}[\tilde{f}], \tag{123}$$

where we have used the fact that $C_{22}[\tilde{f}] = 0$ and

$$\begin{aligned} \tilde{n} &\equiv \int \frac{d\mathbf{p}}{(2\pi)^3}\tilde{f}(\mathbf{p},\mathbf{r},t) \\ \tilde{n}\mathbf{v}_n &\equiv \int \frac{d\mathbf{p}}{(2\pi)^3}\frac{\mathbf{p}}{m}\tilde{f}(\mathbf{p},\mathbf{r},t) \\ \Gamma_{12}[\tilde{f},\Phi] &\equiv \int \frac{d\mathbf{p}}{(2\pi)^3}C_{12}[\tilde{f},\Phi]. \end{aligned} \tag{124}$$

More explicitly, we find [compare with (71)]

$$\begin{aligned}\Gamma_{12}[\tilde{f}] &= \frac{2g^2 n_c}{(2\pi)^5}[e^{-\beta[\tilde{\mu}-\mu_c-\frac{1}{2}m(\mathbf{v}_n-\mathbf{v}_c)^2]}-1] \\ &\times \int d\mathbf{p}_1 \int d\mathbf{p}_2 \int d\mathbf{p}_3 \delta(m\mathbf{v}_c+\mathbf{p}_1-\mathbf{p}_2-\mathbf{p}_3) \\ &\times \delta(\varepsilon_c+\tilde{\varepsilon}_1-\tilde{\varepsilon}_2-\tilde{\varepsilon}_3)(1+\tilde{f}_1)\tilde{f}_2\tilde{f}_3 \\ &\equiv \frac{n_c}{\tau_{12}}\left[e^{-\beta[\tilde{\mu}-\mu_c-\frac{1}{2}m(\mathbf{v}_n-\mathbf{v}_c)^2]}-1\right].\end{aligned} \tag{125}$$

We note that τ_{12} is a collision time which describes the C_{12} collisions between the condensate and non-condensate atoms. Adding (123) to the continuity equation in (60)

$$\frac{\partial n_c}{\partial t} = -\nabla\cdot(n_c\mathbf{v}_c) - \Gamma_{12}[\tilde{f},\Phi], \tag{126}$$

we see that the source term Γ_{12} cancels out to give

$$\frac{\partial(n_c+\tilde{n})}{\partial t} = -\nabla\cdot(n_c\mathbf{v}_c+\tilde{n}\mathbf{v}_n). \tag{127}$$

Thus our theory gives the correct continuity equation for the total local density $n = n_c + \tilde{n}$.

Similarly, one finds

$$\begin{aligned}\int d\mathbf{p}\mathbf{p}\left\{\hat{\mathcal{L}}\tilde{f} = C_{12}[\tilde{f}]\right\} &\rightarrow m\tilde{n}\left(\frac{\partial\mathbf{v}_n}{\partial t}+\frac{1}{2}\nabla\mathbf{v}_n^2\right) \\ &= -\nabla\tilde{P}(\mathbf{r},t) - \tilde{n}\nabla U(\mathbf{r},t) - m(\mathbf{v}_n-\mathbf{v}_c)\Gamma_{12}[\tilde{f}],\end{aligned} \tag{128}$$

where the kinetic pressure is given by

$$\tilde{P}(\mathbf{r},t) = \frac{m}{3}\int\frac{d\mathbf{p}}{(2\pi)^3}(\mathbf{p}-m\mathbf{v}_n)^2\tilde{f}(\mathbf{p},\mathbf{r},t). \tag{129}$$

Finally, the second moment gives

$$\begin{aligned}&\int d\mathbf{p}p^2\left\{\mathcal{L}\tilde{f} = C_{12}[\tilde{f}]\right\} \rightarrow \frac{\partial\tilde{P}}{\partial t} + \nabla\cdot(\tilde{P}\mathbf{v}_n) \\ &= -\frac{2}{3}\tilde{P}\nabla\cdot\mathbf{v}_n + \frac{2}{3}\left[\mu_c+\frac{1}{2}m(\mathbf{v}_n-\mathbf{v}_c)^2-U\right]\Gamma_{12}[\tilde{f}].\end{aligned} \tag{130}$$

The detailed derivation of these equations is not important here. It involves straightforward manipulations using the explicit form of $\tilde{f}$ in (117).

The hydrodynamic equations (123), (128) and (130) describe the non-condensate in terms of three new "coarse-grained" local variables:

$$\tilde{n}(\mathbf{r},t), \mathbf{v}_n(\mathbf{r},t) \text{ and } \tilde{P}(\mathbf{r},t).$$

These are coupled to the two additional local variables which describe the condensate:

$$n_c(\mathbf{r},t), \quad \mathbf{v}_c(\mathbf{r},t).$$

We note that the two condensate equations of motion given by (60) are always "hydrodynamic" in form. In contrast, it is only in the collision-dominated region that the non-condensate dynamics can *also* be described in terms of a few collective variables. Both components exhibit *coupled, coherent* collective motions at the same frequency. This is the essence of two-fluid superfluid behaviour, familiar in liquid ^{4}He studies but still an unexplored frontier in trapped Bose gases.

We will now discuss the linearized version of our condensate and non-condensate equations for local equilibrium as given by (60), (122), (128) and (130). We work to first order in the fluctuations around static equilibrium,

$$\begin{aligned} \tilde{n} &= \tilde{n}_0 + \delta\tilde{n}, \quad \mathbf{v}_n = \delta\mathbf{v}_n, \quad \tilde{P} = \tilde{P}_0 + \delta\tilde{P} \\ n_c &= n_{c0} + \delta n_c, \quad \mathbf{v}_c = \delta\mathbf{v}_c\,. \end{aligned} \tag{131}$$

What is *new* about the two-fluid hydrodynamic equations derived above is the role of the source term $\Gamma_{12}[\tilde{f}]$. In a linearized theory expanded around the *static* equilibrium Bose distribution f_0 (where $\Gamma_{12}[f_0,\Phi_0]$ vanishes), one finds

$$\Gamma_{12}[\tilde{f},\Phi] = \delta\Gamma_{12}[\tilde{f},\Phi] = -\frac{\beta_0 n_{c0}}{\tau_{12}^0}\delta\mu_{diff}, \tag{132}$$

where

$$\mu_{diff}(\mathbf{r},t) \equiv \tilde{\mu}(\mathbf{r},t) - \mu_c(\mathbf{r},t). \tag{133}$$

Here $\tau_{12}^0(\mathbf{r})$ is the C_{12} collision time defined in (125) with both components being in static equilibrium (see also (77)). The linearized coupled hydrodynamic equations for the two components are given by the ZGN′ equations [7,32]:

$$\begin{aligned} \frac{\partial \delta n_c}{\partial t} &= -\boldsymbol{\nabla}\cdot(n_{c0}\delta\mathbf{v}_c) - \delta\Gamma_{12} \\ m\frac{\partial \delta\mathbf{v}_c}{\partial t} &= -\boldsymbol{\nabla}\delta\mu_c \end{aligned} \tag{134}$$

and

$$
\begin{aligned}
\frac{\partial \delta \tilde{n}}{\partial t} &= -\nabla \cdot (\tilde{n}_0 \delta \mathbf{v}_n) + \delta \Gamma_{12} \\
m\tilde{n}_0 \frac{\partial \delta \mathbf{v}_n}{\partial t} &= -\nabla \delta \tilde{P} - \delta \tilde{n} \nabla U_0(\mathbf{r}) - 2g\tilde{n}_0[\nabla \delta n_c + \nabla \delta \tilde{n}] \\
\frac{\partial \delta \tilde{P}}{\partial t} &= -\frac{5}{3}\nabla \cdot (\tilde{P}_0 \delta \mathbf{v}_n) + \frac{2}{3}\delta \mathbf{v}_n \cdot \nabla \tilde{P}_0 - \frac{2}{3} g n_{c0}\, \delta \Gamma_{12}\,, \qquad (135)
\end{aligned}
$$

with $\delta\Gamma_{12}$ given by (132) and (within the TF approximation)

$$\delta \mu_c = g\delta n_c + 2g\delta \tilde{n}. \qquad (136)$$

We note that to lowest order, $\delta\Gamma_{12}$ does not appear in the second equation in (135). In addition, in the last term in the third equation in (135), we have used $\mu_{c0} - U_0 = -gn_{c0}$. The condensate couples into the non-condensate equations in (135) directly via the mean field terms $g\delta n_c$ and, indirectly, through $\delta\Gamma_{12}$ (see Section 5).

Despite appearances, this coupled set of hydrodynamic equations form a closed set. We have 9 scalar equations in (134) and (135), while they appear to involve 10 fluctuating variables

$$\delta n_c, \delta \mathbf{v}_c;\ \delta \tilde{n}, \delta \mathbf{v}_n;\ \delta \tilde{P}, \delta \mu_{diff}. \qquad (137)$$

However one can show [7] that $\delta\mu_{diff}$ can be written as a linear combination of the variables $\delta\tilde{P}, \delta\tilde{n}$ and δn_c and thus we are in fact left with only 9 variables. In Section 5, we prove that (134) and (135) are equivalent to the two-fluid hydrodynamic equations of Landau when applied to a trapped Bose gas, but only when the condensate and non-condensate are in diffusive local equilibrium, ie, $\mu_c(\mathbf{r}, t) = \tilde{\mu}(\mathbf{r}, t)$. It will turn out that the rate at which $\delta\mu_{diff}(\mathbf{r}, t)$ relaxes to zero is controlled by a relaxation time τ_μ related to the C_{12} collisions. The Landau two-fluid hydrodynamics [10] is only valid for low frequency oscillations which satisfy $\omega\tau_\mu \ll 1$, as we discuss in Section 5.

5 Two-fluid hydrodynamics in Bose liquids and gases

To put our new two-fluid equations derived in Section 4 into some sort of context, we first briefly review the more general theory developed by Landau in 1941 to explain superfluidity in liquid ^{4}He.

The original discovery of superfluidity in liquid ^{4}He is associated with the famous 1938 papers of Kapitza in Moscow and Allen and Misener at Cambridge. (I cannot resist remarking that Allen and Misener had been graduate students at the University of Toronto where they carried out (with Burton) the pioneering studies on vanishing viscosity in the period 1935-1937). These and subsequent experiments in the next few years [47] showed that superfluid ^{4}He could exhibit very bizarre behaviour compared to ordinary liquids. This led to the development of a two-fluid theory of the hydrodynamic behaviour of liquid ^{4}He by Landau (1941). An earlier but less complete version of Landau's hydrodynamic equations was developed by Tisza in the period 1938-40. For further discussion of this early history, I refer to a recent article of mine. [3]

In this early work, superfluidity (the term was coined in 1938 by Kapitza) was entirely associated with the *relative* motion of the normal fluid and the superfluid components under a variety of conditions. [47] The main point was that while the normal fluid exhibited finite viscosity and thermal conductivity typical of ordinary fluid, the superfluid component (which exhibited irrotational flow) did not. In more recent times, the aspect of superfluidity which has been emphasized (see Ch.4 of Ref. [21] and Leggett's lectures in this book, for example) are those most directly tied to the fact that the superfluid velocity is associated with the gradient of the phase of the macroscopic wavefunction $\Phi(\mathbf{r},t)$. Given that this point of view is indeed more fundamental, it is still crucial to understand why superfluidity persists *even* in the presence of a dissipative normal fluid. This question can be addressed using two-fluid hydrodynamic equations.

In essence, Landau developed his generic two-fluid hydrodynamics by generalizing the standard theory of classical hydrodynamics to include the equations of motion for a new "superfluid" degree of freedom. We recall that classical fluid dynamics was developed well before one knew about the existence of atoms. Since the work of Maxwell and Boltzmann in the 1880's, we know the "coarse-grained" hydrodynamic description of a fluid in terms of a few quantities like $n(\mathbf{r},t)$ and $\mathbf{v}(\mathbf{r},t)$ is only valid when the collisions between atoms are strong enough to produce local equilibrium. [38] It only describes low frequency phenomena, where the condition in (94) is satisfied.

In his 1941 paper, Landau did not connect the superfluid component with the motion of a "Bose condensate." Indeed, he rejected the efforts by Tisza and F. London [3] to use a Bose-condensed gas to get some insight into superfluid ^{4}He. However, since the period 1957-1965, Landau's superfluid degree of freedom has been understood microscopically in terms of the complex order parameter $\Phi(\mathbf{r},t)$. As noted earlier, the superfluid velocity field $\mathbf{v}_s(\mathbf{r},t)$ is related to the gradient of the phase of $\Phi(\mathbf{r},t)$, as given by (19). In the same period, it was also realized that there are two distinct kinds of quantum fluids:

(a) Bose fluids (associated with a Bose condensate wavefunction Φ)

(b) Normal Fermi fluids (associated with the key role of a Fermi surface).

These two kinds of quantum fluids were magnificently described in two books [37,21] by Nozières and Pines written around 1965, although the one on superfluid Bose liquids was only published in 1990 (it was in wide circulation as a preprint before then). I think the clearest account of the connection between $\Phi(\mathbf{r},t)$ and superfluidity in Bose fluids is still the discussion given in Chapters 4 and 5 of Vol II by Nozières and Pines. [21] I highly recommend it to everyone in the BEC field.

Landau's pioneering work on superfluid ^{4}He has two separate aspects which are logically distinct but sometimes confused with each other:

(a) The two-fluid equations describing hydrodynamic behaviour. These equations are generic and apply (under certain conditions) to trapped Bose gases as well as to superfluid ^{4}He. In the period 1947-1950, Landau and Khalatnikov extended these equations to include hydrodynamic damping of the normal fluid (described by various kinds of viscosities, thermal conductivity, etc). This work is all described in the classic 1965 monograph by Khalatnikov. [10]

(b) A "microscopic" theory of the elementary excitations describing the normal fluid of liquid ^{4}He - the famous phonon - roton spectrum. Within his picture of a weakly interacting gas of *quasiparticles*, Landau and coworkers could calculate the thermodynamic and transport properties of superfluid ^{4}He. These quantities enter into the expressions given by the two-fluid equations for the first and second sound modes (velocity and damping). Of course, the roton part of the spectrum is not valid for a dilute Bose gas.

As we have noted, Landau's original formulation of his two-fluid hydrodynamic equations was phenomenological in that the superfluid component was not given an explicit microscopic basis. The first derivation of the Landau hydrodynamic equations starting simply from the existence of the macroscopic order parameter $\Phi(\mathbf{r},t)$ was given by Bogoliubov in 1963. [17] This derivation built on Bogoliubov's earlier work on deriving hydrodynamic equations for classical liquids without going through the intermediate stage of using Boltzmann-like kinetic equations. While Bogoliubov's derivation is often viewed as being quite general, buried in his complex analysis is the assumption that the normal fluid and the superfluid are in local equilibrium with each other. Even today, probably the definitive account which formulates the various levels of theory for Bose superfluids is the classic paper by Hohenberg and Martin [19] published in 1965. It clearly shows (to the patient reader!) the central unifying role of $\Phi(\mathbf{r},t)$, summarizes the collisionless and hydrodynamic domains, and finally gives criteria for developing and judging various approximation schemes using Green's function techniques.

With the preceding discussion as a preamble, I will now summarize the linearized two-fluid hydrodynamic equations of Landau, following the approach given in Ch. 7 of Ref. [21]. As noted, these equations are valid for both superfluid liquids as well as gases. The differences come in only at the last stage when we evaluate the thermodynamic coefficients appearing in these equations, using the appropriate quasiparticle excitation spectrum. [48,49,50] The first two (linearized) Landau equations are familiar from ordinary fluid dynamics

$$\frac{\partial \delta n}{\partial t} + \nabla \cdot \delta \mathbf{j} = 0 \tag{138}$$

$$m\frac{\partial \delta \mathbf{j}}{\partial t} = -\nabla \delta P - \delta n \nabla V_{ex}, \tag{139}$$

where we have included the effect of an external potential. Landau's work incorporated two components and hence the total mass density and mass current fluctuations are given by

$$\delta\rho \equiv m\delta n = \delta\rho_s + \delta\rho_n \tag{140}$$

$$m\delta\mathbf{j} \equiv \rho_{s0}\,\delta\mathbf{v}_s + \rho_{n0}\,\delta\mathbf{v}_n. \tag{141}$$

The new superfluid component was argued to only exhibit pure potential (irrotational) flow and carry no entropy. Thus Landau's final two hydrodynamic equations were new,

$$m\frac{\partial \delta \mathbf{v}_s}{\partial t} = -\nabla \delta\mu \tag{142}$$

$$\frac{\partial \delta s}{\partial t} + \nabla \cdot (s_0 \delta \mathbf{v}_n) = 0\ . \tag{143}$$

Here the local equilibrium entropy density is $s(\mathbf{r},t) = s_0 + \delta s$, the local equilibrium pressure is $P(\mathbf{r},t) = P_0 + \delta P$ and the local chemical potential is $\mu(\mathbf{r},t) = \mu_0 + \delta\mu(\mathbf{r},t)$. We note that the Landau hydrodynamic equations only involves 8 equations, in contrast to the 9 equations we derived at the end of Section 4. In particular, we see that Landau does not have *separate* continuity equations for the superfluid $\rho_s(\mathbf{r},t)$ and normal fluid $\rho_n(\mathbf{r},t)$ densities. Finally, it is assumed that these two components are always in local equilibrium with each other and hence the fluctuations are related by the thermodynamic identity

$$n_0\delta\mu = -s_0\delta T + \delta P, \tag{144}$$

where n_0 is the equilibrium total density.

For simplicity, we now specialize our analysis to a uniform superfluid, where $\rho_{s0}, \rho_{n0}, s_0$ and other thermodynamic quantities are all position-independent. Multiplying (142) by n_{s0} gives [using (144)]

$$\begin{aligned}\rho_{s0}\frac{\partial\delta\mathbf{v}_s}{\partial t} &= -n_{s0}\boldsymbol{\nabla}\delta\mu \\ &= -\frac{n_{s0}}{n_0}\boldsymbol{\nabla}(-s_0\delta T + \delta P) \\ &= -\frac{n_{s0}}{n_0}\boldsymbol{\nabla}\delta P + \frac{n_{s0}}{n_0}s_0\boldsymbol{\nabla}\delta T.\end{aligned} \tag{145}$$

Using this in (138) gives

$$\begin{aligned}-\boldsymbol{\nabla}\delta P &= \rho_{s0}\frac{\partial\delta\mathbf{v}_s}{\partial t} + \rho_{n0}\frac{\partial\delta\mathbf{v}_n}{\partial t} \\ &= -\frac{n_{s0}}{n_0}\boldsymbol{\nabla}\delta P + \frac{n_{s0}}{n_0}s_0\boldsymbol{\nabla}\delta T + \rho_{n0}\frac{\partial\mathbf{v}_n}{\partial t}\end{aligned}$$

or

$$\rho_{n0}\frac{\partial\delta\mathbf{v}_n}{\partial t} = -\frac{n_{n0}}{n_0}\boldsymbol{\nabla}\delta P - \frac{n_{s0}}{n_0}s_0\boldsymbol{\nabla}\delta T. \tag{146}$$

Combining (138) and (139) gives

$$\frac{\partial^2\delta\rho}{\partial t^2} = -m\boldsymbol{\nabla}\cdot\frac{\partial\delta\mathbf{j}}{\partial t} = \nabla^2\delta P. \tag{147}$$

Finally, (143) gives [using (146)]

$$\begin{aligned}\frac{\partial^2\delta s}{\partial t^2} &= -s_0\boldsymbol{\nabla}\cdot\frac{\partial\delta\mathbf{v}_n}{\partial t} \\ &= \frac{s_0}{\rho_0}\nabla^2\delta P + \frac{s_0^2}{\rho_0}\left(\frac{\rho_{s0}}{\rho_{n0}}\right)\nabla^2\delta T \\ &= \frac{s_0}{\rho_0}\frac{\partial^2\delta\rho}{\partial t^2} + \frac{s_0^2}{\rho_0}\left(\frac{\rho_{s0}}{\rho_{n0}}\right)\nabla^2\delta T.\end{aligned} \tag{148}$$

The last equation gives the local entropy fluctuations in terms of the local mass density $\delta\rho$ and temperature δT fluctuations. Eq.(148) can be re-written [21] in terms of the local entropy per unit mass $\bar{s}(\mathbf{r},t) \equiv s(\mathbf{r},t)/\rho(\mathbf{r},t)$. Using

$$\delta\bar{s} = -\frac{s_0}{\rho_0^2}\delta\rho + \frac{1}{\rho_0}\delta s, \tag{149}$$

(148) takes on the simpler form

$$\frac{\partial^2 \delta\bar{s}}{\partial t^2} = \bar{s}_0^2\left(\frac{\rho_{n0}}{\rho_{s0}}\right)\nabla^2\delta T\,. \tag{150}$$

Expanding δP and δT in terms of $\delta\rho$ and $\delta\bar{s}$ fluctuations

$$\partial P = \left.\frac{\partial P}{\partial\rho}\right|_{\bar{s}}\delta\rho + \left.\frac{\partial P}{\partial\bar{s}}\right|_{\rho}\delta\bar{s}$$

$$\delta T = \left.\frac{\partial T}{\partial\rho}\right|_{\bar{s}}\delta\rho + \left.\frac{\partial T}{\partial\bar{s}}\right|_{\rho}\delta\bar{s}\,, \tag{151}$$

we see that (150) and (147) reduce to two coupled scalar equations for $\delta\rho$ and $\delta\bar{s}$. Inserting the normal mode solutions

$$\delta\rho, \delta\bar{s} \propto e^{i(\mathbf{k}\cdot\mathbf{r}-\omega t)}, \tag{152}$$

one finds these coupled algebraic equations have two phonon solutions $\omega^2 = u^2k^2$, where u^2 is the solution of the quadratic equation:

$$u^4 - u^2\left[\left.\frac{\partial P}{\partial\rho}\right|_T + \frac{T}{\bar{c}_v}\left(\frac{1}{\rho_0}\left.\frac{\partial P}{\partial T}\right|_\rho\right)^2 + \frac{\rho_{s0}}{\rho_{n0}}\frac{T\bar{s}_0^2}{\bar{c}_v}\right] + \frac{\rho_{s0}}{\rho_{n0}}\left(\frac{T\bar{s}_0^2}{\bar{c}_v}\right)\left.\frac{\partial P}{\partial\rho}\right|_T = 0. \tag{153}$$

Here $\bar{c}_v = \left.\frac{\partial\bar{s}}{\partial T}\right|_\rho$ is the equilibrium specific heat per unit mass.

The coefficients in (153) are daunting, but only involve equilibrium thermodynamic quantities which can be calculated for any given superfluid. We emphasize that (153) is valid for both a uniform Bose-condensed gas and superfluid ^{4}He, assuming that the superfluid and normal fluid are in local equilibrium with each other. However, as we shall show, the detailed characteristics and behaviour of the two phonon modes (first and second sound) are quite different in a Bose gas and in superfluid ^{4}He. [48,49]

A key feature about superfluid ^{4}He is that (typical of any liquid) $\partial P/\partial T|_\rho \simeq 0$. This can be shown to be equivalent to $C_p \simeq C_v$, where $C_{p,v}$ are the specific heats at constant pressure or constant volume. In this case, (153) reduces to

$$u^4 - u^2(A+B) + AB = 0, \tag{154}$$

with two solutions $u^2 = A, B$:

$$u_1^2 = \frac{\partial P}{\partial \rho} \text{ : first sound} \tag{155}$$

$$u_2^2 = \frac{\rho_{s0}}{\rho_{n0}} \left(\frac{T\bar{s}_0^2}{\bar{c}_v} \right) \text{ : second sound.} \tag{156}$$

Working out the associated motions, one finds the first sound mode ($\omega = u_1 k$) involves the in-phase motion of the superfluid and normal fluid components. Moreover, one can show that first sound is essentially a pressure wave. In contrast, the second sound mode ($\omega = u_2 k$) involves the out-of-phase motion of the two components, with $\delta \mathbf{j} \simeq 0$ or

$$\rho_{n0}\, \delta \mathbf{v}_n = -\rho_{s0}\, \delta \mathbf{v}_s \, . \tag{157}$$

This corresponds to an almost pure temperature wave. The successful detection in 1946 of a second sound mode (or temperature wave) was of tremendous significance in low temperature physics. [21,47] The good agreement of the measured second sound velocity u_2 with the Landau expression in (156), calculated using the postulated phonon-roton quasiparticle spectrum, vindicated *both* aspects of the Landau theory of superfluid ^{4}He.

In contrast, in a (Bose-condensed) gas, the pressure and temperature fluctuations are strongly coupled and hence $\partial P/\partial T|_\rho$ in (153) plays a significant role. We recall that in a classical gas, one has $C_p/C_v = \frac{5}{3}$. Evaluating all the thermodynamic derivatives using our finite T microscopic model (the single-particle HF spectrum in (59) replaces the roton spectrum of superfluid ^{4}He) and working to lowest order in $gn_c/k_B T$, one finds (after very lengthy calculations)

$$u_1^2 = \frac{5}{3} \frac{k_B T}{m} \frac{g_{5/2}(z_0 = 1)}{g_{3/2}(z_0 = 1)} + O(gn) \tag{158}$$

$$u_2^2 = \frac{gn_{c0}}{m} + \dots \tag{159}$$

It turns out that the first sound mode in a Bose gas is largely an oscillation of the thermal cloud (the normal fluid). The second sound mode is largely an oscillation of the condensate (the superfluid). Both involve density fluctuations and thus will have significant weight in the dynamic structure factor related to the density response function. The quite different features of first and second sound in Bose gases vs superfluid ^{4}He should be remembered when reading standard texts about superfluid ^{4}He. Results equivalent to (158) and (159) are given in Refs. [48,49,50,51].

Comparing (159) with (28), it is clear that in a uniform Bose-condensed gas, the *hydrodynamic* second sound mode at finite T smoothly extrapolates to the Bogoliubov phonon mode in the *collisionless* region. The fact that the first sound velocity u_1 does not depend on the interaction strength g to lowest order is typical of ordinary sound waves in any gas. However, one must remember that interactions (collisons) play a crucial indirect role in enforcing dynamic local equilibrium.

We now turn to a discussion of the generalized two-fluid hydrodynamic equations we derived at the end of Section 4. Again, for simplicity, we limit our analysis to a uniform Bose-condensed gas. If we can ignore vorticity in both fluids, we can introduce two velocity potentials

$$\begin{aligned} \delta \mathbf{v}_c &\equiv \nabla \phi_c(\mathbf{r}, t) \\ \delta \mathbf{v}_n &\equiv \nabla \phi_n(\mathbf{r}, t), \end{aligned} \tag{160}$$

and reduce the equations in (134) and (135) to three equations for ϕ_c, ϕ_n and $\delta\mu_{diff}$ [the latter is defined in (133)]:

$$m\frac{\partial^2 \phi_c}{\partial t^2} = g n_{c0} \nabla^2 \phi_c + 2g\tilde{n}_0 \nabla^2 \phi_n + \frac{\sigma_H}{\tau_\mu}\delta\mu_{diff} \tag{161}$$

$$m\frac{\partial^2 \phi_n}{\partial t^2} = \left(\frac{5}{3}\frac{\tilde{P}_0}{\tilde{n}_0} + 2g\tilde{n}_0\right)\nabla^2\phi_n + 2g n_{c0}\nabla^2\phi_c - \frac{2}{3}\frac{\sigma_H}{\tau_\mu}\delta\mu_{diff} \tag{162}$$

$$\frac{\partial \delta\mu_{diff}}{\partial t} = \frac{2}{3} g n_{c0}\nabla^2\phi_n - g n_{c0}\nabla^2\phi_c - \frac{\delta\mu_{diff}}{\tau_\mu}. \tag{163}$$

Here τ_μ is a new relaxation time governing how $\delta\mu_{diff}$ relaxes to 0 (ie, $\tilde{\mu} \to \mu_c$). It is found to be related to the C_{12} collision time τ' defined in (80) by

$$\frac{1}{\tau_\mu} = \frac{1}{\sigma_H \tau'}, \tag{164}$$

where the dimensionless hydrodynamic renormalization factor σ_H is given by [7,32]

$$\sigma_H \equiv \frac{\frac{5}{2}\tilde{\gamma}_0\tilde{P}_0 - \frac{3}{2}g\tilde{n}_0^2}{\frac{5}{2}\tilde{P}_0(1-\tilde{\gamma}_0) + 2g\tilde{n}_0 n_{c0} + \frac{2}{3}g\tilde{\gamma}_0 n_{c0}^2 + \frac{3}{2}g\tilde{n}_0^2}. \tag{165}$$

For completeness, we recall that

$$\begin{aligned} \tilde{P}_0(z_0) &= \frac{k_B T}{\Lambda_0^3} g_{5/2}(z_0 = e^{-\beta g n_{c0}}) \\ \tilde{n}_0(z_0) &= \frac{1}{\Lambda_0^3} g_{3/2}(z_0) \end{aligned}$$

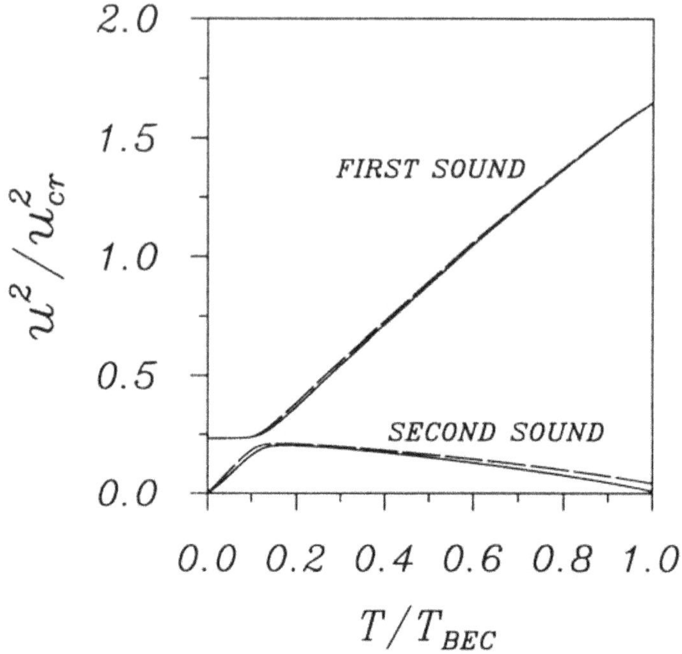

Figure 5. Squares of the first and second sound velocities (normalized by the first sound velocity of the ideal gas at $T = T_{BEC}$) vs. T/T_{BEC}. The solid lines are for the Landau limit $\omega\tau_\mu \ll 1$ while the dashed lines are for $\omega\tau_\mu \gg 1$. The curves were generated for $gn/k_BT_{BEC} = 0.2$. The interesting crossover at low temperatures was first pointed out in Ref. [49]. It is reproduced by our equations, even though this region is outside the region of validity of our model. See Refs. [7,48].

$$\tilde{\gamma}_0 = \frac{g}{k_BT\Lambda_0^3} g_{1/2}(z_0), \tag{166}$$

where the Bose-Einstein functions are $g_n(z) \equiv \sum_{l=1}^{\infty} z^l/l^n$ and Λ_0 is the equilibrium thermal de Broglie wavelength. Calculation shows that σ_H becomes large as $T \to T_{BEC}$ and thus τ_μ can become very large near the superfluid transition. This is not in contradiction with the fact that the present discussion is for the collision-dominated region [$\omega\tau_{22} \ll 1$, where τ_{22} is some relaxation time associated with the C_{22} collision integral in (88)].

It is clear from the fact that we have 3 coupled equations for the 3 variables ϕ_c, ϕ_n and $\delta\mu_{diff}$, we will obtain a new mode, in addition to the usual first and second sound oscillations discussed earlier. It is convenient to first eliminate

$\delta\mu_{diff}$ using the solution of (163),

$$-i\omega\delta\mu_{diff} = \frac{2}{3}gn_{c0}(-k^2)\phi_n - gn_{c0}(-k^2)\phi_c - \frac{\delta\mu_{diff}}{\tau_\mu}$$

or

$$\delta\mu_{diff} = \frac{gn_{c0}\tau_\mu}{1-i\omega\tau_\mu}(\phi_c - \frac{2}{3}\phi_n)k^2. \tag{167}$$

Using this in (161) and (162) gives

$$m\omega^2\phi_c = gn_{c0}\left(1 - \frac{\sigma_H}{1-i\omega\tau_\mu}\right)k^2\phi_c + 2g\tilde{n}_0\left(1 + \frac{\sigma_H}{3(1-i\omega\tau_\mu)}\frac{n_{c0}}{\tilde{n}_0}\right)k^2\phi_n \tag{168}$$

$$m\omega^2\phi_n = \left(\frac{5}{3}\frac{\tilde{P}_0}{\tilde{n}_0} + 2g\tilde{n}_0\left[1 - \frac{2\sigma_H}{9(1-i\omega\tau_\mu)}\frac{n_{c0}^2}{\tilde{n}_0^2}\right]\right)k^2\phi_n$$
$$+ 2gn_{c0}\left(1 + \frac{\sigma_H}{3(1-i\omega\tau_\mu)}\frac{n_{c0}}{\tilde{n}_0}\right)k^2\phi_c \,. \tag{169}$$

These two coupled equations for ϕ_n and ϕ_c are easily solved and we obtain (as expected) first and second sound modes. Clearly the velocities u_1 and u_2 will now depend on the value of $\omega\tau_\mu$, although it turns out that this dependence is not very strong. This is shown by the results in Fig. 5.

The $\omega\tau_\mu \to 0$ limit is of special interest since the equations (168) and (169) can then be shown to be completely equivalent to the predictions of the Landau two-fluid equations discussed earlier in this section. To be precise, the first and second sound velocities $u_{1,2}^2$ obtained from (168) and (169) in the limit $\omega\tau_\mu \to 0$ agree with those given by (153). This equivalence makes physical sense since in the limit of $\tau_\mu \to 0$, one sees that $\delta\mu_{diff} \to 0$ very rapidly [see (163) and (167)]. Thus we have proven that the Landau two-fluid hydrodynamic equations are correct if the condensate and thermal cloud are in diffusive local equilibrium. This, as noted earlier, is an (sometimes implicit) assumption in all previous derivations of the Landau equations. [17,51]

This Landau limit ($\omega\tau_\mu \to 0$) is, in fact, very subtle in the context of our microscopic calculation. We see that in this limit, there are still correction terms in (168) and (169) which are proportional to the hydrodynamic renormalization factor σ_H as defined in (165). These terms are crucial in ensuring that (168) and (169) reproduce the results of the Landau two-fluid equations. Another way of seeing this is that even though $\delta\mu_{diff} \to 0$ when $\tau_\mu \to 0$, we note that $\delta\Gamma_{12}$ in (132) is still finite. Using (167), one finds

$$\delta\Gamma_{12} = -n_{c0}\sigma_H\left(\phi_c - \frac{2}{3}\phi_n\right)k^2. \tag{170}$$

In conclusion, one might say that the hydrodynamic renormalization factor σ_H contains a key part of the physics buried in the hydrodynamic two-fluid equations of Landau.

In the opposite limit $\omega\tau_\mu \gg 1$ (which can arise near T_{BEC}) we see all the terms proportional to σ_H in (168) and (169) are negligible. This domain is missed in the Landau two-fluid equations. It describes situations in which the condensate and thermal cloud are out of diffusive equilibrium with each other. It would be of great interest to look for this kind of phenomenon in trapped Bose gases.

One can work out the frequency of the new mode associated with the dynamics of $\delta\mu_{diff}$, and for a uniform gas it is well approximated by

$$\omega_R \simeq -i/\tau_\mu. \tag{171}$$

Thus, in general, our two-fluid hydrodynamic equations predict the existence of a relaxational mode peaked at zero frequency. One can improve the theory to include deviations from local equilibrium [ie, deviations of f from $\tilde{f}$ in (116)]. This involves a Chapman-Enskog kind of calculation familiar in the theory of classical gases and gives rise to hydrodynamic damping. [38] The normal fluid (non-condensate) equations of motion have new terms corresponding to shear viscosity (η) and thermal conductivity (κ) transport coefficients [51,52], where we recall that κ and η are proportional to some τ_{22} collision time and hence go as $1/g^2$. The damping of first sound, second sound and the relaxational mode due to small deviations from local equilibrium among the thermal atoms ($C_{22}[f,\Phi]$) has been recently worked out in detail by Nikuni, Zaremba and the author. [52] Of particular interest is the effect on the relaxational mode, which is now described by [compare with (171)]

$$\omega_R \simeq -i\left[\frac{1}{\tau_\mu} + A\kappa k^2\right]. \tag{172}$$

Effectively the relaxation mode is strongly coupled into thermal conduction processes (or flow of heat). The expression for the coefficient A in (172) is somewhat complicated [52] but above $T_{\rm BEC}$ ($\frac{1}{\tau_\mu} \to 0$), we find the mode reduces to the well-known thermal diffusion mode,

$$\omega_R = -i\frac{\kappa k^2}{n_0 C_p} = -iD_T k^2 . \tag{173}$$

This result strongly suggests that our new relaxational mode below $T_{\rm BEC}$ is the renormalized version of the usual thermal diffusion mode above $T_{\rm BEC}$.

We recall that in the standard two-fluid hydrodynamic theory of Landau, the thermal diffusion mode disappears below $T_{\rm BEC}$, with its spectral weight

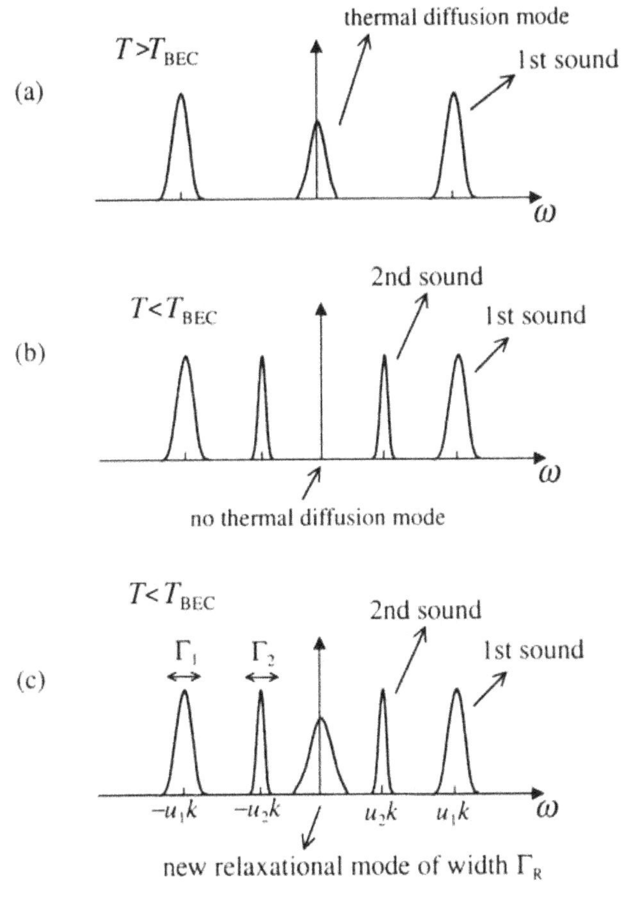

Figure 6. A schematic illustration of the predicted hydrodynamic mode spectrum for a uniform Bose gas. (a) Above T_{BEC}, one has the usual first sound mode and a zero-frequency thermal diffusion mode. (b) Below T_{BEC}, the standard Landau-Khalatnikov two-fluid hydrodynamics predicts first and second sound, but no remnant of any zero-frequency mode. (c) In the ZGN′ theory presented in Sections 4 and 5, one has first and second sound modes as well as a zero-frequency relaxational mode. Based on Ref. [52].

going into the emerging second sound doublet. In Fig. 6, we schematically illustrate the different hydrodynamic mode spectra predicted for above and below T_{BEC}.

To illustrate the essential physics, the detailed discussion in this section has been limited to a uniform Bose-condensed gas. The analysis can, of course, be extended to trapped Bose gas. In particular, the out-of-phase dipole mode [53,7] in the hydrodynamic limit is of special interest. This corresponds to centre-of-mass oscillation of the equilibrium condensate and non-condensate density profiles,

$$\begin{aligned} n_c(\mathbf{r},t) &= n_{c0}(\mathbf{r}-\boldsymbol{\eta}_c(t)) \\ \tilde{n}(\mathbf{r},t) &= \tilde{n}_0(\mathbf{r}-\boldsymbol{\eta}_n(t)), \end{aligned} \tag{174}$$

with

$$\mathbf{v}_n = \frac{d\boldsymbol{\eta}_n}{dt} \; ; \; \mathbf{v}_c = \frac{d\boldsymbol{\eta}_c}{dt} \, . \tag{175}$$

One finds that the hydrodynamic equations give (as expected) an undamped *in-phase* normal mode, with $\mathbf{v}_n = \mathbf{v}_c$ and $\omega = \omega_0$, where ω_0 is the trap frequency. This is a generalized version of the $T = 0$ Kohn mode discussed at the end of Section 2. In addition, however, there is an *out-of-phase* mode satisfying

$$N_c\mathbf{v}_c = -\tilde{N}\mathbf{v}_n, \tag{176}$$

with a frequency different from the trap frequency ω_0. The frequency of this mode is shown in Fig. 7. Calculation shows that this *out-of-phase* mode is only damped by its coupling to the relaxational mode given by (171). There is no hydrodynamic-type damping from the finite transport coefficients of the normal gas. Calculations are in reasonable agreement with the observed damping of such an out-of-phase mode. [31] A careful study of this out-of-phase dipole mode as a function of the temperature would be a nice way of probing the unusual hydrodynamics of a Bose-condensed gas.

In these lectures, I have not made any attempt to analyze the available experimental investigations [11,12,30,31] of collective modes at finite temperatures. These pioneering experimental studies are very promising but we need more systematic investigations, especially as a function of temperature and density. A key reason why such studies are perhaps a unique way of probing the many-body dynamics of superfluid gases [54] is that one can measure collective mode frequencies very accurately (at $T = 0$, with errors of only a few percent).

In this section, we have put emphasis on deriving two-fluid hydrodynamic equations by starting from an approximate but still microscopic model. Such

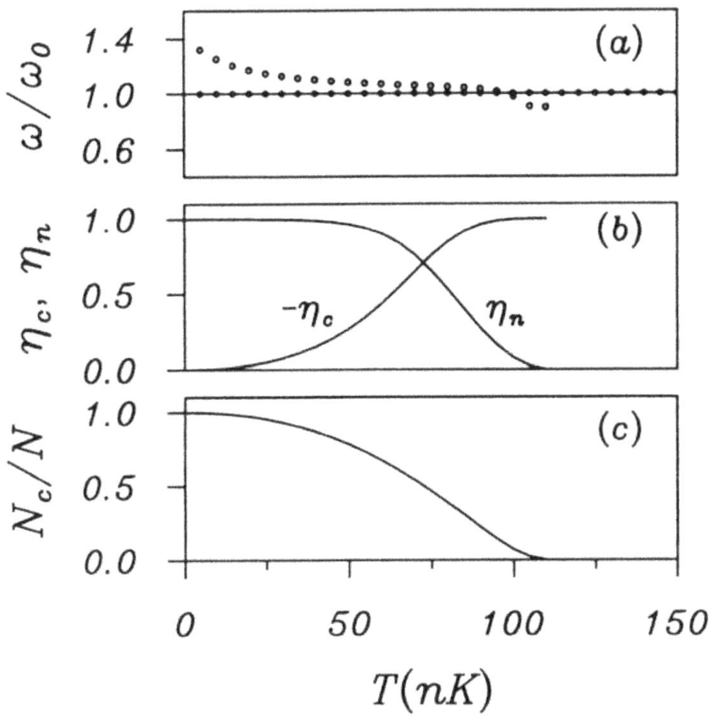

Figure 7. (a) Mode frequencies for the in-phase (solid dots) and out-of-phase (open dots) dipole modes vs. temperature for 2000 Rb atoms in an isotropic parabolic trap (see Ref. 13 for values of the physical parameters used). (b) Condensate (η_c) and non-condensate (η_n) amplitudes for the out-of-phase dipole mode. (c) Fraction of atoms in the condensate as a function of temperature. From Ref. [53].

a derivation has been carried out recently by Zaremba, Nikuni and the author [7] for a trapped Bose gas. The pioneering work on deriving two-fluid hydrodynamic equations for a uniform dilute Bose gas was by Kirkpatrick and Dorfman. [51] In this regard, it is useful to point out that the linearized Landau two-fluid hydrodynamic equations [as given by (138) - (143)] are expected to be *exact* as long as the two components are in local equilibrium. Thus, for a uniform system, the exact first and second sound velocities are given by the solutions of (153). The only question is how to calculate the various thermodynamic functions which are involved in this equation. The analogous Landau equations for a trapped gas allows us to go past the simplified microscopic models we have used in Section 4 to derive equations of this kind. In particular, one should be able to use the Landau two-fluid equations as a direct probe of the superfluid density, just as one does in superfluid ^{4}He. [21,47] One might be able to detect small differences between the magnitude of the superfluid density n_{s0} and the condensate density n_{c0} (at the level of our model in Section 4, n_{s0} and n_{c0} are equal). In this connection, we note that in the formal zero temperature limit, the Landau two-fluid equations reduce to two coupled equations for a *pure* superfluid ($\rho_s = \rho, \rho_n = 0$ at $T = 0$). This limit has been used [54] to find a more accurate version of the $T = 0$ quantum hydrodynamic equations discussed in Section 2.

In conclusion, I hope I have given some insight into why the two-fluid hydrodynamics of trapped Bose gases has so much potential interest. I hope some of the young experimentalists attending this Summer School will take up the challenge to study this new frontier. In a sense, the natural next step after understanding the dynamics of a pure condensate ($T = 0$) is to study the two-fluid hydrodynamics of trapped gases (at finite T) since now one has *two* components which can execute coupled coherent collective motions. In contrast, the collisionless region at finite T as discussed in Section 3 does not seem as interesting since the thermal cloud has always such a low density (see Fig. 2). As a result, mean field effects produced by the non-condensate are never very important [46,55], above or below $T_{\rm BEC}$, and thus no new many-body dynamics emerges which is different than already found at $T = 0$. What is new is the different mechanisms of damping of collisionless condensate motion, as discussed in Sections 3 and 4.

Acknowledgments

I am grateful to Craig Savage and Mukunda Das for organizing this Australian BEC Summer School, for inviting me to participate, and their hospitality and assistance. Much of the work reviewed in these lectures was carried out with Eugene Zaremba, Tetsuro Nikuni, Jamie Williams and Milena Imamović-Tomasović. I would like to thank Jamie Williams for a critical reading and help with preparing the manuscript. I also thank Helen Iyer for texing the manuscript in record time. My research is supported by NSERC of Canada.

References

1. F. Dalfovo, S. Giorgini, L.P. Pitaevskii and S. Stringari, Rev. Mod. Phys. **71**, 463 (1999).
2. *Bose-Einstein Condensation in Atomic Gases*, Proceedings of the International School of Physics "Enrico Fermi", ed. M. Inguscio, S. Stringari and C. E. Wieman (IOS Press, Amsterdam, 1999).
3. F. London, Nature **141**, 643 (1938); see also the historical overview by A. Griffin in Ref. [2], p.1.
4. N.N. Bogoliubov, J. Phys. U.S.S.R. **11**, 23 (1947).
5. S.T. Beliaev, Sov. Phys.—JETP **7**, 289 (1958).
6. S. Stringari, Phys. Rev. Lett. **77**, 2360 (1996).
7. E. Zaremba, T. Nikuni and A. Griffin, Journ. Low Temp. Phys. **116**, 227 (1999).
8. J.E. Williams and A. Griffin, cond-mat/0003481.
9. L.D. Landau, J. Phys. U.S.S.R. **5**, 71 (1941).
10. I.M. Khalatnikov, *An Introduction to the Theory of Superfluidity* (W. A. Benjamin, N. Y., 1965).
11. E.A. Cornell, J.R. Ensher and C.E. Wieman, in Ref. [2], p.15.
12. W. Ketterle, D.S. Durfee and D.M. Stamper-Kurn, Ref. [2], p.67
13. D.A.W. Hutchinson, E. Zaremba and A. Griffin, Phys. Rev. Lett. **78**, 1842 (1997).
14. M. Edwards, P.A. Ruprecht, K. Burnett, R.J. Dodd and C.W. Clark, Phys. Rev. Lett. **77**, 1671 (1996).
15. A.L. Fetter, in Ref. [2], p.201.
16. A. Svidzinsky and A.L. Fetter, Phys. Rev. A **58**, 3168 (1998).
17. N.N. Bogoliubov, *Lectures on Quantum Statistics*, Vol. 2 (Gordon and Breach, N.Y., 1970), p.148.
18. J. Gavoret and P. Nozières, Ann. Phys. (N.Y.) **28**, 349 (1964).
19. P.C. Hohenberg and P.C. Martin, Ann. Phys. (N.Y.) **34**, 291 (1965).
20. See, for example, P.W. Anderson, Rev. Mod. Phys. **38**, 298 (1996); P.W. Anderson in *A Career in Theoretical Physics* (World Scientific, Singapore. 1994).
21. P. Nozières and D. Pines, *The Theory of Quantum Liquids*, Vol. II: *Superfluid Bose Liquids* (Addison-Wesley, Redwood City, 1990).
22. A. Griffin, Phys. Rev. B **53**, 9341 (1996).
23. M. Imamović-Tomasović and A. Griffin, Phys. Rev. A **60**, 494 (1999).
24. S. Giorgini, Phys. Rev. A **57**, 2949 (1998); S. Giorgini, cond-mat/9911377.
25. Hua Shi and A. Griffin, Physics Reports **304**, 1 (1998).

26. A. Minguzzi and M.P. Tosi, J.Phys. Cond. Mat. **9**, 10211 (1997).
27. R.J. Dodd, M. Edwards, C.W. Clark and K. Burnett, Phys. Rev. A **57**, R32 (1998).
28. H.T.C. Stoof, Journ. Low Temp. Phys. **114**, 11 (1999).
29. R. Walser, J. Williams, J. Cooper and M. Holland, Phys. Rev. A **59**, 3878 (1999).
30. D.S. Jin *et al.*, Phys. Rev. Lett. **78**, 764 (1997).
31. D.M. Stamper-Kurn *et al.*, Phys. Rev. Lett. **81**, 500 (1998).
32. T. Nikuni, E. Zaremba and A. Griffin, Phys. Rev. Lett. **83**, 10 (1999).
33. P.O. Fedichev, G.V. Shlyapnikov and J.T.M. Walraven, Phys. Rev. Lett. **80**, 2269 (1998); P.O. Fedichev and G.V. Shlyapnikov, Phys. Rev. A **58**, 3146 (1998).
34. S. Giorgini, L.P. Pitaevskii and S. Stringari, J. Low Temp. Phys. **109**, 309 (1997).
35. C.W. Gardiner and P. Zoller, Phys. Rev. A **61**, 033601 (2000), and references therein.
36. M.J. Bijlsma, E. Zaremba and H.T.C. Stoof, cond-mat/0001323.
37. D. Pines and P. Nozières, *The Theory of Quantum Liquids*, Vol. I: *Normal Fermi Liquids* (W.A. Benjamin, N.Y., 1966).
38. For a nice introduction to kinetic equations in classical gases, see K. Huang, *Statistical Mechanics*, second ed. (J. Wiley, N.Y., 1987).
39. L.P. Kadanoff and G. Baym, *Quantum Statistical Mechanics*, (W.A. Benjamin, N.Y., 1962).
40. S. T. Beliaev, Sov. Phys.—JETP **7**, 299 (1958).
41. M. Guilleumas and L.P. Pitaevskii, Phys. Rev. A **61**, 013602 (2000).
42. S. A. Morgan, cond-mat/9911278.
43. J. Reidl, A. Csordás, R. Graham and P. Szépfalusy, Phys. Rev. A **61**, 043606 (2000).
44. P. Szépfalusy and I. Kondor, Ann. Phys. **82**, 1 (1974).
45. L.P. Pitaevskii and S. Stringari, Phys. Lett. A **235**, 398 (1997), and references therein.
46. G.M. Kavoulakis, C.J. Pethick and H. Smith, Phys. Rev. A **61**, 053603 (2000); U.A. Khawaja, C.J. Pethick and H. Smith, Journ. Low Temp. Phys. **118**, 127 (2000).
47. The classic account is given by J. Wilks, *The Properties of Liquid and Solid Helium*, (Oxford, 1967).
48. A. Griffin and E. Zaremba, Phys. Rev. A **56**, 4839 (1997).
49. T.D. Lee and C.N. Yang, Phys. Rev. **113**, 1406 (1959).
50. V.B. Shenoy and T.L. Ho, Phys. Rev. Lett. **80**, 8895 (1998).
51. T.R. Kirkpatrick and J.R. Dorfman, Phys. Rev. A **28**, 2576 (1983);

Journ. Low Temp. Phys. **58**, 399 (1985).
52. T. Nikuni, A. Griffin and E. Zaremba, cond-mat/9911336.
53. E. Zaremba, A. Griffin and T. Nikuni, Phys. Rev. A **57**, 4695 (1998).
54. L.D. Pitaevskii and S. Stringari, Phys. Rev. Lett. **81**, 4541 (1998).
55. D. Guery-Odelin, F. Zambelli, J. Dalibard, and S. Stringari, Phys. Rev. A **60**, 4851 (1999).

FINITE-TEMPERATURE EFFECTS IN BOSE-EINSTEIN CONDENSATES

KEITH BURNETT

Department of Physics, University of Oxford, Clarendon Laboratory, Parks Road, Oxford, OX1 3PU, United Kingdom

E-mail: k.burnett1@physics.ox.ac.uk

Web: http://www-kbg.physics.ox.ac.uk/

In this article we discuss aspects of fluctuations and correlations that arise in the theory of trapped condensates at finite temperature. This requires us to look at both thermal and quantum fluctuations in the field that describes the condensates. We discuss ways in which these effects can be observed in experiments.

1 Introduction

In these lectures we shall discuss fluctuations, correlations and entanglements in Bose-Einstein Condensed gases. Fluctuations are defined relative to the mean-field description of condensates and arise because of both thermal and quantal fluctuations in the quantum field we use to describe condensed samples of atoms. The thermal fluctuations require us to consider a quantum field at finite temperature: a subject of considerable importance in a wide area of science. Quantum fluctuations are closely related to the correlations between particles that are ignored in the simplest mean-field theories. The description of correlations between particles in condensates links the theory of dilute gases at finite temperature to that of quantum information processing.

In the first section of the notes we will discuss how the binary correlations between atoms define the effective interactions between them. To understand this we shall need to look at the wavefunction of the relative motion of pairs of atoms at ultra-cold energies. These binary correlations occur on much shorter length and time scales than those on which we observe condensates in the lab. They do not, therefore, appear explicitly in the mean-field theory of condensate evolution. They do however effect the strength of effective contact interaction that we use in the mean field equations. We shall show how the correlations between particles on longer length scales are modified in the presence a condensate. This is a small modification in three dimensions but in two it produces a crucial modification to the strength of the effective interaction. The change in the correlation between the particles is due to the fact that they are colliding in a medium, and we shall show how this is included in the theory of condensates.

We shall explain briefly how much of the evolution of a BEC far from equilibrium can be treated using the Gross-Pitaevskii equation (GPE). This approach is complementary to that of quantum kinetics and allows us to study the evolution of structure formed in condensates after they have passed through the BEC transition region and formed non-equilibrium condensates.

The theory of collective excitations in a trapped gas is now at the same level of sophistication as that reached for the homogeneous gas and we shall give an outline of this important work. We now have systematic methods for calculating the shifts and widths of collective modes at finite temperature in a trapped gas. Comparison of the predictions of this theory with experiment will provide crucial tests of finite temperature field theory. This theory will also, in our view, form the basis of thermometry in the nK regime for condensed gases.

The proper definition of the phase of a condensate has been the subject of some debate. We shall explain the role of entanglement in the definition of the phase of a condensate relative to a phase standard. The entanglement we meet in the definition of a phase standard is robust to decoherence and is not a useful "quantum resource". By that we mean it is not useful for quantum information processing and computation: it can only define a classical variable of the condensed system. We shall then look at the entanglement between particles within a condensate and discuss how it may be observed in experiments. This will lead us to consider entanglement properties of condensates that may be a useful resource for quantum information theory.

In the last part of the lectures I will discuss how we can produce useful entanglement through four wave mixing of matter waves. This is produced by the same "Bogoliubov" terms in the Hamiltonian responsible for the damping of collective excitations by mode mixing. We shall link this to the generation of entangled pairs by interactions within a BEC.

2 Effective Interactions between ultracold atoms

In the theory of condensed gases we make a great deal of use of an effective interaction between atoms. This effective interaction describes the interaction between atoms in the limit of s-wave scattering, i.e. when the de Broglie wavelength is much larger than the range of the potential. An important feature of evaporatively cooled gases is that they are so cold that one can attain BEC while the gas is still very dilute on the scale of interatomic forces. Why is this so? BEC is reached when the interparticle separation is comparable with the de Broglie wavelength of the atoms. If the atoms are sufficiently cold, the de Broglie wavelength is big enough for us to reach BEC while keeping the interparticle separation large, large being defined with respect to the range of the potential. In fact, in the case of evaporatively cooled gases at sub-micro-Kelvin temperatures, the de Broglie wavelength of the atoms is enormous compared to the range of the interatomic potential. This means we can use an effective delta-function interaction which is proportional to the scattering length. This has the form:

$$V(\mathbf{r}-\mathbf{r}') = U_0\,\delta(\mathbf{r}-\mathbf{r}'). \tag{1}$$

Here U_0 is given in terms of the binary s-wave scattering length a by $U_0 = 4\pi\hbar^2 a/m$[12]. We shall return to the form of this interaction below and explain how it relates to an effective excluded volume produced by an atom. This interaction gives the exact low energy scattering amplitude $(-a)$ when used in the simplest, first order perturbation theory, approximation for the scattering amplitude i.e. the Born approximation. This effective interaction only describes the effective coupling between low momentum states, but all of the momenta involved in the evaporative cooling process satisfy this criterion. We shall use the contact interaction to construct the mean-field theory of condensates and then examine the nature of the approximations involved in using it. A useful way to see how the contact interaction gets into the energy of the gas is to see how it is related to the scattering wavefunction. By that we mean the wavefunction for the relative motion of a pair of atoms as they collide in a condensate. Because the density is so low we can assume, as a first approximation, that the collision occurs just as it does in vacuo. For zero collision energy the wavefunction has the simple form:

$$\chi[1-\frac{a}{r}] \tag{2}$$

Here a is the scattering length and χ is the asymptotic value of the wavefunction. We compare this with the general form of the wavefunction at finite energies, i.e.

$$\chi[e^{i\mathbf{k}.\mathbf{r}} - \frac{a}{r}e^{ikr}] \tag{3}$$

In this form we see that $-a$ is the scattering amplitude in the limit of low energies and is determined by the position of the node in the zero-energy wavefunction. For the atoms that have been Bose-condensed to date, the scattering length has been determined by a combination of spectroscopic measurement and multi-channel scattering theory. This means that we can really start from first principles in computing properties of the many-body systems; a is not an adjustable parameter, it is an input to the theory. The binary form of the wavefunction given above is only valid, of course, for $r \geq a$: for smaller distances the wavefunction depends on the details of the interatomic potential. For the case of the dilute gas the scattering length gives us all we need to calculate the change in the energy of the gas due to the interactions between the particles.

We now want to show how this form of the wavefunction affects the energy of a gas. The presence of the interatomic potential between atoms gives rise to a new node of their relative wavefunctions at $r = a$. This increases the kinetic energy in much the same way an excluded volume inside $r = a$ would do. This extra kinetic energy in the wavefunction is given by:

$$\int_a^\infty dr 4\pi r^2 \frac{\hbar^2}{m}\{\chi\nabla[1-\frac{a}{r}]\}^2 = U_0\chi^2 \tag{4}$$

If we take χ^2 as the density of "other particles" we obtain an expression for the energy of a given particle in the presence of others in the gas. The expression we get is then exactly the same as that we would obtain by using first order perturbation theory with the contact interaction. The fact that the contact interaction produces the correct scattering amplitude with just first order perturbation theory is the reason for this agreement between the two routes. The form of the wavefunction at small internuclear distances corresponds to purely binary correlations between the particles. One could also speak of the small distance entanglement between the particles. As we go along in these lectures we shall expand on this concept of entanglement between particles produced by interactions.

So we now have the form of the interaction between particles needed to define an effective low energy theory of condensates. This leads to a non-linear Schrödinger equation, the GPE, as we shall discuss below. The presence of the condensate changes this effective interaction in a way that can be calculated directly for a dilute gas. For three dimensional condensates this correction is very small!, i.e. of order $n_0 a^3 = (\frac{a}{l_0})^3$, where l_o is the average interparticle separation. One can see how this arises that by taking the upper limit of the integral on the kinetic energy we found above to be l_0 rather than ∞. This gives an estimate of the effect of other particles on the binary interaction. For two dimensional systems the modification of the strength of the contact interaction due to the presence of the medium is crucial. In spite of the large change in the strength of the interactions it is still valid to use an effective contact interaction. The difference to the three dimensional case comes about because of the form of the relative wavefunction of the particles. In two dimensions this has the following logarithmic dependence on r:

$$\Psi = \chi \frac{\ln(r/a)}{\ln(L/a)} \tag{5}$$

This is obtained by assuming that the wavefunction vanishes at $r = a$ and is equal to χ , at some finite $r = L$. This form of the wavefunction implies that the low energy limit of the scattering amplitude goes to zero with increasing L. We can see the consequences for the effective interaction between particles by finding the kinetic energy in the relative motion as we did above for the three-dimensional case. We find for this logarithmic wavefunction

$$E \sim \int_a^L \frac{\hbar^2}{2m}[\nabla \chi \frac{\ln(r/a)}{\ln(L/a)}]^2 2\pi r dr \tag{6}$$

$$E \sim \frac{\hbar^2}{m} \frac{\chi^2}{\ln(L/a)} \tag{7}$$

This shows that the binary interaction between the particles goes to zero as the size L of the box is increased. We can obtain a more sensible result by arguing that we should cut off this binary correlation on the interparticle distance. When this is included the energy in the two-dimensional gas is found to be

$$E \sim \frac{\hbar^2}{m} \frac{\chi^2}{\ln(l_o/a)}, \tag{8}$$

where l_0 is the interparticle spacing in the gas. A contact interaction with this modified strength can be used to describe the evolution of the momenta involved in the condensation process. This change in the strength of the effective interaction is equivalent to a renormalisation of the interaction strength relative to its value in vacuo. The "renormalisation" of the effective interaction that we use in the GPE can only be carried out if the gas is dilute and fails for a strongly coupled system such as liquid helium In the next section we shall suppose that the contact interaction is a sensible starting point and examine the simplest mean field theory for condensed gases.

3 The Gross-Pitaevskii Equation and Mean-Field Theory

The simplest theory of the trapped gases is Hartree-Fock theory where we assume that the atoms all share the same wavefunction i.e. the condensate wavefunction $\Psi(\mathbf{r}, t)$. This is equivalent to neglecting any quantum fluctuations in the gas which come from longer range correlations or entanglement between the atoms. The short range binary correlations are included implicitly through the use of a contact potential. This description of the gas will, therefore, only be valid for times and length scales that are large compared to collisional ones. This, of course, is what we mean by an effective low energy theory. So we put the effect of collisions into the description using the contact potential above and find a local mean-field interaction proportional to the density of the particles in the gas. The evolution of the condensate is then described by a non-linear Schrödinger equation called the Gross-Pitaevskii equation (GPE) [1]. In free space this has the form:

$$i\hbar\frac{\partial\Psi}{\partial t} = -\frac{\hbar^2\nabla^2\Psi}{2m} + U_0\left|\Psi\right|^2\Psi \tag{9}$$

This equation has to be solved numerically in general for the case of the trapped gases produced in experiments. For a homogeneous gas it has a static solution with all the atoms sharing the mean energy, $n_0 U$, where n_0 is the density of atoms in the condensate. Excitations are defined relative to this uniform solution. For small k this leads to the dispersion relation for longitudinal excitations in the condensate:

$$\omega = \sqrt{\frac{n_o U_o}{m}}k = ck \tag{10}$$

This spectrum of purely longitudinal phonons is characteristic of a system that exhibits superfluidity with a critical speed equal to c. We should caution that the critical speed is determined not only by this sound speed by the generation of vortices and many other issues that we do not have time to tackle here. This equation has been widely used to predict the properties of trapped condensates with near zero temperatures through the simple addition of the trapping potential to the GP equation, to obtain the following:

$$i\hbar\frac{\partial\Psi}{\partial t} = -\frac{\hbar^2\nabla^2\Psi}{2m} + V_{\text{trap}}(\mathbf{r})\Psi + U_0\left|\Psi\right|^2\Psi. \tag{11}$$

The agreement between the predictions of this equation for low temperature excitations and the experimental measurements of them is excellent. To obtain the frequencies and shapes of the excitations one finds the normal modes of oscillation of the condensate given by linearizing the GPE around the ground state. The set of equations one then has to solve is:

$$\left\{-\frac{\hbar^2\nabla_{\mathbf{r}}^2}{2m} + V_{\text{trap}}(\mathbf{r}) - \mu + U_0|\psi(\mathbf{r})|^2\right\}\psi(\mathbf{r}) = 0. \tag{12}$$

$$\hat{\mathcal{L}}(\mathbf{r})u_j(\mathbf{r}) + U_0\psi^2(\mathbf{r})v_j^*(\mathbf{r}) = \epsilon_j u_j(\mathbf{r}) \tag{13}$$

$$\hat{\mathcal{L}}(\mathbf{r})v_j(\mathbf{r}) + U_0\left[\psi^*(\mathbf{r})\right]^2 u_j(\mathbf{r}) = -\epsilon_j v_j(\mathbf{r}) \tag{14}$$

In these equations, the condensate wavefunction has been denoted by $\psi(\mathbf{r})$, μ represents the chemical potential of the assembly, $u_j(\mathbf{r})$ and $v_j(\mathbf{r})$ are the Bogoliubov functions[3,4] and the hermitian operator $\hat{\mathcal{L}}$ is defined by:

$$\hat{\mathcal{L}}(\mathbf{r}) = -\frac{\hbar^2\nabla_{\mathbf{r}}^2}{2m} + V_{\text{trap}}(\mathbf{r}) - \mu + 2U_0|\psi(\mathbf{r})|^2 \tag{15}$$

This method can be generalised to finite temperatures by the inclusion of the effects of thermal atoms on the mean field. We shall return to how this is done later on in the lectures. In mean-field theory we assume that the excitations that it predicts are long lived i.e. their interaction and decay due to processes beyond the linearizing we have described here can be ignored. The full GPE can, in fact, be used to describe this interaction and decay but only for modes that are sufficiently highly occupied. In essence, it describes the evolution of the modes that can be described classically. Quantum fluctuations of the field cannot be treated in this way and we shall also describe aspects of their nature below. The GPE is only valid for dilute gases where these quantum fluctuations simply renormalise the effective interaction between atoms. The diluteness criterion means that we need the scattering length to be much smaller than the interparticle spacing l_o: which certainly holds for the condensates produced using evaporative cooling. It is clearly not true for liquid helium. For a homogeneous gas we usually say that mean field theory only holds outside the region where:

$$\frac{T - T_c}{T_c} \gg \frac{a}{l_o} \tag{16}$$

We need to be precise about what this means. It does not mean that the GPE cannot be used close to the critical point. The fluctuations that affect the critical point are at long wavelengths and correspond to highly occupied, i.e. classical, modes of excitation. To make this clear some say that the state of the gas is a "quasi-condensate". Within the critical region the thermal fluctuations interact strongly and cannot be treated using perturbation theory.

Outside the critical region, below the BEC transition, we can describe the gas by linearizing around the static condensate, finding the excitations and assuming they interact weakly. Inside the critical region in a homogeneous gas the strongly interacting long-wavelength fluctuations can be treated using the Renormalisation Group. These methods still implicitly use the GPE to describe the strongly interacting, effectively classical, thermal excitations. This means that we can use a computational approach to the GPE to describe the evolution of a gas close to the transition point.

We should add the important point that for a trapped gas mean-field theory works right through the BEC transition. For a trapped gas, in the harmonic potentials commonly used in experiments, this comes from the fact that the density of states is proportional to energy squared. This removes the infra-red singularity that produces the breakdown of perturbation near the critical point in the uniform gas case. The cross-over to critical behaviour in a homogeneous gas is a subject of current studies. In the next section we shall show how the evolution of a non-equilibrium gas may be studied using the GPE.

4 Treating Thermal and Non-equilibrium effects using the GPE

In the first generation of experiments on Bose-Einstein condensation in atomic gases the condensates formed directly into the ground state of the trap. They then have the same shape as the self-consistent ground state with no nodes: equivalent to the condensate containing no topological defects. The confinement suppresses the critical fluctuations that are so important in the critical region for the homogeneous case, preventing the formation of defects in the BEC transition. In future experiments the condensates may be much larger and we need to consider the topological structure of the condensates one should expect to produce. These issues are likely be important in atom laser design where we want to produce as large as coherence length as possible.

The formation of condensates is an important theoretical problem that has attracted considerable attention in recent years. A rigorous quantum kinetic formalism was developed by Gardiner and Zoller[5], and led, after some simplifying assumptions, to a modified Quantum Boltzmann Equation (MQBE) which incorporates the effect the condensate has on modifying the single particle levels. Gardiner et al.[6] have shown that this MQBE gives a description of the overall condensate growth which is in good agreement with current experimental results. It also implicitly describes some aspects of the coherence of the condensate formation, because the wavefunction of the lowest level (i.e. the one that becomes macroscopically occupied) is determined by the Gross Pitaevskii equation.

We stated above that domain structures, vortices etc., may be formed in the condensate during the growth process. These features are not part of the ground state, and must in general be described by a coherent superposition of a number of low lying condensate states. The GPE should give a very good description of condensate modes that are significantly populated. Thus a collection of significantly populated low lying quasi-particle levels can be simply treated by a single wavefunction evolving according to the time dependent GPE. Kagan et al.[7] have utilized this idea and employed the GPE on its own to describe condensate growth in its early stages, over a small period of time they termed the "coherent stage". In subsequent work on condensate growth, Damle et al.[8] and Marshall et al.[9] further extended the use of the GPE by assuming the wavefunction described both the thermal gas as well as the condensate. They used a random initial wavefunction, and evolved it until it attained an equilibrium state, and identified part of this as the condensate. We should emphasize that this treatment treats computationally the same problem addressed by the classical Renormalisation group. For the low-lying modes we have a classical system in which all length scales are coupled and perturbation theory fails. In the case of the homogeneous gas we can be sure that it is only modes that are guaranteed to be highly populated that dominate the critical dynamics. This is not necessarily true for the trapped gas. The pure wavefunction approach is not well suited to the description of the higher energy levels, which have small populations, and evolve on a rapid time scale. Furthermore the single particle wavefunction misrepresents the correlations that develop between the various levels. It is natural to apply the kinetic approach to the evolution of these levels.

In the rest of this section we shall give a brief derivation of a kinetic equation from the Gross-Pitaevskii equation. This will show how the GPE equation describes the coherent evolution domain. It is valid in the realm of perturbation theory, where the non-interacting basis is a good one. The full GPE equation is valid as long as the occupation of the modes being dealt with is much greater than one, and does not rely on perturbation theory. We begin with the Gross-Pitaevskii equation and decompose the wave-function $\psi(\mathbf{r},t)$ as

$$\psi(\mathbf{r},t) = \sum_k c_k(t) e^{-i\omega_k t} \phi_k(\mathbf{r}) \tag{17}$$

This gives us the GPE in the following form:

$$i\hbar\frac{dc_p}{dt} = U_0 \sum_{qmn} \langle pq|mn\rangle c_q^* c_m c_n e^{i(\omega_p+\omega_q-\omega_m-\omega_n)t}, \tag{18}$$

where

$$\langle pq|mn\rangle = \int d^3\mathbf{r}\phi_p^*(\mathbf{r})\phi_q^*(\mathbf{r})\phi_m(\mathbf{r})\phi_n(\mathbf{r}). \tag{19}$$

We can then write down the equation for the population $n_p(t) \equiv |c_p(t)|^2$ in level p:

$$i\hbar\frac{dn_p}{dt} = U_0 \sum_{qmn} \langle pq|mn\rangle c_p^* c_q^* c_m c_n e^{i(\omega_p+\omega_q-\omega_m-\omega_n)t} + c.c. \tag{20}$$

Up to this point we have not made any approximations to the evolution process described by the GPE. In order to show how this equation produces evolution towards an equilibrium state we shall treat the right hand side of this equation to second order in the interaction. To do this we approximate the terms that appear on the right hand side of the equation by:

$$\begin{aligned}
c_p^* c_q^* c_m c_n = {} & c_q^* c_m c_n \frac{iU_0}{\hbar}\int_{-\infty}^{t} dt' \sum_{rkj} \langle kj|pr\rangle c_k^* c_j^* c_r e^{i(\omega_k+\omega_j-\omega_p-\omega_r)t'} \\
& + c_p^* c_m c_n \frac{iU_0}{\hbar}\int_{-\infty}^{t} dt' \sum_{rkj} \langle kj|qr\rangle c_k^* c_j^* c_r e^{i(\omega_k+\omega_j-\omega_q-\omega_r)t'} \\
& + c_p^* c_q^* c_n \frac{-iU_0}{\hbar}\int_{-\infty}^{t} dt' \sum_{rkj} \langle mr|kj\rangle c_r^* c_k c_j e^{i(\omega_m+\omega_r-\omega_k-\omega_j)t'} \\
& + c_p^* c_q^* c_m \frac{-iU_0}{\hbar}\int_{-\infty}^{t} dt' \sum_{rkj} \langle nr|kj\rangle c_r^* c_k c_j e^{i(\omega_n+\omega_r-\omega_k-\omega_j)t'}
\end{aligned} \tag{21}$$

We want to consider the long time behaviour of these quantities, so we assume that the c's under the integrals on the right hand side are slowly varying. Ignoring the principal parts of the integrals that correspond to energy shifts, which we presume are small, we find:

$$\begin{aligned}
c_p^* c_q^* c_m c_n = & \frac{i2\pi U_0}{\hbar} \sum_{rkj} \langle kj|pr\rangle c_q^* c_k^* c_j^* c_r c_m c_n \delta(\omega_k + \omega_j - \omega_p - \omega_r) \\
& + \frac{i2\pi U_0}{\hbar} \sum_{rkj} \langle kj|qr\rangle c_p^* c_k^* c_j^* c_r c_m c_n \delta(\omega_k + \omega_j - \omega_q - \omega_r) \\
& + \frac{-i2\pi U_0}{\hbar} \sum_{rkj} \langle mr|kj\rangle c_p^* c_q^* c_r^* c_k c_j c_n \delta(\omega_m + \omega_r - \omega_k - \omega_j) \\
& + \frac{-i2\pi U_0}{\hbar} \sum_{rkj} \langle nr|kj\rangle c_p^* c_q^* c_r^* c_k c_j c_m \delta(\omega_n + \omega_r - \omega_k - \omega_j) \quad (22)
\end{aligned}$$

Now we use the random phase approximation i.e.

$$c_m^* c_n = |c_m|^2 \delta_{mn} = n_m \delta_{mn} \quad (23)$$

This is equivalent to Wick's theorem in the quantum field theory version of our procedure. For each set of six c's, there are six independent contractions that can be made. Substituting into the equation for dn_p/dt we find the final result:

$$\begin{aligned}
\frac{dn_p}{dt} = & \frac{8\pi U_0^2}{\hbar^2} \sum_{qmn} |\langle pq|mn\rangle|^2 \delta(\omega_m + \omega_n - \omega_p - \omega_q) \times \\
& \{(n_p + n_q) n_m n_m - n_p n_q (n_m + n_n)\} \\
& + \frac{16\pi U_0^2}{\hbar^2} \sum_{qmn} \langle pq|qn\rangle \langle mn|pm\rangle n_q n_m (n_n - n_p) \delta(\omega_n - \omega_p) \\
& + \frac{16\pi U_0^2}{\hbar^2} \sum_{qmn} \langle pq|pn\rangle \langle mn|qm\rangle n_p n_m (n_n - n_q) \delta(\omega_n - \omega_q) \quad (24)
\end{aligned}$$

In the homogeneous limit, we can calculate the matrix elements of the basis functions. The basis is that of plane waves i.e.

$$\phi_p(\mathbf{r}) = (2\pi V)^{-d/2} e^{-i\mathbf{k}_p \cdot \mathbf{r}} \quad (25)$$

where $d \equiv$ number of dimensions. Thus

$$\begin{aligned}
\langle pq|mn\rangle &= (2\pi V)^{-2d} \int_V d^d\mathbf{r} e^{i(\mathbf{k}_p + \mathbf{k}_q - \mathbf{k}_m - \mathbf{k}_n)\cdot\mathbf{r}} \\
&= (2\pi V)^{-d} \delta(\mathbf{k}_p + \mathbf{k}_q - \mathbf{k}_m - \mathbf{k}_n) \quad (26)
\end{aligned}$$

This result means that in the homogenous limit, the second two terms of Eq. (24) are zero, as the momentum conservation forces the quantum numbers in the difference of populations to be the same. So we have

$$\frac{dn_p}{dt} = \frac{8\pi U_0^2}{\hbar^2}(2\pi V)^{-2d}\sum_{qmn}\delta(\omega_p+\omega_q-\omega_m-\omega_n)\,\delta(\mathbf{k}_p+\mathbf{k}_q-\mathbf{k}_m-\mathbf{k}_n)$$
$$\times\{(n_p+n_q)n_m n_m - n_p n_q(n_m+n_n)\} \tag{27}$$

This is the same as the quantum Boltzmann equation for bosons, but without the spontaneous terms. It is correct in the limit that perturbation theory is valid, and also where $n_p \gg 1$. The equilibrium populations for the levels are *equipartition* rather than Bose-Einstein — in the limit of large occupation numbers this takes the form:

$$\begin{aligned} n_p &= \frac{1}{e^{(\varepsilon_p-\mu)/kT}-1} \\ &\approx \frac{1}{1+(\varepsilon_p-\mu)/kT+\cdots-1} \\ &= \frac{kT}{\varepsilon_p-\mu} \end{aligned} \tag{28}$$

If we calculate the amount of energy at equilibrium we find

$$\begin{aligned} E_{tot} &= \sum_p(\varepsilon_p-\mu)\frac{kT}{\varepsilon_p-\mu} \\ &= \sum_p kT \\ &\to \infty \end{aligned} \tag{29}$$

This diverges as the number of states included in the sum, and means some type of momentum cutoff must be made. This is consistent with the fact that the effective interaction we use in the GPE can only represent coupling between low lying states.

We have seen that the GPE reduces to a kinetic equation if we assume mean field effects on the basis states are negligible and the mean free time is long. This kinetic equation is a good approximation to the quantum Boltzmann equation for when the modes are highly occupied, having an equipartition distribution function in equilibrium. This suggests that the Gross-Pitaevskii equation could be a good method of simulating the low lying states of the quantum Bose field that are sufficiently highly occupied in the presence of a large mean field and outside the domain of perturbation theory. This is the subject of continuing work at Oxford.

5 Correlations between particles in a condensate

When collisions take place in a condensate the other atoms affect the long range wavefunction of the colliding pair. We should, in fact, expect the pair wavefunction to be drastically changed when the distance apart is comparable to the average distance between atoms. A given collision will not have "finished" before another one starts up. We can estimate the consequences for the effective interaction between atoms by cutting off the integral we use to compute it at the mean interparticle distance. This is what we did above to see how the effective interaction in a two-dimensional gas is changed by the presence of the particles around it. The systematic approach to this problem involves the calculation of the pair correlation in the gas beyond mean-field theory. To do this we need to take into account the correction to mean-field theory produced by the quadratic terms in the approximate Hamiltonian i.e. terms of the form:

$$n_0 U_0 \Sigma_p [a_p^\dagger a_{-p}^\dagger + a_p a_{-p}] \tag{30}$$

These terms in the Hamiltonian promote pairs of atoms from the condensate into excited states. These pairs deplete the condensate and represent the quantum fluctuations in the mean field from that given by the condensate alone. The quadratic terms in the Hamiltonian also change the excited states from particles to quasi-particles. This is represented by the following Bogoliubov transformation:

$$\alpha_p = u_p a_p^\dagger + v_p a_{-p} \tag{31}$$

where,

$$u_p = \left[\frac{E_p + \epsilon_p + n_0 U_0}{2E_p}\right]^{1/2}, \quad v_p = -\left[\frac{\epsilon_p + n_0 U_0 - E_p}{2E_p}\right]^{1/2}. \tag{32}$$

and the spectrum of quasi-particles is that for phonons given above, i.e.

$$E_k = \sqrt{\epsilon_p^2 + 2n_0 U_0 \epsilon_p}, \tag{33}$$

This means that we should consider colliding quasi-particles rather that particles in determining the effective interaction between atoms. This change in the energy spectra of particles is equivalent to changing the "free" Hamiltonian of the particles and produces a characteristic change in the wave-function of a pair in the condensate. The effective interaction between particles that includes the change in the pair correlation function can be written in the form:

$$\tilde{\Gamma}_0(\mathbf{r}) = U_0 \left[1 + \tilde{m}(\mathbf{r})/\psi^2(\mathbf{r})\right] \tag{34}$$

The second term in the effective interaction describes the change in the correlation between atoms because of the Bogoliubov terms in the Hamiltonian. These produce correlations, i.e. pairing between particles of equal and opposite momenta in the gas. Pairs of this sort become bound in the case of electrons in a superconductors (or Helium three atoms) below their BCS transition temperature[1]. For the case of a dilute Bose gas they are just an admixture in the ground state of the gas. They are the correlated or configuration interaction part of the ground state wavefunction. $\tilde{m}(\mathbf{r})$ is also termed the "anomalous average" of treatments based on the Green function or variational approaches to the theory of many body systems. This part of the wavefunction is locked to the condensate and contributes to the superfluid density. $\tilde{m}(\mathbf{r})$ for a homogeneous gas may be written, after the subtraction of the ultra-violet divergent part, in the form:

$$\tilde{m}(\mathbf{r}) = \tilde{m} = \int \frac{d^3\mathbf{k}}{(2\pi)^3} \left(u_k v_k [2N(E_k) + 1] - \lim_{k \to \infty} u_k v_k \right) \tag{35}$$

Here ϵ_k is the bare particle energy, and E_k is the self-consistent energy of a quasi-particle given by:

$$E_k = \sqrt{\epsilon_k^2 + 2n_0\tilde{\Gamma}_0\epsilon_k}, \tag{36}$$

u_k and v_k are the corrected Bogoliubov transformation factors and are given by:

$$u_k = \left[\frac{E_k + \epsilon_k + n_0\tilde{\Gamma}_0}{2E_k} \right]^{1/2}, \quad v_k = - \left[\frac{\epsilon_k + n_0\tilde{\Gamma}_0 - E_k}{2E_k} \right]^{1/2}. \tag{37}$$

The fact that $\tilde{\Gamma}_0$ is the effective interaction including the effect of the correlations, rather than just the bare U_0 makes the calculation self-consistent. The presence of the Bogoliubov transformation in the effective interaction is due to the fact that we have to consider colliding quasi-particles, rather than particles, due to the presence of a condensate around the collision. A simple calculation then gives $u_k v_k = -n_0\tilde{\Gamma}_0/2E_k$ enabling us to write $\tilde{m}$ as

$$\tilde{m} = -n_0\tilde{\Gamma}_0 \int \frac{d^3\mathbf{k}}{(2\pi)^3} \left(\frac{2N(E_k) + 1}{2E_k} - \frac{1}{2\epsilon_k} \right). \tag{38}$$

The term with $-(1/2\epsilon_k)$ corresponds to the pair correlation, or pair wavefunction in vacuo. It is the momentum space version of the low energy wavefunction as a function of interparticle separation that we gave above. It is subtracted from the full correlation function in the medium so that we calculate the change in the effective interaction from that in vacuo due to the presence of the condensate. This is equivalent to expressing the effective interaction in the medium in terms of the observed scattering in vacuo. This is formally a renormalisation process and has the welcome effect of removing the ultraviolet divergence. This renormalisation process only works because the gas is dilute. Diluteness means that the effects of the mean field do not affect the short range of the collision dynamics, i.e. the effective interaction for higher momenta goes over to the binary form. If it didn't we couldn't simply subtract the binary term in the equation above to obtain a finite change in the effective interaction in going from collisions in a vacuum to collisions within a condensate.

We can also obtain this effective interaction from the many-body T-matrix for particles colliding in a condensate (denoted here by $\tilde{\Gamma}_0$ as in[10])[11,12]. It can be shown that $\tilde{\Gamma}_0$ is related to the zero-momentum limit of the vacuum scattering amplitude U_0 via

$$\tilde{\Gamma}_0 = \frac{U_0}{1+\alpha(T)U_0}\ , \tag{39}$$

where $\alpha(T)$ depends both on the energies and occupation of the intermediate states in a collision[10], and is given by:

$$\alpha(T) = \int \frac{d^3\mathbf{k}}{(2\pi)^3}\left(\frac{1}{2E_k}\coth\frac{\beta E_k}{2} - \frac{1}{2\epsilon_k}\right), \tag{40}$$

where n_0 is the condensate density. $\alpha(T)$ can be rewritten using the following identity: $\coth(\beta E_k/2) = (e^{\beta E_k}+1)/(e^{\beta E_k}-1) = 2N(E_k)+1$, where $N(E_k) = [e^{\beta E_k}-1]^{-1}$ is the Bose-Einstein distribution. We then get the following expression for $\tilde{\Gamma}_0$

$$\tilde{\Gamma}_0 = U_0\left[1 - \tilde{\Gamma}_0\int \frac{d^3\mathbf{k}}{(2\pi)^3}\left(\frac{2N(E_k)+1}{2E_k} - \frac{1}{2\epsilon_k}\right)\right]. \tag{41}$$

Since $n_0 = \psi^2$, comparison with Eq. (38) clearly shows that:

$$\tilde{\Gamma}_0 = U_0\left[1+\frac{\tilde{m}}{\psi^2}\right] \tag{42}$$

As we said above this expression corresponds formally to the many-body T-matrix which takes into account the modification imposed by the medium on interactions. It is expected to be a very good approximation for most temperatures, except for a very small region near $T=0$. We have seen, therefore, that the effective interaction in a medium is changed from that in free space. This change can be pictured in terms of the change in the pair correlation function or equivalently in terms of quantum fluctuations in the mean field. Detailed calculations have shown that their effect on the predictions for excitations in a trapped gas are modest but observable[13]. This change in the correlation between atoms is also a function of temperature through the Bose occupation factors. These indicate that correlations that come from promotion to excited pair states are enhanced by their being populated at finite temperature. The extra pair correlations between particles can also be thought of as the entangled portion of the gas. This entanglement can be seen in the output of atom lasers for the right choice of output coupler. In the next section we shall outline the full second order theory of the trapped gas. This includes the effect of dynamical correlations and thermal excitations along with the static correlation effects discussed here.

6 Second-Order Theory of Widths and Shifts at Finite Temperature

The first predictions for the frequencies of collective excitations at low temperatures were remarkably successful. This success was clear evidence that the atoms in atomic condensates share the same wavefunction. The subsequent measurements at higher temperatures of the frequencies and damping of collective modes was a great challenge to theorists in the field. They are a tremendous opportunity to test quantum field theory at finite temperatures.[14,15,16]

In order to produce a quantitative theory of collective excitations of trapped gases we have to develop more fully the quantum field theory of the dilute Bose gas. This theory has to go beyond the simplest theory based on the GPE or the Bogoliubov approximation. We can, however, use the quadratic Bogoliubov Hamiltonian as the starting point for perturbation theory. We then treat the terms ignored in the Bogoliubov approximation systematically to second order in the effective interaction. This has to be done with care as we encounter ultra-violet and infra-red divergences on the way. These issues have now been fully resolved and implemented in quantitative predictions for trapped gases. Before going into the details we'll give a brief account of previous developments.

The first second order theory of the dilute Bose gas was developed by Beliaev for homogeneous systems at $T = 0$ [17]. This approach was based on the Green function approach to many-body systems. The theory of excitations that is closest to that developed in our group is that by Mohling *et al.* [18,19]. These are important and clear papers and deserve careful reading. Later on work by Popov and Fadeev[20,21] recast the theory of the dilute Bose gas in a new perturbation scheme based on density and phase fluctuations. This work is closest to the spirit of the renormalisation group and avoids the potential infra-red divergence problems of the Bogoliubov route. As mentioned above, the infra-red issues need not be a problem for either the homogeneous or trapped gas case. We now have well posed second order theories that describe the collective modes of trapped condensates[22,23]. We shall describe the version of the theory developed by Morgan as it is the one most easily adapted to the use of the computational methods we have developed in recent years.

These computational methods for describing condensates' properties have, so far, been based on the finite temperature mean field theory. These could not include decay processes as the thermal component of the trapped gas was assumed to be static[24]. In the last section we saw how static correlations modify the effective interaction between the particles. Including the effect of evolving correlations is more of a challenge. They also ignored damping of excitations due to Beliaev wave-mixing. More recently theories have developed that treat the decay using various approximations[25]. Calculations of lifetimes have been performed using an extension of the Green-function method to a quasi-homogeneous system[26].

In our method we studied the evolution of excited states using a resolvent formalism in order to avoid approximations that may limit the relevance to experimental measurements[16]. Our starting point is the second quantised Hamiltonian for a system of structureless Bosons with pairwise interactions as used by Bogoliubov[3]. For system with large ground state population N_0 the dominant terms of the Hamiltonian are those up to quadratic order in the field operators. Neglecting all higher order terms leaves us with an approximate Hamiltonian, which we can diagonalise exactly. This is equivalent to finding the normal modes or quasi-particles as discussed above. They can also be found using a mean-field[3] or a number conserving approach[23].These equations describe the condensate and its collective excitations for a system with negligible thermal component, and are in excellent agreement with the experiment[27]. At finite temperatures, however, the terms left out in the Bogoliubov theory have to be included to account for excited quasi-particles and their effect on the condensate. We deal with the non-quadratic terms using second order perturbation theory in the quasi-particle basis[23]. The set of equations we obtain satisfy the Hugenholtz-Pines theorem[28,4], requiring the excitation energy for homogeneous systems to tend to zero in the long wavelength limit. They are also free of infra-red divergences that can cause considerable problems in second order theories. The theory is well suited to computational implementation for trapped Bose gas at finite temperature. The condition for a perturbative treatment to be valid is, for homogeneous systems,

$$\left(\frac{k_B T}{N_0 U_0}\right)(N_0 a^3)^{1/2} \ll 1, \tag{43}$$

We are, of course, always working in the dilute gas limit, $N_0 a^3 \ll 1$. This is closely related to the Ginzburg criterion for the validity of mean-field theory[2]. For trapped condensates one finds that the perturbative shifts indeed are small in comparison to the leading order terms even up to T_C and that perturbation theory is valid.

In the experiments the frequency and decay rate of the mode initially excited is measured. In the theory we can look at the response to a driving field in the region of the bare quasi-particle state or equivalently the Fourier transform of the quasi-particles evolution. This is most easily related to the resolvent i.e. $\hat{G}(z) = (z - \hat{H})^{-1}$, $z \in C$. This resolvent when restricted to the evolution of a quasi-particle gives the single particle Green function and is related to the time evolution operator through the following;

$$\hat{U}(t) = -\frac{1}{2\pi i}\int_{-\infty}^{\infty} e^{-i\omega t}[\hat{G}(\omega + i0) - \hat{G}(\omega - i0)]d\omega. \tag{44}$$

The Hamiltonian $\hat{H}$ is hermitian and, for a finite and trapped condensate, has a discrete and real spectrum. Thus the singularities of $\hat{G}(z)$ are single poles on the real z-axis. The projection of the resolvent, $G_p(z) = \langle p|\hat{G}|p\rangle$, onto the quasi-particle[29] is given by:

$$G_p(z) = [z - \epsilon_p - \Delta E_3(z) - \Delta E_4(p) - \Delta E_\mu(p) - \Delta E_{shape}(p)]^{-1}. \tag{45}$$

Note the z dependence of $\Delta E_3(z)$. For a trapped system the spectrum of available decay channels, given by sums and differences of excitation energies, is discrete. For anisotropic parabolic traps it has been argued that the spectrum of decay channels can be approximated by a continuous one. The time evolution of an excitation then is dealt with as a decay into a continuum of states[26]. This is valid if we can treat the second order perturbative term as a slowly varying function, and we get a constant width and shift i.e. $\Delta E_3(z) \simeq \delta_{E3}(\epsilon_p) + i\Gamma_{E3}(\epsilon_p)$. In our explicit computations for a spherical trap we find this is not the case. We, therefore, calculate the spectrum of an excitation to second order by summing over the contributions from individual couplings. The full second order theory is rather complex and we shall not be able to go through the whole development. Instead we shall quote the results of the analysis and explain their physical significance in the limit of a large enough trap to obtain a width and shift. The detailed analysis shows that the energy of an excitation in the continuum limit can be written in the form

$$E(p) = \epsilon_p + \Delta E_4(p) + \Delta E_3(p) + \Delta E_\lambda(p) + \Delta E_{shape}(p). \qquad (46)$$

The full expressions for the energy shifts are as follows:

$$\Delta E_4(p) = U_0 \int d\mathbf{r}\, [2\rho(\mathbf{r})\Delta\rho_p(\mathbf{r}) + \kappa(\mathbf{r})\Delta\kappa_p(\mathbf{r})], \qquad (47)$$

$$\begin{aligned}\Delta E_3(p) = &- \sum_{i,j\neq 0} \frac{|A^P_{pij}|^2}{2(\epsilon_p + \epsilon_i + \epsilon_j)}[1 + n_i + n_j] \\ &+ \sum_{i,j\neq 0} \frac{|B_{pij} + B_{pji}|^2}{2(\epsilon_p - \epsilon_i - \epsilon_j)}[1 + n_i + n_j] \\ &- \sum_{i,j\neq 0} \frac{|B_{ijp} + B_{ipj}|^2}{(\epsilon_p - \epsilon_i + \epsilon_j)}[n_j - n_i], \end{aligned} \qquad (48)$$

$$\Delta E_\lambda(p) = (\Delta\mu) \int d\mathbf{r}\, \Delta\rho_p(\mathbf{r}), \qquad (49)$$

$$\begin{aligned}\Delta E_{shape}(p) = -N_0U_0 \int d\mathbf{r}\, (2\Delta\rho_p(\mathbf{r}) + \Delta\kappa_p(\mathbf{r})) \\ (\psi_0(\mathbf{r})^2 - \tilde{\psi}_0(\mathbf{r})^2).\end{aligned} \qquad (50)$$

$\Delta E_4(p)$ denotes the first order energy shift due to excited particles. In the expression for it $\rho(\mathbf{r}) = \sum_p[\Delta\rho_p(\mathbf{r})n_p + |v_p(\mathbf{r})|^2]$ is the density of excited particles, and $m(\mathbf{r}) = \sum_p[\Delta\kappa_p(\mathbf{r})(1+2n_p)]$ is the anomalous average. Furthermore $\Delta\rho_p(\mathbf{r}) = (|u_p(\mathbf{r})|^2 + |v_p(\mathbf{r})|^2)$ and $\Delta m_p(\mathbf{r}) = u_p(\mathbf{r})v_p^*(\mathbf{r})$. n_p is the population of the respective quasi-particle level.

ΔE_3 is the second order contribution to the width and shift. Its first two terms describe Beliaev-type damping processes in which an excitation mixes into two others. This can also be viewed in terms of an excited atom in one state colliding with an atom in the condensate. This leads to the production of population in two other states: $p+0 \rightarrow i+j$. The third term corresponds to Landau damping, where the excitation energy is transferred to another excited particle: $p + i \rightarrow 0 + j$. The matrix elements A^P_{ijk} and B_{ijk} are given by sums of overlap integrals of the ground state with the relevant quasi-particle amplitude. The energy of the quasi-particle given by the quadratic theory ϵ_p, depends only the condensate density. At finite temperature the thermal atoms will affect the condensate's energy and shape, giving new values denoted above by $\mu + \Delta\mu$ and $\tilde{\psi}_0$. The terms $\Delta E_\mu(p)$ and $\Delta E_{shape}(p)$ account for the resulting energy shifts due to these effects. For the derivation and the details of the above equations we refer to the work of Morgan[23].

For the actual calculation of $G_p(z)$ in a trap, we obtain a set of quasi-particle energies, ϵ_i, and amplitudes u_i and v_i as well as the ground state $\psi_0(\mathbf{r})$ for an appropriate value of the condensate population N_0. For a given temperature T we can calculate the population of quasi-particle levels using the Bose distribution function, $n_i = [e^{\epsilon_i/k_B T} - 1]^{-1}$, and find $\rho(\mathbf{r})$ and $m(\mathbf{r})$ from the quasi particle amplitudes. The condensate depletion, N_{ex}, is given by the integral over $\rho(\mathbf{r})$ and thus our choice of N_0 and T must be such that it is consistent with total number of particles, $N_{tot} = N_0 + N_{ex}$. The finite temperature values of $\tilde{\psi}_0$ and μ are found solving the generalised GPE[23]. We then proceed calculating the perturbative corrections using our quadratic theory wave functions and energies. This theory should soon provide quantitative predictions for experiments and give an alternative way of calibrating the temperatures of condensates through the lifetimes of quasi-particles.

7 Entangled Pairs in Condensates and Atom Laser Output

Bose-Einstein condensates of trapped atoms are ideal sources[30,31,32] of coherent atomic beams. At the moment it is only possible to produce pulsed sources of this sort but a great deal of effort will go into making cw versions of these atom-lasers[33]. Various authors have addressed the output coupling of a single-mode non-interacting condensate[34] or an interacting trapped condensate at $T = 0$ [35,36]. Under these conditions the output beam is a coherent matter wavepacket described by a single complex function of space and time. However, at finite temperatures one expects thermal excitations to play a major role. It is important to quantify and optimize the coherence properties of the output coupled atomic beam, for both continuous wave and pulsed optical laser sources. To do this we need to understand the factors that influence the nature of output beams. Determining the properties of the output may also serve as a probe of the nature of the ground-state and excitations in the atomic gas, just as quantum evaporation from the surface of ^{4}He can be used to probe its internal dynamics[37]. In our group we have studied weak output coupling of a trapped atomic gas at finite temperatures below the critical temperature T_c. We find that there are three components that contribute to the output: (i) coherent coupling of the condensate fraction; (ii) stimulated quantum evaporation of the thermal excitations; (iii) the simultaneous appearance of an output atom and an internal excitation, indicating the existence of pair correlations in the ground state. The last process predicted here occurs even at $T = 0$ and may serve as a probe of the deviation of the ground-state from a direct product of single atom states.

To see how these processes arise, we consider output coupling from a trapped atomic level $|t\rangle$ to a free state $|f\rangle$. Typical mechanisms are a direct (one-photon) radio-frequency (rf) transition and an indirect (two-photon) stimulated Raman transition. This coupling can be represented by the following Hamiltonian.

$$H_{\text{couple}} = \hbar \int d^3\mathbf{r}\lambda(\mathbf{r},t)\hat{\psi}_f^\dagger(\mathbf{r})\hat{\psi}_t(\mathbf{r}) + h.c. \tag{51}$$

Here $\hat{\psi}_t(\mathbf{r})$ describes the annihilation of a trapped atom at $\mathbf{r}$, while $\hat{\psi}_f^\dagger(\mathbf{r})$ describes the creation of a free atom with amplitude $\lambda(\mathbf{r},t)$. The coupling amplitude can be written as $\lambda(\mathbf{r},t) = \tilde{\lambda}(\mathbf{r},t)e^{i(\mathbf{k}_{em}\cdot\mathbf{r}-\Delta_{em}t)}$ where $\tilde{\lambda}$ is slowly varying in space and time. Here $\hbar\mathbf{k}_{\text{em}}$ and $\hbar\Delta_{\text{em}}$ are the momentum and energy transferred to an output atom. For Raman output coupling, where two laser beams are used to induce a transition from $|t\rangle$ to $|f\rangle$ through an intermediate level $|i\rangle$, $\lambda = \Omega_1\Omega_2/\Delta_i$ where $\Omega_{1,2}$ are the Rabi frequencies corresponding to the intermediate transitions and Δ_i is their detuning from resonance with the two beams. In this case Δ_{em} and $\mathbf{k}_{\text{em}}$ are the differences between the frequencies and momenta associated with the two laser beams. We consider a small rate of output coupling, so that the output atoms are dilute enough to neglect the interactions between them outside the trap. The coupled equations for the field operators $\hat{\psi}_t$ and $\hat{\psi}_f$ then take the form

$$i\hbar\dot{\hat{\psi}}_t(\mathbf{r}) = \mathcal{L}_t\hat{\psi}_t(\mathbf{r}) + U_0\hat{\psi}_t^\dagger(\mathbf{r})\hat{\psi}_t(\mathbf{r})\hat{\psi}_t(\mathbf{r}) + \hbar\lambda^*(\mathbf{r},t)\hat{\psi}_f(\mathbf{r}), \tag{52}$$

$$i\hbar\dot{\hat{\psi}}_f(\mathbf{r}) = \mathcal{L}_f\hat{\psi}_f(\mathbf{r}) + \hbar\lambda(\mathbf{r},t)\hat{\psi}_t(\mathbf{r}). \tag{53}$$

Here $\mathcal{L}_{t,f} \equiv -\hbar^2\nabla^2/2m + V_{t,f}(\mathbf{r})$, where $V_t(\mathbf{r})$ is the magnetic trapping potential, $V_f(\mathbf{r}) = U_1\langle\hat{\psi}_t^\dagger(\mathbf{r})\hat{\psi}_t(\mathbf{r})\rangle$ is the effective potential induced by the collisions with the trapped gas, and $U_i = \frac{4\pi\hbar^2}{m}a_i$ are proportional to the s-wave scattering lengths a_i for collisions between the atoms. Equations (52) and (53) have been previously treated[35,36] in the Gross-Pitaevskii approximation, where the field operators are replaced by their mean values. This approximation neglects the existence of elementary excitations at finite temperatures and the fluctuations around the mean value at $T = 0$.

We can find the operator for the stream of output atoms by formally solving Eq. (53) for $\hat{\psi}_f$ in terms of $\hat{\psi}_t$, i.e.

$$\hat{\psi}_f(\mathbf{r},t) = \hat{\psi}_f^{(0)}(\mathbf{r},t) - i\int_0^t dt' \int d^3\mathbf{r}' \times \\ \times G_f(\mathbf{r},\mathbf{r}',t-t')\lambda(\mathbf{r}',t')\hat{\psi}_t(\mathbf{r}',t'), \tag{54}$$

Here, $\hat{\psi}_f^{(0)}$ satisfies the time-dependent Schrödinger equation $i\hbar\dot{\hat{\psi}}_f^{(0)} = \mathcal{L}_f\hat{\psi}_f^{(0)}$ and the "free" Green's function $G_f(\mathbf{r},\mathbf{r}',t-t')$ can be written in terms of the solutions $\varphi_\mathbf{k}(\mathbf{r})$ of the corresponding time-independent equation $\hbar\omega_\mathbf{k}\varphi_\mathbf{k}(\mathbf{r}) = \mathcal{L}_f\varphi_\mathbf{k}(\mathbf{r})$: This gives $G_f(\mathbf{r},\mathbf{r}',t-t') = \sum_\mathbf{k}\varphi_\mathbf{k}(\mathbf{r})\varphi_\mathbf{k}^*(\mathbf{r}')e^{-i\omega_\mathbf{k}(t-t')}\theta(t-t')$. The field operator for the trapped atoms $\hat{\psi}_t(\mathbf{r},t)$ in thermal equilibrium contains the effects of the thermal excitations within the trap. The presence of these excitations effects the nature of the output field. The full field operator can be written in terms of the annihilation operators $\hat{\alpha}_0$ and $\hat{\alpha}_j$ of the condensate atoms and elementary excitations respectively. These are obtained using the finite temperature form of the Bogoliubov approximation. This gives the many-body Hamiltonian of the interacting system in diagonal form[38]. The field operator for the trapped atoms can be written in the form:

$$\hat{\psi}_t(\mathbf{r},t) = e^{-i\mu t/\hbar}\left\{\Psi_0^t(\mathbf{r})\hat{\alpha}_0 + \sum_j[u_j^t(\mathbf{r})\hat{\alpha}_j - v_j^{t*}(\mathbf{r})\hat{\alpha}_j^\dagger]\right\}. \tag{55}$$

To remind the reader, μ is the chemical potential, $\Psi_0^t(\mathbf{r})$ is the wavefunction of condensate atoms and $u_j^t(\mathbf{r}), v_j^t(\mathbf{r})$ are the steady-state solution of the equations obtained from Eq. (37) in the absence of output coupling[38]. They then reduce to the finite temperature equations called variously the Hartree-Fock Bogoliubov or Bogoliubov de Gennes equations. In thermal equilibrium and using the Bogoliubov approximation $\hat{\alpha}_0$ is replaced by $\sqrt{N_0}$, N_0 being the macroscopic mean number of condensate atoms. The population of the excited modes is given by the Bose-Einstein distribution $N_j^{eq} \equiv \langle\hat{\alpha}_j^\dagger\hat{\alpha}_j\rangle_{eq} = [e^{\hbar\omega_j/T} - 1]^{-1}$. The interpretation of the various terms in this expression for the quantum field is important in the interpretation of the form of the output field. Eq. (55) implies that the annihilation of a trapped particle at $\mathbf{r}$ is equivalent to the annihilation of a particle in the condensate, (with amplitude $\Psi_0^t(\mathbf{r})$), or the annihilation of an elementary excitation (with amplitude u_j^t) or the *creation* of a new excitation (with amplitude $-v_j^{t*}$). The last amplitude is associated with the non-condensate component of the ground-state of the system, which contains entangled pairs of correlated atoms in excited single-particle states, formed by binary collisions. We saw above how this contributes to the anomalous average and changes the pair correlations in the gas. Its role in the output spectrum, as shown below, is the simultaneous creation of an excitation within the trap along with an output atom. The Bogoliubov functions $u_j^t(\mathbf{r})$ of a dilute weakly interacting Bose gas are nearly the same as the energy eigenfunctions $\phi_n(\mathbf{r})$ of a single particle in the effective trapping potential $V_t(\mathbf{r}) + 2U_0N_0|\Psi_0^t(\mathbf{r})|^2$. A formal solution of the Bogoliubov de Gennes equations yields an expression for the functions $v_j^t(\mathbf{r})$ in terms of $u_j^t(\mathbf{r})$ e.g. ref. [38]

$$v_j(\mathbf{r}) = U_0N_0 \sum_n \frac{\int d^3\mathbf{r}' \phi_n^*(\mathbf{r}')[\Psi_0^*(\mathbf{r}')]^2 u_j^t(\mathbf{r}')}{E_n + \hbar\omega_j} \phi_n(\mathbf{r}), \tag{56}$$

This shows that the function $-v_j^*(\mathbf{r})$ describes collisional scattering of two atoms in the condensate into two excited states. Such virtual processes are the essence of pairing effects in the ground state of a Bose gas. We described them above in terms of the anomalous average and the change in the pair correlation function for atoms colliding in a condensate. In this part of the lectures we want to show how they can be observed in the spectrum of the output atoms.

We assume that the output rate is small compared to the energy separation $\delta\omega_j$ between the trap states. In this case the energies and wavefunctions of the condensate and elementary excitations do not vary significantly from their initial equilibrium values and the main contribution to the time-dependence of $\hat{\psi}_t(\mathbf{r},t)$ comes from the operators $\hat{\alpha}_0, \hat{\alpha}_j$. Moreover, for short times we may assume that the system stays close to equilibrium and the population of the trap states decays exponentially with decay coefficients γ_0, γ_j. For $\delta\omega_j^{-1} \ll t \ll \gamma_0^{-1}, \gamma_j^{-1}$ we thus assume $\hat{\alpha}_0(t) = \hat{\alpha}_0(0)e^{-\gamma_0 t/2}$ and $\hat{\alpha}_j(t) = \hat{\alpha}_j(0)e^{-i\omega_j t}e^{-\gamma_j t/2}$, where γ_0, γ_j are estimated below.

We suppose that all the atoms are in the trap at $t = 0$ and, using Eq. (54) with the free term $\hat{\psi}_t^{(0)}$ set to zero, we express the field operator $\hat{\psi}_f(\mathbf{r},t)$ of the free atoms at $t > 0$ in terms of the operators $\hat{\alpha}_0, \hat{\alpha}_j, \hat{\alpha}_j^\dagger$ appearing in Eq. (55). $\hat{\psi}_f$ then has the form $\hat{\psi}_f = \Psi_0^f\hat{\alpha}_0 + \sum_j [u_j^f\hat{\alpha}_j - v_j^{f*}\hat{\alpha}_j^\dagger]$, where the components ξ^f of the vector $\vec{\xi}^f \equiv (\Psi_0^f, u_j^f, -v_j^{f*})$, are obtained from Eqs. (54) and (55) in terms of the corresponding components of $\vec{\xi}^t(\mathbf{r}) \equiv (\Psi_0^t, u_j^t, -v_j^{t*})$. An expansion in eigenfunctions $\varphi_\mathbf{k}(\mathbf{r})$ of free atoms then gives

$$\xi^f(\mathbf{r},t) = \sum_\mathbf{k} \tilde{\xi}_\mathbf{k}^f \varphi_\mathbf{k}(\mathbf{r}) e^{-i\omega_\mathbf{k} t} \frac{e^{i(\omega_\mathbf{k} - \omega^f - i\gamma^f/2)t} - 1}{\omega_\mathbf{k} - \omega^f - i\gamma^f/2}, \tag{57}$$

where

$$\tilde{\xi}^{f}_{\mathbf{k}} = \int d^3\mathbf{r}\varphi^*_{\mathbf{k}}(\mathbf{r})\tilde{\lambda}(\mathbf{r})e^{i\mathbf{k}_{em}\cdot\mathbf{r}}\xi^{t}(\mathbf{r}). \tag{58}$$

In Eq. (57) the frequencies ω^f are the components of the vector $\vec{\omega}^f = (\omega^f_0, \omega^f_{j+}, \omega^f_{j-})$, where

$$\begin{aligned}\hbar\omega^f_0 &= \mu + \hbar\Delta_{\rm em},\\ \hbar\omega^f_{j\pm} &= \mu + \hbar\Delta_{\rm em} \pm \hbar\omega_j.\end{aligned} \tag{59}$$

After some time t the mean number of output atoms in a state with momentum $\mathbf{k}$ is given by $n_{\mathbf{k}} = \langle \hat{b}^{\dagger}_{\mathbf{k}}\hat{b}_{\mathbf{k}}\rangle$, where $\hat{b}_{\mathbf{k}} \equiv \int d^3\mathbf{r}\varphi^*_{\mathbf{k}}(\mathbf{r})\hat{\psi}_f(\mathbf{r})$. From the form of the output operator $\hat{\psi}_f(\mathbf{r})$ and Eq. (57) we obtain

$$\begin{aligned}n_{\mathbf{k}}(t) &= |\tilde{\Psi}^f_{0\mathbf{k}}|^2 N^{eq}_0 \mathcal{D}(\omega_{\mathbf{k}} - \omega^f_0, \gamma_0/2) +\\ &+ \sum_j \Big[|\tilde{u}^f_{j\mathbf{k}}|^2 N^{eq}_j \mathcal{D}(\omega_{\mathbf{k}} - \omega^f_{j+}, \gamma_j/2)\\ &+ |\tilde{v}^{f*}_{j\mathbf{k}}|^2 (N^{eq}_j + 1)\mathcal{D}(\omega_{\mathbf{k}} - \omega^f_{j-}, \gamma_j/2)\Big].\end{aligned} \tag{60}$$

Here the time-dependent spectral lineshapes $\mathcal{D}(\omega, \gamma/2) = \frac{|1-e^{i\omega t}e^{-\gamma t/2}|^2}{\omega^2+\gamma^2/4}$ tend to Lorentzians of width $\gamma/2$ in the limit $t \gg \gamma^{-1}$.

The first term in Eq. (60) describes a coherent output component generated when atoms with energy μ in the trapped condensate are excited to a free momentum state with kinetic energy $\hbar\omega_{\mathbf{k}} = \mu + \hbar\Delta_{\rm em}$ by absorbing energy $\hbar\Delta_{\rm em}$ from the output coupling field. The second term describes a thermal output component generated by stimulated quantum evaporation, where an atom initially populating an elementary excitation with energy $\mu + \hbar\omega_j$ leaves the trap with kinetic energy $\hbar\omega_{\mathbf{k}} = \mu + \hbar\omega_j + \hbar\Delta_{\rm em}$. The third term describes a process where the energy $\hbar\Delta_{\rm em}$ is sufficient to excite a ground state atom to a free momentum state with kinetic energy $\mu + \hbar\Delta_{\rm em} - \hbar\omega_j$, leaving a counterpart atom in an excited trap state with energy $\mu + \hbar\omega_j$. While a quantum of elementary excitation is annihilated from the trap in the second process of quantum evaporation, the third term describes a process where a new quantum of elementary excitation (a quasi-particle) is created. The factor $N^{eq}_j + 1$ in this term implies that this process occurs also in the absence of excitations (at $T = 0$), but it is amplified by the presence of excitations. According to the interpretation following Eq. (55) we may anticipate that this process points to the existence of pairs of correlated atoms in the ground state in the trap, which are broken when one atom is forced out. This process occurs also in the limit of weak (adiabatic) coupling and therefore it should be differentiated from non-adiabatic processes that may contribute to the generation of new elementary excitations[39].

Energy conservation requires $\hbar\omega_{\mathbf{k}} = \mu \pm \omega_j + \Delta_{\rm em}$ and momentum conservation follows from Eq. (58). In the first two processes the condition $\mathbf{k}_f = \mathbf{k}_t + \mathbf{k}_{\rm em}$, holds. This means that the momentum $\mathbf{k}_t$ of an initially trapped atom plus the momentum $\mathbf{k}_{\rm em}$ supplied by the output coupling field equals the momentum $\mathbf{k}_f$ of a free output atom at the trap, before it is accelerated by the repulsive interaction with the atoms left in the trap. Far from the trap its momentum is given by $|\mathbf{k}| \approx \sqrt{k_f^2 + mU_1N/V}$, $U_1N/2V$ being the effective repulsive potential. For the third process we have $\mathbf{k}_f + \mathbf{k}_t = \mathbf{k}_{\rm em}$, This shows that the total momentum of the resulting output atom-quasi-particle pair is equal to the momentum supplied by the field. The contribution of a specific term in Eq. (60) to the output rate of atoms with a corresponding asymptotic kinetic energy $k^2/2m = \mu \pm \hbar\omega_j + \hbar\Delta_{\rm em}$ is proportional to the number of atoms with momentum $\hbar\mathbf{k}_t$ satisfying $\hbar^2(\mathbf{k}_t \pm \mathbf{k}_{\rm em})^2/2m + U_1N/2V = \mu \pm \hbar\omega_j + \hbar\Delta_{\rm em}$ in the initial momentum distribution of the corresponding trap state and to the density of free states with the corresponding output momentum. The main contribution of the condensate wavefunction Ψ_0^* is near $\mathbf{k}_t = 0$, corresponding to an EM field detuning $\hbar\Delta_{\rm em} = \hbar\omega_{\mathbf{k}} - \mu \sim U_1N/2V + \hbar^2k_{\rm em}^2/2m - \mu$. The width of this distribution as $\Delta_{\rm em}$ is varied is governed by the width Δk_t of the initial atomic distribution and the additional momentum $\mathbf{k}_{\rm em}$. The contribution of processes revealed in the second and third terms in Eq. (60) is dominant at field detuning $\hbar\Delta_{\rm em} \sim U_1N/2V + \hbar^2k_{\rm em}^2/2m - \mu \pm \hbar\omega_j$. By tuning the value of $\Delta_{\rm em}$ we can select one of the three processes discussed above to be dominant. Low (or negative) values of $\Delta_{\rm em}$ give rise to a thermal output leading to an evaporative cooling of the trapped gas. High values of $\Delta_{\rm em}$ give rise to an increase in the number of excitations. Output atomic beams with the best coherence properties are expected for medium values of $\Delta_{\rm em}$, where the condensate atoms alone are coupled efficiently out of the trap.

This theory shows that the measurement of the spectrum of a weakly coupled output as a function of the output coupling parameters may be used to study the quantum state of a trapped atomic Bose gas at a finite temperature. The contribution to the output from pair breaking in the ground state shows it contains features that diagnose the quantum fluctuations in the condensate.

8 The Phase of a Condensate

This is an issue that has produced a great deal of discussion, and some novel insight into spontaneous symmetry breaking in mesoscopic systems. In the case of infinite systems the concept of spontaneous symmetry breaking has a rigorous basis. The broken symmetry corresponds to states with a mean value of the atomic field operator, with well defined amplitude and phase. This is analogous to the mean value of a laser field and corresponds to the quantum field being in a coherent state. In the case of an infinite system, states with different mean values of the atomic field operator become orthogonal and correspond to different macroscopic states. For the case of the trapped gas we have to consider what it means for a finite number of particles.

If we look at a region where two classical fields overlap we expect to see a steady interference pattern as long as their relative phase does not vary too rapidly. We know, however, from the quantum theory of phase that the relative phase between two laser fields (or condensates), with definite numbers of particles in them, is not defined. The uncertainty principle for phase takes the form $\Delta N \Delta \phi \gtrsim 1$. Does this mean that we should not expect to see an interference pattern if we start with condensates in Fock states i.e. for $\Delta N = 0$? It doesn't because as atoms are detected the measurement process itself builds up the relative phase between the two condensates. The way this happens is fairly straightforward to understand. If we make measurements of the arrival of atoms on a detector from a pair of condensates producing an interference pattern we cannot know from which condensate a given atom comes. This means that after a few measurements we cannot know the number of atoms in each condensate: precisely what we need to produce a state with a well defined relative phase! One gets for all practical purposes the same interference pattern one would predict on the basis of an assumption of a pre-existing phase of each of the condensates before the measurement. This is the same as taking both to initially be in coherent states.

In recent work we have argued that this procedure is a most sensible way to set up a phase standard for condensates. In fact it only makes sense to speak of the phase of a condensate relative to another one. Mølmer [40], has shown that isolated atomic fields cannot have non-vanishing mean amplitudes and the concept of an absolute phase for a condensate has no meaning. Such a conclusion gives special importance to the role of a 'phase standard': a reference condensate with which the phases of other condensates can be compared and the arbitrariness eliminated. Recent work [41,42] has shown in detail how making measurements entangles condensates and generates a relative phase. The question arises whether a condensate can be used as a consistent phase reference.

This concept has been discussed in previous work both in the context of Josephson junctions[43,44], and atomic condensates [45]. This shows, in particular, that the interference patterns observed are indistinguishable from those obtained if the condensates start in coherent states, i.e. states of well defined phase. From this observation, they proposed that if relative phases were generated between each of two condensates (A and B) and a third one (C), the relative phase between A and B can be predicted from the results of the other two measurements. Such a proposal has C acting in precisely the role of a phase standard.

We should bear in mind that Leggett has shown that such a phase standard cannot exist if the system is allowed to reach equilibrium with the environment[43] but might work for shorter time-scales [46]. This, in effect, is what we proposed: a phase standard for times short compared to phase-diffusion times.

The entanglement between the three modes is crucial to such a scheme, and we need to keep a full record of its evolution. Previous methods which use conditional probabilities to determine the position (or time) of detection of the next atom in interference schemes are not sufficient[41,42]. We choose to keep track of the quantum state of the system throughout the measurements. To do this we follow the quantum jumps due to atom detection (see, for example,[47]) of a system of two condensates incident on a beam splitter. This enables us to keep track of the entanglement throughout a particular realization and may be viewed as a representative history of an experiment[48]. The set-up is analogous to the theoretical work of Javanainen and Yoo[41], who analysed the case of two initially spatially separated condensates that are allowed to overlap and the positions of atoms in the interference pattern recorded. An entanglement between the two 'modes' is established due to the fact that one cannot know from which condensate the atoms have come. This entanglement leads to a relative phase.

There have been various proposals for measuring the relative phase between condensates. These are based on recording interference patterns [41], or by inducing Raman transitions which couple the condensates[49,50]. The precise way in which the measurement is made does not matter as all of these schemes rely on detecting a superposition of the two fields. We have chosen to use a beam splitter to produce this superposition but a variant of the same analysis would work for any of these schemes. We consider a 50:50 beam splitter with a condensate as the input to each of the two ports. The two condensate modes are identified with the operators a and b. If we transform to a frame rotating at the frequency of mode a, ω_a, the field operators at the two output ports of the beam splitter are

$$C_1 = \sqrt{\kappa/2}\left(a + ibe^{-i\Omega t}\right) \tag{61}$$

$$C_2 = \sqrt{\kappa/2}\left(ia + be^{-i\Omega t}\right), \tag{62}$$

where κ is the rate of detection of atoms and $\Omega = \omega_b - \omega_a$.

We follow the quantum state $|\psi_c\rangle$ of the system as it evolves under the influence of the detection process. The procedure for simulating this measurement process is as follows:

(i) We assume some mean rate of detection of atoms and find the probability that an atom is detected at port i in a time interval, Δt, thus

$$P_i(t, \Delta t) = \Delta t \langle \psi_c(t) | C_i^\dagger C_i | \psi_c(t) \rangle. \tag{63}$$

We then generate random numbers, r_i, from a flat distribution between zero and one, and compare it with P_i.

(ii) If $r_i < P_i$, a detection is made in port i during Δt and the system jumps to the renormalized form,

$$|\psi_c(t)\rangle \longrightarrow \frac{C_i|\psi_c(t)\rangle}{\sqrt{\langle \psi_c(t)|C_i^\dagger C_i|\psi_c(t)\rangle}}. \tag{64}$$

(iii) If we find $r_1 > P_1$ and $r_2 > P_2$, we assume that there is no detection in the interval, Δt, and the system evolves in the presence of the following effective Hamiltonian i.e.

$$|\psi_c(t)\rangle \longrightarrow \exp\left[-iH_{\text{eff}}\Delta t\right] |\psi_c(t)\rangle. \tag{65}$$

Here, we have taken $\hbar \equiv 1$, and the effective Hamiltonian, H_{eff} is given by, $H_{\text{eff}} = H_0 - i\kappa(a^\dagger a + b^\dagger b)/2$, where H_0 is the system Hamiltonian. The state needs to be renormalized after each step since H_{eff} is non-Hermitian.

(iv) We then repeat these steps to propagate the state in time until the desired number of atoms has been detected.

We study the phase build up between the condensates as atoms are detected at the output ports of the beam splitter. The system remains in a state of definite number since we detect and record every atom that escapes from the traps. Throughout the simulation, we find the relative phase between a and b, ϕ_{ab}. This is given by

$$\phi_{ab}(t) = \arg\left\{\langle \psi_c(t)|a^\dagger b|\psi_c(t)\rangle\right\}. \tag{66}$$

The measurement scheme serves the dual role of both creating and measuring the relative phase. This is a rather simple but effective definition of the phase and a fuller analysis is certainly possible. In analogy with the position of the spatial fringes in interference experiments, we can use the 'time-position' of the temporal fringes as a measurement of the relative phase.

On average, the difference in the number of atoms detected at each port per time interval is given by

$$D(t)/\Delta t = \langle\psi_c(t)|C_1^\dagger C_1 - C_2^\dagger C_2|\psi_c(t)\rangle \tag{67}$$

$$= i\kappa\left(\langle a^\dagger b\rangle e^{i\Omega t} - \langle b^\dagger a\rangle e^{-i\Omega t}\right). \tag{68}$$

We can write

$$\langle a^\dagger b\rangle = |\langle a^\dagger b\rangle| e^{i\phi_{ab}(t)}, \tag{69}$$

where $\phi_{ab}(t)$ contains the non-deterministic part of the relative phase due to the randomness of each trajectory. With this substitution, we get

$$D(t)/\Delta t = -2\kappa \sin\left(\Omega t + \phi_{ab}(t)\right). \tag{70}$$

The relative phase is thus given by the argument of the sinusoidal plot of the difference in the number of atoms detected at each port per time interval. As atoms are detected, it fluctuates for a while before settling down to a fixed constant value, Φ_{ab}. This value is random and varies for measurements made on identically prepared systems. At the end of this detection process, modes a and b are entangled and mode c is unaffected. The state vector for the system is given by

$$|\psi_c\rangle = \left(\sum_{i=N-l}^{N} c_i|2N-l-i\rangle_a|i\rangle_b\right)|N\rangle_c, \tag{71}$$

where N is the initial number of atoms in each trap, l is the number of atoms detected ($l < N$), and $\{c_i\}$ are the coefficients determined by the particular realisation.

In the second part of the simulation, we take mode b (now entangled with a) as the input to one port of a beam splitter, and mode c as the other input. As before, we record the times that atoms are detected at the two output ports. As in the first part, this eventually settles down to a fixed constant value, Φ_{cb}, and this varies randomly for measurements made on identically prepared systems.

An important observation is that the entanglement of b and c does not affect the existing entanglement between a and b, so long as the modes are not too severely depleted in the process. This is a very desirable property for a phase standard since it means that the standard is not corrupted in the process of comparing it with another condensate.

In this simulation, we consider that there is no independent loss from the traps. Such a process will corrupt the entanglements and so degrade the phase standard. This explains why the scheme will not work if the system is allowed to reach equilibrium with the environment[43]. We investigate the system on a different time-scale: long before it reaches equilibrium. The effect of interactions and loss are treated by Sinatra and Castin[51].

At the end of this second measurement stage, modes a and c are each entangled with b. The state vector for the system is

$$|\psi_c\rangle = \sum_{j=N-m}^{N} \sum_{i=2N-m-j}^{N} d_{i,j} |3N-i-j-m\rangle_a |j\rangle_b |i\rangle_c, \tag{72}$$

where m is the total number of atoms detected in both measurement stages ($m < N$), and $\{d_{i,j}\}$ are the coefficients determined by the numerical simulation.

Finally, in the third part, we calculate the relative phase between a and c, ϕ_{ac}. This is given by

$$\begin{aligned}\phi_{ac} &= \arg\left\{\langle\psi_c|a^\dagger c|\psi_c\rangle\right\} \\ &= \arg\left\{\sum_{j=N-m}^{N} \sum_{i=2N-m-j}^{N} d_{i,j} d^*_{(i-1),j} \sqrt{i(3N-i-j-m+1)}\right\}. \end{aligned} \tag{73}$$

Proposals have been made for how this could be measured[49]. For this scheme to work successfully as a phase standard, we require that $\phi_{ac} = \Phi_{ab} - \Phi_{cb}$. and this is found to be the case.

Of course, we could carry on and measure the relative phase between another condensate and our phase standard. This measurement will not destroy the entanglements with the other condensates, so we could then predict the relative phases between this condensate and the other two. The only limitation on how many times this phase reference can be used is the number of atoms it contains. Provided that the initial number of atoms is very large, many measurements can be made before the standard is depleted. Such an arrangement demonstrates all the properties that we require in a phase standard.

The relative phase becomes better defined as we become more and uncertain of the relative number distribution between the two condensates in the usual $\Delta N \Delta\phi$ sense. The distribution over relative numbers between the condensates will settle down to a Poissonian one after many atom detection events. This is consistent with the conventional field theoretic way of treating symmetry breaking in condensates i.e. taking the atomic field to have a mean value, $\langle\hat{\Psi}\rangle \neq 0$. This is equivalent to the assumption that a condensate is in a superposition of number states with a Poissonian distribution. It is then in a coherent state analogous to the state of the photon field in a laser cavity. We have seen that the uncertainty in the number of particles can be associated with entanglement with another condensate. In the conventional treatment this point is glossed over and we just suppose that the number of particles in a condensate is uncertain. The mean field is then given by

$$\langle \hat{\Psi} \rangle \equiv \sum_N c^*_{N+1}(t)\, c_N(t)\,.$$

Where $|c_N|^2$ is a Poissonian distribution, The mean value of the field then has the following time dependence:

$$\sum_N c^*_{N+1}(t)\, c_N(t) \exp\left[i\left(E_{N+1} - E_N\right) t/\hbar\right].$$

We can now expand the energy around its value for the mean number of particles N_0 in the superposition thus:

$$E_N \approx E_{N_0} + \left.\frac{\partial E}{\partial N}\right|_{N_0} (N - N_0) + \frac{1}{2}\left.\frac{\partial^2 E}{\partial N^2}\right|_{N_0} (N - N_0)^2$$

In the limit of very large numbers we can ignore the quadratic variation; all the components of the order parameter then have the same time dependence: $e^{-i\mu t/\hbar}$, where $\mu = \partial E/\partial N$. In that limit, i.e., a truly macroscopic condensate, the condensate continues to be described by a coherent state as it evolves. If, however, we look at some finite system then there will be a time t_0 for which the components of the coherent state will get out of step[52]. Whether this can be observed in an experiment is currently being studied. This analysis can be extended to relative phase evolution. It shows that the conventional picture of a slowly diffusing mean value will go over to a fixed order parameter in an infinite system.

9 Entanglement in four wave mixing

So far we have seen how the interactions between atoms in a condensate produce entanglement between them. In the description of output coupling we saw how it may be possible to probe this pairing through the choice of the right parameters for the output coupling field. In this section I want to discuss briefly the production of entangled beams of particles using four-wave mixing of matter waves. The crucial experiment in the field was recently carried out at NIST[53] and the background to this work is discussed in the lectures by Helmerson. The principle of the experiments can be explained by analogy with the optical case. The non-linearity in the matter wave case comes from the atomic interaction rather than the nonlinear response of the medium that the wave is travelling in. The resultant nonlinear Schrödinger equation is very similar to the optical case[36]. The properties of the GPE and the soliton solutions that it supports have been discussed in the lectures by Drummond. In this section we want to discuss the quantum nature of the output beams produced in the four-wave mixing process. We have already seen the sort of terms in the Hamiltonian that are important for generating the entangled beams:

$$\kappa a^{+} b^{+} + \kappa^{*} ab \tag{74}$$

In the case of the quadratic terms that promoted pairs from the condensate the coupling is not resonant and we just find an admixture of pairs of atoms into the ground state. This is the term in the Hamiltonian that we had to include in our description of the gas to go beyond the simplest version of mean-field. It was also responsible for the Beliaev damping process i.e. down-conversion of a single quasi-particle into a pair: the only process that can damp an excitation at zero temperature.

To see resonant coupling from condensates we need to start with colliding coherent beams, where energy and momentum is available to produce real pairs. In that case we can think of the quadratic Hamiltonian as representing the transfer of two particles in the colliding beams into two other states labelled by a and b. This process will produce appreciable transfer of pairs to the a and b beams when we have energy and momentum conservation holding for the process. This is why it only produced an admixture of excited states into the ground state of the system in the case of a static condensate.

In the case of colliding condensates we can get entangled pairs of atoms scattered in all directions. This is the analogue of parametric downconversion in a nonlinear crystal. One can stimulate the process through the use of a third beam in a four wave mixing process. This produces beams where a portion of the atoms is entangled[54]. The maximum entanglement that can be achieved in experiments of this sort is the subject of current investigations.

Acknowledgements

We would like to thank EPSRC, the EU and NIST for supporting this work. Discussions and help from: Rob Ballagh, Charles Clark, Stephen Choi, Matthew Davis, Jacob Dunningham, Mark Edwards, Chris Foot, Kris Helmerson, David Hutchinson, Yoni Jahpa, Mark Lee, Sam Morgan, Bill Phillips, Martin Rusch are gratefully acknowledged.

References

1. E.M. Lifshitz and L.P. Pitaevskii, *Statistical Physics, Part 2* (Pergamon Press, Oxford, 1980).
2. K. Huang, *Statistical Mechanics* (Wiley, New York, 1987).
3. N.N. Bogoliubov, J. Phys. U.S.S.R. **11**, 23 (1947).
4. A. Griffin, Phys. Rev. B **53**, 9341 (1996).
5. C.W. Gardiner and P. Zoller, Phys. Rev. A **58**, 536 (1998).
6. C.W. Gardiner, P. Zoller, R.J. Ballagh, and M.J. Davis, Phys. Rev. Lett. **79**, 1793-1796 (1997); C.W. Gardiner, M.D. Lee, R.J. Ballagh, M.J. Davis, and P. Zoller, Phys. Rev. Lett. **81**, 5226 (1998).
7. Yu. Kagan and B. V. Svistunov, Phys. Rev. Lett. **79**, 3331 (1997).
8. K. Damle, S.N. Majumbdar and S. Sachdev, Phys. Rev. A **54**, 5037 (1996).
9. R. J. Marshall, G.H.C. New, S. Choi and K. Burnett Phys. Rev. A **59**, 2085(1999).
10. H. Shi and A. Griffin, Physics Reports, **301**, 1 (1998).
11. M. Bijlsma and H.T.C. Stoof, Phys. Rev. A **55**, 498 (1997).
12. W.E. Parry, *The Many-Body Problem* (Oxford University Press 1974).
13. D.A.W. Hutchinson, R.J. Dodd, and K.Burnett, Phys. Rev. Lett **81**, 2198 (1998).
14. D.S. Jin, *et al.*, Phys. Rev. Lett. **77**, 420 (1996).
15. M.O. Mewes *et al.*, Phys. Rev. Lett. **77**, 988 (1996).
16. D.S. Jin *et al.*, Phys. Rev. Lett. **78**, 764 (1997).
17. S.T. Beliaev, Sov. Phys. JETP **7**, 289 (1958); *ibid.* **7**, 299 (1958).
18. F. Mohling and A. Sirlin, Phys. Rev. **118**, 370 (1960).
19. F. Mohling and M. Morita, Phys. Rev. **120**, 681 (1960).
20. V.N. Popov and L.D. Fadeev, Sov. Phys. JETP **20**, 890 (1965).
21. V.N. Popov, Sov. Phys. JETP **20**, 1185 (1965).
22. F.D. Fedichev and G.V. Shlyapnikov, Phys. Rev. A **58**, 3146 (1998).
23. S.A. Morgan, cond-mat/0011278.
24. D.A.W. Hutchinson *et al.*, to appear in J. Phys. B, and references therein.
25. J. Reidel, A. Csordas, R. Graham and P. Szepfalusy, cond-mat/9911018.
26. F.D .Fedichev, G.V. Shlyapnikov, and J.T.M. Walraven, Phys. Rev. Lett. **80**, 2269 (1998).
27. M. Edwards, P.A. Ruprecht, K. Burnett, R.J. Dodd, and C.W. Clark, Phys. Rev. Lett. **77**, 1671 (1996).
28. N.M.H. Hugenholtz and D. Pines, Phys. Rev. **116**, 489 (1959).
29. C. Cohen-Tannoudji, J. Dupont-Roc, and G. Grynberg, *Atom-Photon Interactions* (Wiley, New York, 1992).

30. M.H. Anderson, J.R. Ensher, M.R. Matthews, C.E. Wieman, E.A. Cornell, Science **269**, 193 (1995).
31. C.C. Bradley, C.A. Sackett, J.J. Tollett, R.G. Hulet, Phys. Rev. Lett. **75**, 1687 (1995).
32. K. Davis, M.O. Mewes, M R. Andrews, N.J. van Druten, D.S. Durfee, D. M. Kurn, and W. Ketterle, Phys. Rev. Lett. **75**, 3969 (1995).
33. M.O. Mewes *et al.*, Phys. Rev. Lett. **78**, 582 (1997).
34. G.M. Moy and C.M. Savage, Phys. Rev. A **56**, R1087 (1997).
35. R.J. Ballagh, K. Burnett and T.F. Scott, Phys. Rev. Lett. **78**, 1607 (1997); H. Steck, M. Naraschewski and H. Wallis, Phys. Rev. Lett. **80**, 1 (1998).
36. Y.B. Band, M. Trippenbach and P.S. Julienne, Phys. Rev. A, (submitted).
37. F. Dalfovo *et al.*, Phys. Rev. Lett. **75**, 2510 (1995); J. Low. Temp. Phys. **104**, 367 (1996); A.F G. Wyatt, Nature **391**, 56 (1998); A. Griffin, Nature **391**, 25 (1998).
38. D.A.W. Hutchinson, E. Zaremba and A. Griffin, Phys. Rev. Lett **78**, 1842 (1997); R.J. Dodd, M. Edwards, C.W. Clark and K. Burnett, Phys. Rev. A **57**, R32 (1998).
39. Y. Castin and R. Dum, Phys. Rev. Lett. **77**, 5315 (1996); **79**, 3553 (1997); *ibid.*, Phys. Rev. A **57**, 3008 (1998).
40. Klaus Mølmer, Phys. Rev. A **55**, 3195 (1997).
41. Juha Javanainen and Sung Mi Yoo, Phys. Rev. Lett. **76**, 161 (1996).
42. M.W. Jack, M.J. Collett, and D.F. Walls, Phys. Rev. A **54**, R4625 (1996); M. Naraschewski *et al.*, Phys. Rev. A **54**, 2185 (1996); Yvan Castin and Jean Dalibard, Phys. Rev. A **55** 4330 (1997).
43. A.J. Leggett, Found. Phys. **25**, 113 (1995).
44. A.J. Leggett, in *Bose-Einstein Condensation*, edited by A. Griffin, D.W. Snoke, and S. Stringari (Cambridge University Press, Cambridge, England, 1995).
45. J.I. Cirac *et al.*, Phys. Rev. A **54**, R3714 (1996).
46. A.J. Leggett and Fernando Sols, Found. Phys. **21**, 353 (1991).
47. H.J. Carmichael, *An Open Systems Approach to Quantum Optics* (Springer-Verlag, Berlin, 1993).
48. H.M. Wiseman and G.J. Milburn, Phys. Rev. A **47**, 1652 (1993).
49. J. Ruostekoski and D.F. Walls, Phys. Rev. A **56**, 2996 (1997); A. Imamoḡlu and T.A.B. Kennedy, Phys. Rev. A **55**, R849 (1997); J. Javanainen, Phys. Rev. A **54**, R4629 (1996).
50. D.S. Hall *et al.*, Phys. Rev. Lett. **81**, 1539 (1998).
51. A. Sinatra and Y. Castin, Eur. Phys. J. D **4**, 247 (1998).

52. E.M. Wright, D.F. Walls, and J.C. Garrison, Phys. Rev. Lett. **77**, 2158 (1996).
53. L. Deng *et al.*, Nature **398**, 218 (1999).
54. G. Lens *et al.* Phys. Rev. Lett. **71**, 3271 (1993).

THE THEORY OF ATOM LASERS

R.J. BALLAGH
Physics Department, University of Otago, Dunedin, New Zealand
E-mail: ballagh@physics.otago.ac.nz
Web: www.physics.otago.ac.nz/research/bec

C.M. SAVAGE
Physics Department, The Australian National University,
Canberra, ACT 0200, Australia.
E-mail: craig.savage@anu.edu.au
Web: www.anu.edu.au/physics/Savage

We review the current theory of atom lasers. A tutorial treatment of second quantisation and the Gross-Pitaevskii equation is presented, and basic concepts of coherence are outlined. The generic types of atom laser models are surveyed and illustrated by specific examples. We conclude with detailed treatments of the mechanisms of gain and output coupling.

1 Introduction

An atom laser is a device which produces a bright, directed, and coherent beam of atoms. An ideal atom laser beam is a single frequency de Broglie matter wave, approximating a noiseless sinusoidal wave; in particular it will have a well defined frequency, phase, and amplitude. Quantum mechanically, such a wave is a mode of the quantised field, and the properties of brightness and well defined phase require the mode to be highly occupied. This is equivalent to requiring Bose degeneracy, and rules out the possibility of a fermionic atom laser[1].

Atom lasers are a recent concept, and to our knowledge the earliest scientific description appeared in 1993 [2]. The first comprehensive papers on matter wave amplification[3] and atom lasers[4,5,6] appeared in 1995, and the first experimental demonstration of an atom laser in 1997 [7]. The essential components of an atom laser have been identified by analogy with optical lasers and are illustrated in Fig. 1. They are; a reservoir of atoms from which the laser is pumped, a laser mode in which atoms are accumulated, a pumping (or gain) mechanism which transfers atoms from the reservoir into the laser mode by stimulated transitions, and an output coupler which produces an output beam from the laser mode while preserving its coherence.

Many of the possible applications of atom lasers are probably still to be imagined, nevertheless atom lasers have the potential to revolutionise atom

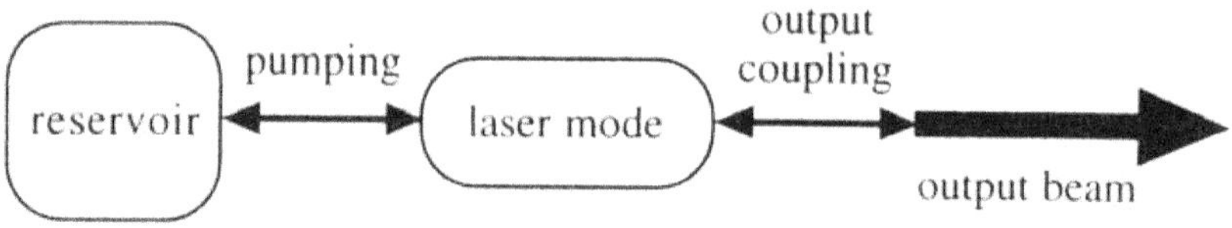

Figure 1. Schematic diagram of a generic atom laser.

optics just as optical lasers did conventional optics[8,9]. A bright coherent beam greatly enhances interferometric capabilities, for example allowing fringes to be obtained on a small time scale, and the use of unequal path length interferometers. The wavelength of the atoms is typically much smaller than light, and an obvious potential technological application is the precision deposition of matter for nanofabrication. While the similarities to optical lasers are important, the fundamental differences between atoms and photons are also significant. Atoms interact with each other, are subject to gravity, have complex internal structure, and may travel at any speed, not just c. Their interactions (collisions) make atom optics intrinsically nonlinear, as illustrated by a recent experiment showing self-induced four-wave mixing of atom laser beams[11]. Their interaction with the gravitational field gives the opportunity for atom laser beams to be used in high precision gravitational and inertial measurements[10], and very sensitive atom interferometers may be able to detect changes in space-time[8]. Eventually we expect that atom lasers will provide a convenient source for greatly improved atomic clocks, or for manipulation of the quantum mechanical state of the matter wave, analogous for example to optical squeezing[12].

As of April 2000 five experimental groups have demonstrated atom lasers experimentally, in either pulsed operation[7,14,15] or with quasi-continuous output[16,17]. In each case, the laser mode consists of a Bose-Einstein condensate confined in the ground state of a magnetic trap, gain is achieved by evaporatively cooling a reservoir of thermal atoms, and the output coupling is accomplished by using an external electromagnetic field to transfer the internal state of the atom from a trapped to an untrapped state. In these experiments the output coupling and the pumping processes are separated in time. Q-switching in pulsed optical lasers also separates the pumping and output coupling, but the cycle time is much shorter. Hence, a closer analogy is to a cavity loaded with coherent light which then leaks out through a mirror. The use of the name "atom laser" is justified by the coherence of the output beam, which was first demonstrated by the MIT group[7,13], and is due fundamentally to the stimulated emission of bosonic atoms into the ground state of the trap. It remains an outstanding experimental goal to achieve simultaneous pumping and output coupling to make a truly continuous atom laser.

Despite the parallels with optical laser theory, the fundamental differences between photons and atoms remain important. They arise principally because atoms have rest mass, and because they interact with each other. Unlike photons, atoms cannot be created or destroyed, rather they are transferred into the laser mode from some other mode. The interactions produce phase dynamics, which degrades the coherence of the atom laser, and self repulsion which spreads the output beam and limits the possible focussing. Even when these interactions are neglected, atoms have dispersive propagation in vacuum.

Various critical parameters for atom and optical lasers are radically different. For example the characteristic frequency in the optical case is the frequency of the light, of order 10^{15} Hz, while for atoms the de Broglie frequency is closer to that of the trapping potential, of order 10^3 Hz. And, whereas it is so far only practical to get atoms to accumulate into the ground mode of a trap, optical lasers typically operate on high modes of the optical cavity. Finally, a practical consideration is that atom lasers can operate only in ultra high vacuum.

In this paper we review the current state of theory of atom lasers. We begin, in sections 2-4, with a brief tutorial review of second quantisation and the Gross-Pitaevskii equation, and the concepts of coherence. In section 5 we survey the main types of theoretical models for atom lasers, and illustrate these with specific examples. Finally, in sections 6 and 7 we provide more detailed treatments of the mechanisms of gain and output coupling.

2 Second Quantization

In this section we review the theory of quantum fields: the most fundamental theoretical framework for treating atom lasers. As long as their internal atomic structure is not disrupted, atoms may be approximated as field quanta. However the internal state of the atom may change, e.g. between trapped and untrapped orientational spin states, and then each relevant internal state is described by its own field.

Identity of particles arises naturally in quantum field theory, as all particles of the same type are indistinguishable quantised excitations of the same field. We introduce quantum fields by second quantization of the wavefunction. For a more complete treatment see *Quantum Mechanics* by E. Merzbacher. The Hamiltonian for N particles interacting via the pairwise potential $V(\mathbf{r},\mathbf{r}')$ is

$$H=\sum_{i=1}^{N}H_0(\mathbf{r}_i)+\frac{1}{2}\sum_{i,j}V(\mathbf{r}_i,\mathbf{r}_j), \tag{1}$$

where the $\mathbf{r}_i$ are the particle position vectors. The factor of a half in the 2-body interaction term corrects for double counting of particle pairs. The single particle Hamiltonian $H_0(\mathbf{r})$ typically represents the kinetic energy and an external potential $V_E(\mathbf{r})$, such as a trap. In the position representation it is

$$H_0=-\frac{\hbar^2}{2m}\nabla^2+V_E(\mathbf{r}). \tag{2}$$

Second quantization allows us to automatically account for the identity of the particles, and is achieved by introducing the operator field $\Psi(\mathbf{r})$ which annihilates a particle at position $\mathbf{r}$. It obeys the following canonical commutation rules with its hermitian conjugate particle creation field $\Psi^\dagger(\mathbf{r})$,

$$\left[\Psi(\mathbf{r}),\Psi^\dagger(\mathbf{r}')\right]=\delta(\mathbf{r}-\mathbf{r}'),\qquad [\Psi(\mathbf{r}),\Psi(\mathbf{r}')]=0. \tag{3}$$

The second quantized Hamiltonian has the form of an "expectation value" of the Hamiltonian (1) with respect to these fields;

$$H=\int\Psi^\dagger(\mathbf{r})H_0\Psi(\mathbf{r})\,d^3r+\frac{1}{2}\int\int\Psi^\dagger(\mathbf{r})\Psi^\dagger(\mathbf{r}')V(\mathbf{r},\mathbf{r}')\Psi(\mathbf{r}')\Psi(\mathbf{r})\,d^3r'd^3r. \tag{4}$$

In the Heisenberg picture, operators obey the usual Heisenberg equation of motion. For example the field operator obeys

$$i\hbar\frac{d\Psi(\mathbf{r},t)}{dt}=[\Psi(\mathbf{r},t),H]. \tag{5}$$

Using the commutation rules (3) and the Hamiltonian (4) this becomes

$$i\hbar\frac{d\Psi(\mathbf{r})}{dt}=\left\{H_0(\mathbf{r})+\int\Psi^\dagger(\mathbf{r}')V(\mathbf{r},\mathbf{r}')\Psi(\mathbf{r}')\,d^3r'\right\}\Psi(\mathbf{r}). \tag{6}$$

It is often useful to expand the field operators in terms of products of operators a_i and the elements of a basis set of orthonormal mode functions $\{\phi_i(\mathbf{r})\}$,

$$\Psi(\mathbf{r})=\sum_i a_i\phi_i(\mathbf{r})\Longrightarrow a_i=\int\phi_i^*(\mathbf{r})\Psi(\mathbf{r})\,d^3r \tag{7}$$

This expansion is the basis of the Fock, or particle number, representation of the field. The commutation relations for the annihilation and creation operators follow from Eqs.(3) and the mode function orthonormality,

$$\left[a_i, a_j^\dagger\right] = \int\int \phi_i(\mathbf{r})\,\phi_j^*(\mathbf{r}')\left[\Psi(\mathbf{r}), \Psi^\dagger(\mathbf{r}')\right] d^3r d^3r' = \delta_{ij} \tag{8}$$

These commutators are those for a set of harmonic oscillators. Hence we can identify a_i as a harmonic oscillator type annihilation operator for the mode ϕ_i, and its Hermitian conjugate $a_i^\dagger$ as the corresponding creation operator. In quantum field theory these are interpreted as annihilating and creating field quanta, which in our case are atoms, in the mode. The quantum mechanical state of the system is that of these oscillators. In the Fock representation the second quantised Hamiltonian (4) becomes

$$\begin{aligned} H = &\sum_{i,j} a_i^\dagger a_j \int \phi_i^*(\mathbf{r})\,H_0(\mathbf{r})\,\phi_j(\mathbf{r})\,d^3r \\ &+\frac{1}{2}\sum_{i,j,k,l} a_i^\dagger a_j^\dagger a_k a_l \int\int \phi_i^*(\mathbf{r})\,\phi_j^*(\mathbf{r}')\,V(\mathbf{r},\mathbf{r}')\,\phi_k(\mathbf{r}')\,\phi_l(\mathbf{r})\,d^3r d^3r' \end{aligned} \tag{9}$$

If the mode functions are chosen to be eigenstates of the single particle free Hamiltonian, $H_0\phi_i = E_i\phi_i$, the free part of the Hamiltonian can be expressed in terms of the oscillator number operators $\{a_i^\dagger a_i\}$. Denoting the integrals in the interaction part of the Hamiltonian by V_{ijkl} we then find

$$H = \sum_i a_i^\dagger a_i E_i + \frac{1}{2}\sum_{i,j,k,l} a_i^\dagger a_j^\dagger a_k a_l V_{ijkl}. \tag{10}$$

Instead of solving for the full dynamics one may find the quasi-particles: the weakly interacting excitations of the system. The aim is to find an (approximately) diagonal form of H. The field is expressed in terms of the quasi-particle creation $b_i^\dagger$ and annihilation b_i operators by

$$\Psi(\mathbf{r}) = \sum_i \left(u_i(\mathbf{r})\,b_i - v_i^*(\mathbf{r})\,b_i^\dagger\right) + \psi_C(\mathbf{r}). \tag{11}$$

ψ_C is called the condensate, or macroscopic, wavefunction. It is normalised so that the integral over all space of its squared modulus is N_C, the number of particles in the condensate. Assuming a non-zero condensate wavefunction is the basis of the "spontaneous symmetry breaking" method[28] for analysing BECs, which is discussed further in the next section. The functions $u_i(\mathbf{r})$ and $v_i(\mathbf{r})$ are determined such that the Hamiltonian (10) has the approximate and noninteracting form

$$H \approx \sum_i b_i^\dagger b_i E_i' + \text{constant} \tag{12}$$

The quanta of these modes are the quasi-particles. The transformation to a quasi-particle representation is known as the Bogoliubov transformation.

3 The Gross-Pitaevskii Equation

In an atom laser the atoms will interact, both in the laser itself and in the output atomic beam. The interactions between cold atoms are usefully described as collisions. Elastic collisions conserve the total external energy of the colliding atoms. Inelastic collisions do not, and energy may be exchanged between the external and internal degrees of freedom. Changes in internal state due to inelastic collisions are an important source of losses of atoms from magnetic traps. We consider a simple s-wave scattering model of elastic collisions. This approximation is valid for low temperatures and low densities, such that both the atomic de Broglie wavelength and the inter-particle separation exceeds the characteristic range of the interaction potential. For further discussion see the article by Burnett in this volume.

In this limit the atom-atom potential may be approximated by a delta function potential with the same interaction energy

$$V(\mathbf{r},\mathbf{r}') \approx U_0 \delta(\mathbf{r}-\mathbf{r}'), \qquad U_0 = \frac{4\pi\hbar^2 a}{m}. \tag{13}$$

a is called the "scattering length". It is positive for repulsive interactions and negative for attractive ones. For indistinguishable particles the scattering cross section (the ratio of scattered to incident particle flux) is $\sigma = 8\pi a^2$, rather than $4\pi a^2$, as found for distinguishable particles. U_0 is the atom-atom interaction energy per atom. Using this approximation in the field operator equation (6) gives

$$i\hbar\frac{d\Psi(\mathbf{r})}{dt} = \left\{H_0(\mathbf{r}) + U_0\Psi^\dagger(\mathbf{r})\Psi(\mathbf{r})\right\}\Psi(\mathbf{r}). \tag{14}$$

We next approximate this operator equation by a non-linear Schrodinger equation for the condensate wavefunction.

The condensate wavefunction ψ_C is defined by Eq.(11) to be the contribution to the expectation value of the field operator Ψ due to the quasi-particle vacuum state. According to the "spontaneous symmetry breaking" assumption, above the BEC transition temperature $\psi_C = 0$, while below this temperature it is nonzero[28]. The name arises because the Hamiltonian (4) is symmetric, or invariant, under a global phase change $\Psi(\mathbf{r}) \to \Psi(\mathbf{r})e^{i\theta}$, whereas the condensate wavefunction must have a particular phase, breaking the global phase symmetry. This symmetry is a consequence of the pairing of particle creation and annihilation operators, and hence of particle number conservation, in the Hamiltonian (4). Particle number conservation in fact implies that $\Psi(r) = 0$ always[18,19], but spontaneous symmetry breaking is a useful heuristic assumption, which used with due caution, gives experimentally verified results. In fact, a number of calculations have been made without this assumption, and obtained the same results. Griffin's article in this volume further explores the spontaneous symmetry breaking assumption.

At zero temperature there are no quasi-particle excitations and the field operator Ψ in Eq.(14) may be approximated by the single particle condensate wavefunction ψ_C. We further assume that products of field operators may be approximated by the corresponding products of condensate wavefunctions. The condensate wavefunction then obeys the Gross-Pitaevskii equation[20]

$$i\hbar \frac{\partial \psi_C(\mathbf{r})}{\partial t} = \left\{ -\frac{\hbar^2}{2m}\nabla^2 + V_E(\mathbf{r}) + U_0 \left|\psi_C(\mathbf{r})\right|^2 \right\} \psi_C(\mathbf{r}) . \qquad (15)$$

Mathematically, this equation is a non-linear Schrödinger equation about which much is known. Nevertheless, in general a numerical solution is required[21].

4 Coherence

An atom laser is a device which emits a coherent beam of atoms. Coherence and noise are closely related. A noiseless classical wave, that is a perfect single frequency sinusoid, is defined to be perfectly coherent or perfectly correlated. For such a wave the field amplitude at any space-time event $A(x,t)$ completely determines the amplitude at any other event. In practice the field amplitude fluctuates due to classical or quantum mechanical noise, and such determination is not possible. Classical noise may be technical, such as vibrating components, or dynamical, such as relaxation oscillations in optical lasers[22]. Quantum mechanical noise may be interpreted as a consequence of the uncertainty relations.

Optical lasers operating far above threshold have a well stabilised intensity. Semiclassical theories describe optical lasers by quantising the atoms only, not the light. Introducing a phenomenological noise term allows spontaneous emission to be modelled without quantising the light. For such a theory, using the dimensionless intensity I defined by Mandel and Wolf[23], far above threshold

$$\langle (\Delta I)^2 \rangle \approx 2, \quad \langle (\Delta I)^2 \rangle^{1/2} / \langle I \rangle \approx (\sqrt{2}) / \langle I \rangle \qquad (16)$$

The angle brackets indicate the average over the noise which models spontaneous emission. The inverse dependence of the relative intensity fluctuations on the intensity is characteristic of the optical laser. It is expected to be true for the atom flux from atom lasers. In contrast, for thermal beams the relative intensity fluctuations are constant.

The phase of an optical laser far above threshold is also stabilised. This is measured by the electric field amplitude temporal coherence function $\langle E^*(t_1)E(t_2)\rangle$. Its Fourier transform is the spectral density, which is approximately Lorentzian[23]. Its width is the laser bandwidth, which is found to be inversely proportional to the intensity[24]. This is another characteristic optical laser property which we would expect to be true for atom lasers.

We now describe the coherence functions which are used to quantify the coherence of fields. The amplitude coherence function is:

$$\langle A^*(\mathbf{x}_1)A(\mathbf{x}_2)\rangle, \tag{17}$$

where the angle brackets denote an average over noise, and a bold $\mathbf{x}_k$ denotes the spacetime event $\mathbf{x}_k = (x_k, t_k)$. This coherence function is unambiguously referred to as a two-time amplitude correlation function. In the literature it may be referred to as either a first[25] or second order[23] amplitude coherence function. We use the former convention. If this function factorises

$$\langle A^*(\mathbf{x}_1)A(\mathbf{x}_2)\rangle = \langle A^*(\mathbf{x}_1)\rangle\,\langle A(\mathbf{x}_2)\rangle \tag{18}$$

the amplitude is said to be first order amplitude coherent. This definition captures the concept that a coherent field has no more noise in amplitude products than that due to the noise in the amplitude itself.

The definition of coherence as the degree to which coherence functions factorise is retained for quantum fields. Coherence functions (also called correlation functions) of order n were defined by Glauber[26] in terms of the field operators by

$$G^{(n)}(\mathbf{x}_1\ldots,\mathbf{x}_n,\mathbf{x}_{n+1}\ldots\mathbf{x}_{2n}) = \langle\Psi^\dagger(\mathbf{x}_1)\ldots\Psi^\dagger(\mathbf{x}_n)\Psi(\mathbf{x}_{n+1})\ldots\Psi(\mathbf{x}_{2n})\rangle, \tag{19}$$

where the angle brackets denote the quantum mechanical average. If $G^{(n)}$ factorises

$$G^{(n)}(\mathbf{x}_1\ldots\mathbf{x}_{2n}) = \langle\Psi^\dagger(\mathbf{x}_1)\rangle\ldots\langle\Psi^\dagger(\mathbf{x}_n)\rangle\langle\Psi(\mathbf{x}_{n+1})\rangle\ldots\langle\Psi(\mathbf{x}_{2n})\rangle \tag{20}$$

the field is said to be nth order coherent. This particular form of the coherence functions arose from photodetection theory. The pairing of creation and annihilation operators is a result of practical photodetectors measuring photon number. The normal ordering, in which all creation operators are to the left of all annihilation operators, arises because detection relies on absorption rather than stimulated emission, which would be affected by spontaneous emission. An all orders coherent quantum state is the annihilation operator eigenstate, the so called "coherent state". This is the quantum state that most closely approximates (superpositions of) noiseless classical waves.

One difficulty with this definition of coherence is that it relies on the spontaneous symmetry breaking assumption that the field expectation $\langle\Psi(\mathbf{x})\rangle$ value is not zero. Normalised coherence functions overcome this problem:

$$g^{(n)}(\mathbf{x}_1\ldots,\mathbf{x}_{2n}) = \frac{G^{(n)}(\mathbf{x}_1\ldots,\mathbf{x}_{2n})}{\left(G^{(1)}(\mathbf{x}_1,\mathbf{x}_1)\ldots G^{(1)}(\mathbf{x}_{2n},\mathbf{x}_{2n})\right)^{1/2}}. \tag{21}$$

For a maximally coherent field the numerator factorises and $g^{(n)} = 1$.

In a simple spatial interference experiment two sources at x_1 and x_2 propagate for the same time to a detection point. If the sources are coherent, interference results. The relevant coherence function is first-order: $g^{(1)}(x_1,x_2)$, because the intensity at the detection point contains the cross term $\langle\Psi^\dagger(x_1)\Psi(x_2)\rangle$ in the field amplitudes[25]. (Note that when either the spatial or temporal arguments are suppressed, as are the temporal arguments here, they are assumed to be equal.) If this cross term is zero the field fluctuations wash out the interference. The visibility of the interference fringes is maximal when $g^{(1)}(x_1,x_2) = 1$. From the definition Eq.(21), this is true for any single mode field, as for $g^{(1)}$ the only relevant expectation value is $\langle a^\dagger a\rangle$. In a BEC a large number of atoms are cooled into a single mode. Therefore a BEC with output coupling that preserves coherence is a prime candidate for an atom laser.

The first order spatial coherence of two freely expanding Bose-Einstein condensates has been observed experimentally by the detection of interference fringes[27]. In this experiment two condensates separated by 40μm were created in a double-well potential. The potential was switched off and the condensates left to expand over a period of 40ms after which time they overlapped. High contrast matter-wave interference fringes were observed indicating that the BECs were first order spatially coherent.

Ideally one would want a continuous output beam which is stationary, i.e. for which the first order temporal coherence, $g^{(1)}(t+\tau,t) = g^{(1)}(\tau)$ does not depend on t. The first order temporal coherence function for an atom laser beam, at some particular point, is defined as

$$g^{(1)}(\tau) \equiv \frac{\langle\Psi^\dagger(t+\tau)\Psi(t)\rangle}{\sqrt{\langle\Psi^\dagger(t)\Psi(t)\rangle}\sqrt{\langle\Psi^\dagger(t+\tau)\Psi(t+\tau)\rangle}}. \tag{22}$$

This is a useful measure of phase fluctuations, which we require to be small for a laser like source. For a laserlike, first order temporally coherent source,

$g^{(1)}(\tau) \approx 1$ over a time called the coherence time. This is in contrast to a noisy source for which $g^{(1)}(\tau)$, decays quickly to zero. According to Wiseman[1] an atom beam might be regarded as laser-like if the coherence time is much greater than the inverse atomic flux.

The second order coherence function relates two separate field quantum (atoms or photons) detection events. Unlike the first order coherence function. it is sensitive to the quantum mechanical state of the field. For a single mode thermal state $g^{(2)} = 2$ and for a BEC or single mode coherent state $g^{(2)} = 1$.

The second order spatial coherence of a matter wave source is related to the atom-atom interaction energy of the condensate[29]. This is because for a short-range potential, the interaction energy is proportional to the probability that two atoms are nearby, which is in turn proportional to $g^{(2)}$. The experimental evidence[30] is consistent with $g^{(2)} = 1.0 \pm 0.2$ for BECs, confirming that condensates suppress local density fluctuations.

As well as first order temporal coherence, an atom laser beam is expected to have higher order temporal coherence. The second order temporal coherence function (for a stationary system) is

$$g^{(2)}(\tau) \equiv \frac{\langle : I(t+\tau)I(t) : \rangle}{\langle I \rangle^2}, \tag{23}$$

where $I = \Psi^\dagger \Psi$, and the colons denote normal ordering, i.e. all creation operators to the left of all annihilation operators. $g^{(2)}$ is approximately unity for a coherent source. A filtered thermal beam, which is first order coherent, has $g^{(2)}(\tau) = 2$ for short times, approaching 1 for long times. In terms of counting field quanta $g^{(2)}(\tau)$ is determined by the distribution of arrival times. A field with $g^{(2)}(\tau) = 1$ has arrival times distributed according to Poissonian statistics, while $g^{(2)}(\tau) > 1$ corresponds to superpoissonian statistics, or bunching of arrival times. Only purely quantum mechanical correlations, such as those associated with squeezed states, allow $g^{(2)}(\tau)$ to be less than one.

Experimental evidence of third order coherence in BECs has been given by Burt *et al.*[31]. They compared the three-body recombination rate constant in condensed and non-condensed Bose gases. They found that the ratio of $g^{(3)}$ for a thermal cloud to that for a BEC was 7.4 ± 2, in agreement with the predicted value [32] of $3! = 6$. For a BEC $g^{(3)} = 1$, and for the non-condensed fraction $g^{(3)} = 6$. It is notable that $g^{(3)} = 1$ is the same as for classical distinguishable particles. This is because BEC particles occupy the same field mode and are therefore already indistinguishable in principle. Hence state symmetrization. which is used to enforce indistinguishability, thereby generating the quantum mechanical factor of 3!, is neither necessary nor possible.

For light only the Glauber type coherence functions that we have discussed are relevant, because all light detectors use photon absorption. However atoms may be detected in a variety of ways, and different coherence functions may be measured[33,34]. For example the spectrum of off resonant light scattered from a BEC is related to the density correlation function, rather than the Glauber normally ordered coherence function[33]. Normally ordered coherence functions are relevant for destructive detection, such as by ionisation detectors. We have focussed on them because they allow the most straightforward comparison with the optical laser.

5 Atom Laser Models

In this section we discuss the types of theoretical models which have been used to describe the atom laser. The starting point is usually an analogy with the optical laser. However despite the similarities, the differences between atom and optical lasers are also of considerable interest. We consider the advantages and disadvantages of three major types of atom laser models. In order of increasing complexity: rate equation, mean field, and fully quantum mechanical.

Rate equation models only consider state populations, ignoring the dynamics of coherences between quantum states. Hence they are unable to properly describe the quantum mechanical coherence properties of the atom laser. Nevertheless they usefully describe important physics such as the role of Bose-enhancement in producing the laser threshold.

The mean-field approach uses Gross-Pitaevskii type equations, Eq.(15). Since the atom laser is then described by a macroscopic wavefunction, quantum coherences may be included in the models. Additional terms may be introduced to describe the pumping and output coupling. A limitation of these models is that they assume that the field is in a coherent state, as discussed at the end of section 3. This means that the quantum coherence of the output cannot be calculated. However the nonlinear dynamics of the BEC and output beam are fully modelled, as is the spatial propagation of the output beam.

The most common fully quantum mechanical approach uses the quantum optical master equation[25]. This is an equation for a reduced density operator, obtained by averaging (strictly tracing) over part of the system, usually the output and/or pump modes. The Born-Markov approximation is commonly applied during the trace over the output modes to ensure that a differential equation, rather than an integro-differential equation, is obtained[35]. Because this approximation is not necessarily valid we shall also discuss more general quantum operator models.

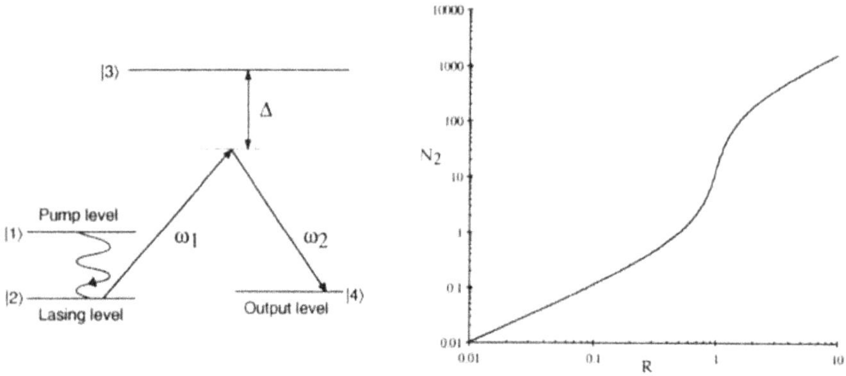

Figure 2. Schematic level diagram for the atom laser model of Moy *et al.*[36] (left). Plot of the steady state number of atoms in the lasing mode, as a function of the dimensionless pumping rate, R (right). Threshold occurs at $R = 1$.

In the rest of this section we give brief descriptions of these three approaches and the kinds of physics they treat. Pumping and output coupling are considered in detail in sections 6 and 7.

5.1 Rate equations

The first rate equation based atom laser models were presented by Olshanii *et al.*[4] and Spreeuw *et al.*[5]. We discuss a later model due to Moy *et al.*[36] which is based on implementation in a hollow optical fibre.

The model consists of atoms with four energy levels, as indicated in Fig. 2. Level $|1\rangle$ is the input pump level, level $|2\rangle$ is the lasing level and level $|4\rangle$ is the output level. Level $|3\rangle$ mediates the output coupling Raman transition and is not populated. Level $|1\rangle$ atoms irreversibly emit a photon to populate the lasing level $|2\rangle$ at the Bose-enhanced rate $r_{12}(N_{21}+1)$, where N_{21} is the number of atoms in level $|2\rangle$ and the ground state of the laser trap. A Raman transition couples atoms from level $|2\rangle$ into level $|4\rangle$ with rate constant r_{24}. The resulting rate equations for the numbers of atoms N_k in level $|k\rangle$ are

$$\frac{dN_1}{dt} = r_1 - \sum_j g_j(N_{2j}+1)r_{12}N_1 - (1 - \sum_j g_j)r_{12}N_1,$$

$$\frac{dN_{2j}}{dt} = g_j r_{12} N_1 (N_{2j}+1) - r_{24}N_{2j} + G_j(N_{2j}+1)N_4 r_{24},$$

$$\frac{dN_4}{dt} = \sum_j N_{2j} r_{24} - \sum_j G_j (N_{2j} + 1) N_4 r_{24} - N_4 \frac{1}{t_0}. \tag{24}$$

The N_{2j} are the numbers of atoms in level $|2\rangle$ in the jth mode of the lasing level trap. g_j is the overlap between the pump mode and the jth mode of the lasing trap. $G_j = N_{4j}/N_4$ is the proportion of atoms with electronic level $|4\rangle$ that have the same spatial mode as the jth mode of the lasing trap. t_0 is the time scale on which atoms are irreversibly lost from the system due to the Raman momentum kick. For a more complete description of these equations refer to Moy *et al.*[36].

The solutions of these rate equations show behaviour familiar from optical lasers. The steady state population of the laser mode N_{21} as a function of a suitably scaled pump rate $R = r_1 g_1 / r_{24}$ shows a rapid increase at the threshold value of $R = 1$, as shown in Fig. 2. As a function of time the above threshold values of the lasing cavity mode populations N_{2j} show initial growth, and then all but the lasing mode N_{21} decay. This is the familiar mode competition in which the "winner takes all" due to runaway Bose stimulation.

The rate equation model of Olshanii *et al.*[4] is similar in form. However they included photon reabsorption and found that lasing is inhibited by small amounts of reabsorption.

5.2 *Mean-field models*

In this approach, the atom laser is modelled using the Gross-Pitaevskii (GP) type equations discussed in section 3. The "mean-field" is the GP wavefunction which determines the effective potential felt by an atom due to its 2-body interactions with all of the other atoms in the condensate. Indeed one of the advantages of GP models is this straightforward treatment of atom-atom interactions.

Kneer *et al.*[42] describe a GP model of an atom laser including pump and loss terms. The modified GP equation for the laser mode macroscopic wavefunction is

$$i\hbar \frac{\partial \psi}{\partial t} = -\frac{\hbar^2}{2m}\frac{\partial^2 \psi}{\partial x^2} + V_E(r)\psi + U_0|\psi|^2\psi + H_{\text{gain}}\psi + H_{\text{loss}}\psi. \tag{25}$$

The first three terms are standard. The gain and loss are represented by phenomenological terms with rate constants Γ and γ_c respectively,

$$H_{\text{gain}}\psi = \frac{i\hbar}{2}\Gamma N_u \psi, \quad H_{\text{loss}}\psi = -\frac{i\hbar}{2}\gamma_c \psi. \tag{26}$$

In addition to the GP equation (25), a rate equation is used to describe the number of uncondensed atoms, N_u,

$$\frac{dN_u}{dt} = R_u - \gamma_u N_u - \Gamma N_c N_u. \tag{27}$$

R_u describes a source which pumps the uncondensed state at rate R_u. The term $\gamma_u N_u$ describes atoms lost from the system, but not trapped in the condensed state. The final term describes the transfer of atoms into the condensed state. The condensate loss is spatially homogeneous, and undamped collective excitations are found. To avoid this they introduce a spatially dependent decay rate, so that γ_c is a function of position, $\gamma_c(x)$. This produces a single lasing mode as the spatially dependent loss leads to mode selection.

Robins *et al.*[43] present a GP model which combines the pumping model of Kneer *et al.*, with an explicit treatment of the output coupling and of the output beam. Including the output beam in the model allows one to investigate its spectrum and spatial distribution. Furthermore, non-Markovian effects in the output, such as Rabi oscillations between the condensate and beam, can be explored. The equations which describe the laser mode $\psi^a(x,t)$, and the output beam $\psi^b(x,t)$ are

$$\begin{aligned} i\hbar\frac{\partial\psi^a}{\partial t} &= -\frac{\hbar^2}{2m}\frac{\partial^2}{\partial x^2}\psi^a + \frac{1}{2}m\omega^2x^2\psi^a + U_0^a|\psi^a|^2\psi^a + U_0^a|\psi^b|^2\psi^a \\ &\quad -i\gamma_r(|\psi^a|^4 + |\psi^b|^4)\psi^a - \hbar\Gamma_R e^{ik_0x}\psi^b(x) + \frac{i\hbar}{2}\,\Gamma\,N_u\psi^a, \\ i\hbar\frac{\partial\psi^b}{\partial t} &= -\frac{\hbar^2}{2m}\frac{\partial^2}{\partial x^2}\psi^b + mgx\psi^b + U_0^b|\psi^b|^2\psi^b + U_0^a|\psi^a|^2\psi^b \\ &\quad -i\gamma_r(|\psi^a|^4 + |\psi^b|^4)\psi^b - \hbar\Gamma_R e^{-ik_0x}\psi^a. \end{aligned} \tag{28}$$

The number of uncondensed atoms N_u obeys Eq.(27). The Raman output coupling is described by the terms containing the momentum kick generators $e^{\pm ik_0x}$. Three body recombination has been included as a condensate loss mechanism, by the terms proportional to γ_r. This is found to have a stabilising effect on the non-linear dynamics of the atom laser. Two-body losses have not been included.

Typical output from a one dimensional numerical solution of this model is shown in Fig. 3. The oscillations represent nonlinear dynamical noise, which will determine the laser's effective linewidth, or first order coherence. This is an example of physics which cannot be treated by the rate equation approach.

How mean field models may be extended to incorporate finite temperature effects, such as the population of excited levels, is described in subsequent sections, and by Burnett and Griffin in this volume.

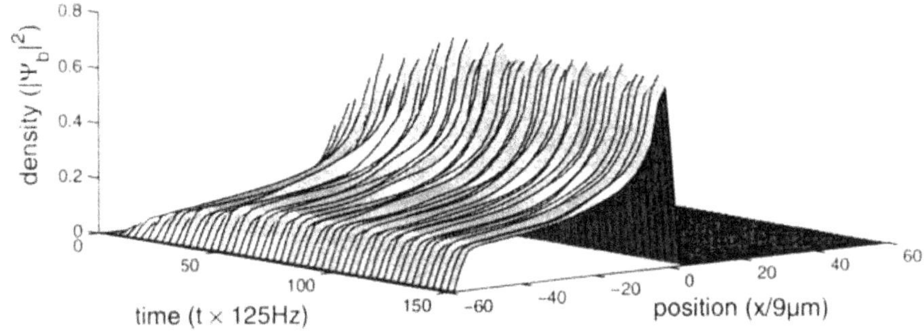

Figure 3. **Atom laser beam density as a function of position and time, showing a quasi-steady state. Atoms are absorbed at the spatial boundaries. Parameters are given by Robins *et al.*[43].**

5.3 *Fully quantum mechanical models*

A number of quantum optical master equation based models of the atom laser have been discussed in the literature[6,37,38,39]. The model of Holland *et al.*[6] is based on evaporative cooling and incorporates: the creation of an atom in the trap from the pump field, the loss of an atom from a high lying trap state, collisions which scatter atoms into or out of the ground state, and the loss of atoms from the ground state through output coupling.

A three level model is sufficient to describe these features. Atoms are injected with rate κ_1 in level one and are lost with rates κ_0 and κ_2 from level zero and level two respectively. Only the energy conserving 2-body interaction terms are considered. Level two is adiabatically eliminated by assuming large damping due to the evaporative cooling. This leads to the master equation for the density operator ρ,

$$\frac{\partial \rho}{\partial t} = \frac{1}{i\hbar}[V_i, \rho] + \kappa_0 \mathcal{D}[a_i]\rho + \kappa_1 \mathcal{D}[a_i^\dagger]\rho + \Omega_r \mathcal{D}[a_0^\dagger a_1^2]\rho \tag{29}$$

where

$$V_i = \hbar(a_0^\dagger a_0 - 2n_0)(V_{0000} a_0^\dagger a_0 + V_{0101} a_1^\dagger a_1) + \hbar V_{1111} a_1^{\dagger 2} a_1^2, \tag{30}$$

and n_0 is the mean population of level zero. The loss terms are assumed to be of the standard (Born-Markovian) Lindblad form, defined by

$$\mathcal{D}[c]\rho = 2c\rho c^\dagger - (c^\dagger c\rho + \rho c^\dagger c) \tag{31}$$

The effective redistribution rate Ω_r is proportional to $|V_{0211}|^2/\kappa_2$.

Equations for the populations, $\langle nk|\rho(t)|nk\rangle$, where $|nk\rangle$ describes a state with n atoms in level zero and $k = 0$ or $k = 1$ atoms in level one, are obtained. Level one is rapidly depleted by two atoms scattering into levels zero and two. The steady state population of the level zero lasing level is found to be Poissonian with mean atom number $2\kappa_1/(3\kappa_0)$. Thus when loss out of the lasing mode is sufficiently small $\kappa_0 << \kappa_1$, a large population can build up in the lasing mode.

In order to produce a state with a well defined phase they consider parameter regimes where Ω_r is much larger than the frequency shifts in V_i. This inhibits the dispersive nature of the latter from destroying phase coherence. They find that, as for an optical laser, line narrowing occurs as the population of the ground state increases. However, in a refinement of the model due to Wiseman *et al.*[39] the linewidth of the laser output is broader than the bare linewidth of the laser mode.

The crucial refinement is in the description of the pumping of level 1. The approximation that Holland *et al.* make allows atoms to leak into, but not out of, the trap. To allow the latter Wiseman *et al.* replace the term $\kappa_1\mathcal{D}[a_1^\dagger]\rho$ with

$$\kappa_1(N+1)\mathcal{D}[a_1^\dagger]\rho + \kappa_1 N\mathcal{D}[a_1]\rho. \tag{32}$$

The N is the mean population of the pump reservoir modes. In fact, there is no limit in which the extra terms can be ignored. When $\kappa_0 << \kappa_1$, there are two distinct parameter regimes depending on the relative size of a parameter proportional to $\kappa_2|V_{0211}|^2$, which they call Γ, and κ_0. These they call the strong and weak collision regimes: $\Gamma >> \kappa_0$ and $\Gamma << \kappa_0$ respectively. In the weak collision regime, $\Gamma << \kappa_0$ there is a large population in the laser mode if $N >> 1$, i.e. with strong pumping. The power spectrum reveals a linewidth larger than the bare linewidth of the lasing mode. Nevertheless, the phase diffusion rate may still be slow in the sense of being much less than the output flux of atoms from the lasing mode.

Wiseman and Collett[44] proposed an atom laser model using optical dark state cooling. The laser is found to have a linewidth narrower than the bare linewidth. This is an improvement on models based on evaporative cooling discussed above. The output is also found to be second order coherent.

The master equation is one means of describing the full quantum statistics of an atom laser. It has the great advantage of drawing on the wealth of techniques that have been developed in quantum optics[25]. However, a fully quantum theory of the atom laser can be developed without using the standard master equation approximations[45,46,47]. In particular, assuming a

Born-Markov form for the output coupling ignores the possibility of the output beam acting back on the laser mode. This may happen, for example, due to Rabi cycling between the output and laser modes.

Hope[45] developed a general theory of the input and output of atoms from an atomic trap. Moy *et al.*[46] used this theory to find an analytic expression for the spectrum of a beam of atoms output from a single mode atomic cavity. For very strong coupling it is just the spectrum of the trapped lasing mode, and is therefore broad. As the strength of the coupling is reduced, the linewidth is reduced. For small coupling strengths the lineshape approaches a Lorentzian, centred about the energy of the cavity with a linewidth proportional to the coupling strength. Hope *et al.*[47] include a pumping mechanism and they find that the output spectrum is substantially wider than that obtained using the Born-Markov approximation. However an important limitation of these theories is that they do not include 2-body interactions.

6 Atomic Gain

We have seen in sections 1 and 5 that a mechanism for atomic gain, to provide coherent amplification of the atomic field in the lasing mode, is an essential element of an atom laser. A number of gain mechanisms have been proposed, (e.g. see section 5.1) but to date only one, evaporative cooling, has been demonstrated experimentally. The process of evaporative cooling is based on the removal of high energy atoms from the gas, so that subsequent collisional rethermalisation of the remaining atoms results in a distribution of lower temperature (i.e. lower average energy). Under appropriate conditions, bosonic stimulation of transitions into the ground state of the confining potential leads to the formation of a Bose condensate: in an atom laser this would constitute the lasing mode.

The kinetics of condensate growth has long been a subject of theoretical study[47–49], but the first quantitative calculations were carried out only recently by Gardiner. Together with coworkers he has given a comprehensive treatment of condensate formation by bosonic stimulation, in a series of papers[50–55]. In this section we review the major features and results of this theory.

6.1 Quantum Kinetic theory

The starting point of the treatment is the third in a series of papers on Quantum Kinetic theory developed by Gardiner and Zoller[53], which we shall refer to as QKIII. We note that an alternative formulation of a Quantum Kinetic

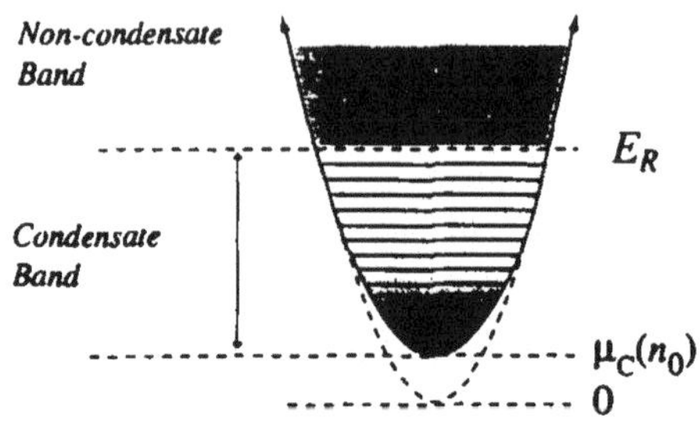

Figure 4. Schematic representation of modification of trap energy levels by condensate meanfield. Levels above E_R are unaffected.

theory has been given by Griffin and coworkers (see this volume, and references therein). The gas of Bose atoms in a harmonic trap of frequency ω is described by a second-quantized field, as in Eqs.(1 - 4). Two-body collisions are included but three-body collisions are neglected. Guided by the fact that the energy spectrum above a certain value E_R is unaffected by the presence of the condensate, we can divide the field into the condensate band (all energy levels below E_R) and the non-condensate band (all the energy levels above E_R), as represented in Fig. 4.

The noncondensate band (which contains the majority of the atoms) is taken to be fully thermalised, in the sense that for any point $\mathbf{r}, \mathbf{k}$ in phase space the single particle distribution function $F(\mathbf{r}, \mathbf{k})$ (in principle a Wigner function) is localised, and has a well defined temperature T and chemical potential μ. Thus the noncondensate band can be treated as a time independent bath, and a master equation is derived for the density matrix of the condensate band, (Eqs. (50a)-(50f) QKIII). It is worth emphasizing that in general the condensate band is not in thermal equilibrium, and the master equation provides a description of its dynamical evolution. The level structure of the condensate band is well described by the Bogoliubov spectrum, and Gardiner[58] has shown that appropriate states for this band are $|N, \{n_m\}\rangle$ which have a definite number N of total particles in the band, and a set of numbers $\{n_m\}$ representing the occupation numbers of the condensate ($m = 0$) and quasiparticle (excited Bogoliubov) levels. In this treatment. the condensate wavefunction and energy eigenvalue (the condensate chemical potential $\mu_C(N)$) are given by the time-independent Gross-Pitaevskii equation with n_0 atoms. The quasiparticles may be particle-like excitations (at higher energies) or phonon-like collective excitations (at low energies). The phonon excitations are not eigenstates of the number operator and thus the total population of the quasiparticles need not be conserved, in contrast to N which is strictly conserved.

By taking a representation of the master equation in terms of the states $|N, \{n_m\}\rangle$, and then taking the diagonal elements, a stochastic master equation for the occupation probabilities $p(N, \{n_m\})$ is obtained. The transitions appearing in this equation can be divided into those which change the number of atoms N in the condensate band, and those which do not. The former, which we call *growth processes*, arise predominantly from collisions of two noncondensate atoms where one of the outgoing atoms goes into the condensate band. The reverse collision is also included in the growth processes. The transitions which do not change N are called *scattering processes*, and arise predominantly from collisions between a condensate band atom and a noncondensate band atom which leave one atom in each band. Other collision processes, in which the pair of colliding atoms either begins or ends in the condensate band, will in principle also contribute to growth or scattering, but can be neglected since the condensate band contains such a small fraction of all the atoms. We note that collisions between pairs of atoms that begin and remain in the noncondensate band play the role of thermalising the bath and do not appear in this master equation.

6.2 *'Particle-like' rate equations*

Using standard methods, a deterministic rate equation for the mean population $\langle n_m \rangle$ of the mth quasiparticle level (which we will subsequently write n_m) can be obtained from the master equation and we write it as

$$\frac{dn_m}{dt} = \dot{n}_m|_{\text{growth}} + \dot{n}_m|_{\text{scatt}} \tag{33}$$

where the forms for the growth transition rates $\dot{n}_m|_{\text{growth}}$ and the scattering transition rates $\dot{n}_m|_{\text{scatt}}$ will be discussed below. A similar deterministic rate equation can be written for the mean population of the condensate band $\langle N \rangle$ (which we will subsequently write N).

The growth transition rates $\dot{n}_m|_{\text{growth}}$ can be obtained in explicit form from the population master equation given in Eq.(189) QKIII, and comprise all the processes where N changes by 1, and one of the quasiparticle populations, n_j say, changes by 1 or 0, (with all other n_k unchanged). Most of these processes (e.g. $N \to N+1$ and $n_j \to n_j + 1$, or $N \to N+1$ and $\{n_m\}$ unchanged) admit a particle-like interpretation. However, those processes where N and n_j change in opposite directions have to be interpreted as phonon-like processes, and for example could represent a process where the condensate is initially oscillating, and a thermal atom is added to the condensate band

and causes the condensate oscillation to be reduced. All such phonon-like processes are neglected in this treatment, since only a small fraction of quasiparticle wavefunctions have phonon-like character, and we expect they will play an unimportant part in the condensate growth. The growth terms for the quasiparticle levels thus simplify to the form

$$\dot{n}_m|_{\text{growth}} = 2W_m^{++}(N)\,(n_m+1) - 2W_m^{--}(N)n_m \tag{34}$$

and the equation for the mean number in the condensate band becomes

$$\frac{dN}{dt} = 2W^+(N)\,(N+1) - 2W^-(N)N \tag{35}$$

where $W_m^{++}(N)$ and $W^+(N)$ are forward rates and $W_m^{--}(N)$ and $W^-(N)$ are backward rates. It is shown in QKIII that when the condensate wavefunction $\xi_N(\mathbf{r})$ is sharply peaked compared to the spatial dependence of $F(\mathbf{r},\mathbf{k})$, then

$$W^+(N) = \frac{4a^2}{m^2(2\pi)^3}\int d^3\mathbf{K}_1 \int d^3\mathbf{K}_2 \int d^3\mathbf{K}_3 \int d^3\mathbf{k}\,\delta(\mathbf{K}_1+\mathbf{K}_2-\mathbf{K}_3-\mathbf{k})$$
$$\delta(\Delta\omega_{123}-\mu_C(N)/\hbar)F(0,\mathbf{K}_1)F(0,\mathbf{K}_2)\,(1+F(0,\mathbf{K}_3))\,|\tilde{\xi}_N(\mathbf{k})|^2 \tag{36}$$

The integrand represents a collision in which a pair of noncondensate atoms of momenta $\mathbf{K}_1$ and $\mathbf{K}_2$ produces one noncondensate atom ($\mathbf{K}_3$) and one condensate band atom of momentum $\mathbf{k}$. Energy conservation is expressed in the second δ function, in which $\Delta\omega_{123} = \omega_{\mathbf{K}_1} + \omega_{\mathbf{K}_2} - \omega_{\mathbf{K}_3}$, where $\hbar\omega_{\mathbf{K}_i} = \hbar^2\mathbf{K}_i^2/2m + V_T(0)$ is the particle energy at the centre of the trap potential V_T. $\tilde{\xi}_N(\mathbf{k})$ is the momentum-space ground-state wavefunction. The expression for $W_m^{++}(N)$ is similar to Eq.(36), and the forward and backward rates are related by

$$W^+(N) = e^{[\mu-\mu_C(N)]/kT}W^-(N) \tag{37}$$
$$W_m^{++}(N) = e^{[\mu-e_m]/kT}W_m^{--}(N) \tag{38}$$

where e_m is the energy of the mth quasiparticle level. Equations (33)-(38) are derived within the approximation that the number of particles in the condensate, n_0, is large enough that we can write $n_0 \approx N$, (which is required for the Bogoliubov spectrum to be accurate). However, in the early stages of condensate growth when n_0 is small, the unperturbed spectrum should be used, and will be accurate provided none of the condensate band levels are too highly populated. This means that by making the replacement $N \to n_0$ in Eqs.(33)-(38) those equations will provide a good description of the dynamical evolution of the levels in the condensate band throughout the formation of the condensate. Values for the rates W^+ can be obtained in varying levels of approximation, and the simplest expression, used in the first quantitative description of growth[52], is obtained by assuming a classical Maxwell-Boltzmann distribution for $F(\mathbf{r}, \mathbf{k})$, and allowing the integrals in $\mathbf{K}_{1,2,3}$ in Eq.(36) to extend over all energies, rather than just the noncondensate band. This leads to the explicit expression

$$W^+(n_0) \approx \frac{4m(akT)^2}{\pi\hbar^3} e^{2\mu/kT} \tag{39}$$

in which we note the prefactor to the exponential is independent of n_0 and is essentially the elastic collision rate $\rho\sigma v$ with quantities evaluated at the critical point for condensation. More accurate evaluations[54,56] of W^+ which take into account the Bose-Einstein nature of the non-condensate distribution and exclude the condensate region in the $\mathbf{K}$ integrals, give a value about three times larger than that of Eq.(39). $W_m^{++}(n_0)$, which has a form similar to Eq.(36), is more difficult to evaluate but is of the same order of magnitude as $W^+(n_0)$, and in the first instance we set $W_m^{++}(n_0) \approx W^+(n_0)$. The sensitivity of the results to this approximation will be discussed in section 6.3. Finally, in order to reduce the number of equations to a tractable set, an ergodic assumption is made which allows us to group together levels of similar energy into small sub-bands, with only the ground state being described by a single level. The sub-bands have (central) energy ϵ_m and degeneracy g_m, and the equations for the growth processes of these bands are

$$\dot{n}_m|_{\text{growth}} = 2W_m^{++}(n_0)\left\{\left[1 - e^{[\epsilon_m - \mu]/kT}\right] n_m + g_m\right\}, \tag{40}$$

$$\dot{n}_0|_{\text{growth}} = 2W^+(n_0)\left\{\left[1 - e^{[\mu_C(n_0) - \mu]/kT}\right] n_0 + 1\right\}. \tag{41}$$

These equations exhibit the most important features of condensate growth, or atomic gain, as we shall discuss shortly in section 6.3. The remaining terms required for Eq.(33), the scattering terms $\dot{n}_m|_{\text{scatt}}$, describe processes in which a condensate band atom and a noncondensate band atom collide and leave one atom in each band. Gardiner and Lee[56] have evaluated the scattering terms within well defined approximations (for the case of an isotropic trap), to give

$$\dot{n}_m|_{\text{scatt}} = \frac{8ma^2\omega^2}{\pi\hbar} e^{\mu/k_BT}\Gamma(T) \times$$
$$\left\{\sum_{k<m} \frac{1}{g_m}\left[n_k(g_m + n_m)e^{-\hbar\omega_{mk}/k_BT} - n_m(g_k + n_k)\right]\right.$$

$$+\sum_{k>m}\frac{1}{g_k}\left[n_k(g_m+n_m)-n_m(g_k+n_k)e^{-\hbar\omega_{km}/k_BT}\right]\Bigg\}. \tag{42}$$

where $\Gamma(T)=\sum_{e_m>E_R}e^{-e_m/kT}$, $\hbar\omega_{nq}=e_n-e_q$, and the notation $k>m$ means $e_k>e_m$.

6.3 Results

Equations (40 - 42) provide an explicit form for the set of deterministic rate equations for populations in the condensate band, with the size of the set determined by the amount of level grouping. Much of the behaviour that occurs in the full set of equations is captured by a simple implementation in which the condensate band is represented by only one level[52], i.e. the condensate. The equations then reduce to a single equation (41), which has the laser-like features of stimulated transition rates (the terms proportional to n_0) and a spontaneous transition rate (the term +1 in the curly braces) which initiates growth. In addition, this so-called *simple growth equation* has thermodynamic features, in that the forward and backward transition rates become equal (i.e. equilibrium is reached) when the chemical potential of the condensate, $\mu_C(n_0)$, becomes equal to μ of the thermal bath (to order $1/N$). A crucial aspect of the growth process is that as the condensate grows. the meanfield causes $\mu_C(n_0)$ to increase, beginning from $\mu_C(0)=3\hbar\omega/2$ (the trap zero-point energy), and rising as $(n_0)^{2/5}$ (in the Thomas-Fermi approximation) at large n_0. It is clear that a condensate can only grow if the thermal bath has $\mu>0$, and this is achieved in the process of evaporative cooling by removing the highest energy atoms. The higher energy levels quickly come to their equilibrium distributions, but the lower energy levels remain far from equilibrium, and the ensuing tendency of collisions of lower energy bath atoms to feed the condensate band can be characterised by a bath chemical potential $\mu>0$.

A typical result for the simple growth equation is shown as the dotted line in Fig.5 where we have chosen a case corresponding to the conditions of the MIT growth experiment[57] in which sodium atoms of scattering length $a=2.75$ nm are contained in a trap of geometrical mean frequency $\omega=2\pi\times 50$ Hz. Although we do not expect quantitative accuracy from this simple model, the characteristic S shape of the growth curve exhibits the key aspects of the growth process. In the initial stages the growth is slow and arises from spontaneous transitions; then as n_0 increases, stimulated transitions dominate and the growth is steep. Finally, as the condensate chemical potential approaches μ, there is a slow approach to equilibrium.

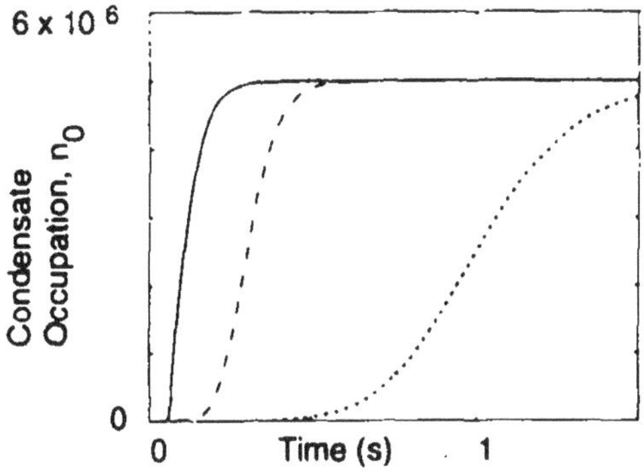

Figure 5. Typical condensate growth curves for different implementations of the model: simple growth equation (dotted line); full model (solid line); with no scattering (dashed line). Parameters are appropriate to the regime of the MIT experiment: $\mu = 43.3\hbar\omega$, $T = 900$ nK, $n_0(t=0) = 100$.

Of course for a quantitatively accurate treatment, more condensate band levels must be included, along with the more careful calculation for W^+ (see Eq.(10) ref[54]). Typically 20-50 sub-bands are used, each of width $\approx \hbar\omega$. The energy e_m of these sub-bands increase as n_0 grows, with the lower levels most affected, and the levels at the top of the condensate band not affected at all. Although the increase of the condensate energy level is crucial to the growth process and must be treated as accurately as possible, it is found that there is little sensitivity to the scheme used to alter the quasiparticle energy levels. Typically, the levels between $\mu_C(n_0)$ and $2\mu_C(n_0)$ are compressed from the bottom in a linear fashion, and higher levels are unaltered. The solid line in Fig.5 shows the effect of using a set of sub-bands in the growth process. The physical parameters are identical to those used in the simple growth model (the dotted curve) but there has been a speed up of about an order of magnitude. The effects of the scattering terms can be separated from the growth terms, as we show with the dashed line in Fig.5, which is a calculation using the same number of sub-bands as for the solid curve, but with all scattering terms put to zero. It is clear that the main effect of the scattering processes is to cause the initiation of condensate growth to occur much more sharply. This is due to the fact that the growth processes can initially populate a wide number of sub-bands, and then scattering allows these populations to be redistributed directly into the condensate. The steep part of the growth curve is essentially unchanged by scattering, since it is dominated by the stimulated terms in the growth processes.

A number of sensitivity studies were carried out by Gardiner and Lee[56], who showed that the scattering transition rates could be varied by two orders of magnitude, the relative size of W^{++} to W^{+} could be varied by a factor of 2, and a wide variety of initial conditions for the rate equations could be used, with only moderate effects on the growth curves.

6.4 Full dynamical solution

The master equation approach described above, which assumes a nondepletable bath of fixed thermodynamic properties, can be expected to be inaccurate when high fraction condensates form. Under those circumstances the full dynamics of the system must be considered, and Davis *et al.* [59] have treated this problem using a modified Quantum Boltzmann formulation. In their treatment the single particle distribution function f_n for the nth energy sub-band evolves according to

$$g_n \frac{\partial f_n}{\partial t} = \frac{8ma^2\omega^2}{\pi\hbar} \sum_{p,q,r} \delta_{e_r+e_n, e_p+e_q} g_{\min(p,q,r,n)} \times \{f_p f_q (1+f_r)(1+f_n) - (1+f_p)(1+f_q) f_r f_n\} . \quad (43)$$

This equation appears similar to the ergodic form of the well-known Quantum Boltzmann equation (e.g. see ref [60]) but the crucial modification from Quantum Kinetic theory is that the energy levels e_m and the degeneracies g_m alter as the condensate population changes. A division of the levels into condensate and noncondensate bands is now made only for computational purposes, and the evolution of all sub-band levels is explicitly followed. In addition to the collisional processes treated in the master equation method, Eq.(43) includes thermal scattering processes where the pair of colliding atoms begin and remain in the noncondensate band, as well as collisions which either begin or end with a pair of atoms in the condensate band. As before, all levels are treated as particle-like, which allows an analytic expression to be obtained for the density of states and its dependence on the condensate occupation. The numerical simulation of Eq.(43), which we shall refer to as the *full dynamical solution*, requires vastly more computation than the rate equation model described in section 6.2, and considerable care must be taken with ensuring energy conservation in the collisional calculations and in rebinning the sub-bands at each time step. Details are given in ref [59].

For small condensate fractions (of less than about 10%) the full dynamical solution agrees well with the rate equation solution. However for larger

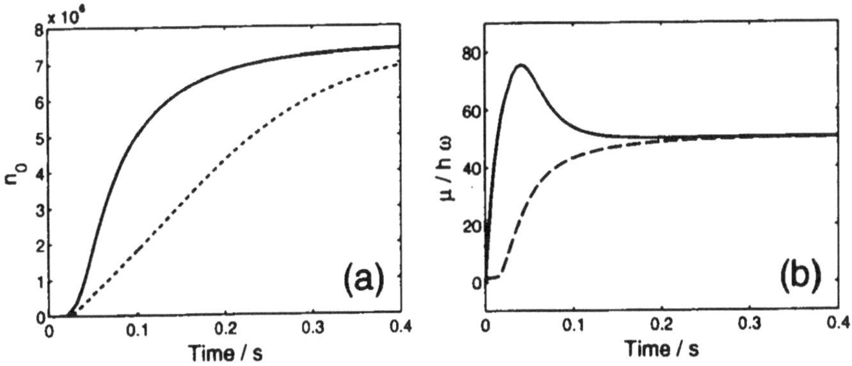

Figure 6. (a) Comparison of condensate growth models for a final condensate fraction of 24%. Solid line, simulation of Eq.(43) [with $\mu = 0$ before evaporative cut]. Dotted line, rate equation model. Final condensate parameters $n_0 = 7.5 \times 10^6, T = 590\text{nK}$. (b) Chemical potentials μ_{eff} (solid line) and $\mu_C(n_0)$ (dashed line).

fractions, as illustrated in Fig. 6(a) we find that the bath dynamics have an appreciable effect. For this case, where parameters are chosen to be appropriate for the MIT experiment and the final condensate fraction is 24%, the full dynamical solution (solid line) grows faster than the rate equation solution (dotted line). This can be most easily understood by introducing the concept of an effective chemical potential μ_{eff} for the thermal cloud (obtained by fitting a Bose-Einstein distribution to the distribution function above E_R) and using this in the simple growth equation Eq.(41). Figure 6(b) shows the evolution of both μ_{eff} and $\mu_C(n_0)$ during the simulation, and we can see that during the stimulation-dominated period of growth, μ_{eff} substantially *exceeds* the final equilibrium value, and thus from Eq.(41), the backward rate from condensate to thermal cloud is greatly inhibited. Ultimately this is due to the very severe cut that is required on the initial equilibrium distribution in order to allow evaporative cooling to proceed to such a high condensate fraction. We note that Bijlsma *et al.*[61] have treated the full dynamical problem using a somewhat different formalism, but obtain essentially the same numerical results.

The theory reviewed in this section can be compared to the experimental results of the MIT group[57]. It is found [54,56,59] that there is very good agreement in some parameter regimes, but significant discrepancies in others, which may be attributable to the difficulty of fully characterising the experimental parameters. A definitive comparison is handicapped by the unavailability of a wide range of primary experimental data, and further experiments will be required to comprehensively test the theoretical predictions.

7 Output Coupling

An output coupler for an atom laser is a device which extracts atoms from the laser mode while preserving their coherence. All current coupler mechanisms are based on applying an electromagnetic field to change the internal state of the atom from a trapped to an untrapped state[2]. The first demonstration of an atom laser, by the MIT group[62], used an RF pulse to transfer atoms into untrapped zeeman substates, where they then fell freely under gravity. A refinement of this technique, using a precisely controlled trap and cw-RF fields has been used to produce a quasi-cw coherent output beam[63]. The use of Raman coupled output fields offers some advantages, principally in providing a momentum kick to the outgoing atom beam[46], but also in that the optical fields can be focussed into a small volume, achieving a measure of spatial selectivity for the coupling. Recently the NIST group[17] demonstrated Raman output coupling in an experiment where the output beam was ejected in the horizontal direction.

A number of mean field theoretical treatments have been given of RF[40,64] and Raman couplers[66], and elsewhere[65] the effects of gravity have been included. Recently the Oxford group[67,68], have given a detailed treatment of output coupling from a finite temperature condensate. In this section we review the theory of output couplers, beginning with a mean field treatment, which establishes a framework for both RF and Raman output couplers. We then briefly summarize the theory of finite temperature couplers.

7.1 Mean field treatment of output coupling

The basis of the mean field treatment of an output coupler is a pair of coupled GP equations for the mean fields ψ_1 and ψ_2 of atoms in internal states $|1\rangle$ and $|2\rangle$, as discussed previously in section 5.2. For convenience, we present a dimensionless form of the equations

$$\frac{\partial\psi_1}{\partial t} = i\nabla^2\psi_1 - iV_1\psi_1 + i\frac{\lambda^*(\mathbf{r},t)}{2}\psi_2 - iC\left[|\psi_1|^2 + w|\psi_2|^2\right]\psi_1 - iGy\psi_1, \quad (44)$$

$$\frac{\partial\psi_2}{\partial t} = i\nabla^2\psi_2 - iV_2\psi_2 + i\frac{\lambda(\mathbf{r},t)}{2}\psi_1 - iC\left[|\psi_2|^2 + w|\psi_1|^2\right]\psi_2 - iGy\psi_2 . \quad (45)$$

We shall assume that state $|1\rangle$ is the trapped state, with a harmonic potential V_1 while state $|2\rangle$ is untrapped ($V_2 = 0$) or anti-trapped ($V_2 = -V_1$), so that ψ_2 represents the output field. In Eqs.(44) and (45) the units of time ($\tilde{t}_0$) and distance ($\tilde{x}_0$) are respectively ω^{-1} and $(\hbar/2m\omega)^{1/2}$ where ω is the trap frequency, and the fields are normalised such that $\int\left[|\psi_1|^2+|\psi_2|^2\right]d^3r = 1$. The nonlinearity constant C is $4\pi\hbar aN/(m\omega\tilde{x}_0^3)$ (in 3D), and the relative scattering length w of the inter- to intra-species collisions is typically close to 1. Gravity acts in the $-y$ direction, and the scaled acceleration is $G = mg\tilde{x}_0/\hbar\omega$. The coupling term λ can describe either a single photon coupling or a two photon (Raman) coupling, and can be written in terms of a slowly varying envelope $\Omega(\mathbf{r},t)$

$$\lambda(\mathbf{r},t) = \Omega(\mathbf{r},t)\exp i(\mathbf{k}_{em}\cdot\mathbf{r}-\Delta_{em}t) \tag{46}$$

where $\mathbf{k}_{em}$ and Δ_{em} measure the net momentum and net energy transfer from the EM field to the output atom (in units $\hbar/\tilde{x}_0$ and $\hbar\omega$ respectively). For the case of a single photon transition, $\Delta_{em} = \omega_1 - \omega_{21}$, where ω_1 is the photon frequency, ω_{21} is the energy difference $E_2 - E_1$ between states $|2\rangle$ and $|1\rangle$ (in the absence of the traps), and $\Omega(\mathbf{r},t)$ is the Rabi frequency. For a Raman transition connecting $|1\rangle$ to $|2\rangle$ (through a far detuned intermediate state) via the absorption of a photon ω_1 and the emission of photon ω_2 then $\Delta_{em} = \omega_1 - \omega_2 - \omega_{21}$, and $\Omega(\mathbf{r},t)$ is equal to $\Omega_2^*\Omega_1/\Delta$ where Ω_j is the Rabi frequency associated with the ω_j radiation field, and Δ is the detuning from the intermediate state. Our main interest here is in cw coupling, and so $\Omega(\mathbf{r},t)$ will be independent of time. The key difference between RF coupling and Raman coupling is that for RF coupling, the momentum $\mathbf{k}_{em}$ is negligible and can be ignored. Typically RF coupling (which may extend into the microwave region) is single photon, but Raman coupling necessarily involves two photons. We note that in Eqs.(44) and (45) we have neglected the light shifts that would occur for the two photon case, under the assumption that they would be small compared to the potentials V_j or the collisional mean fields near the trap.

The coupler equations (44) and (45) can be solved numerically, and we illustrate in Fig. 7 a case for an RF coupler in two dimensions, where the system begins in an eigenstate of the trap V_1, the RF field is plane wave, and $V_2 = 0$ [69]. The first two frames show $|\psi_1|^2$ and $|\psi_2|^2$ soon after the coupler has been turned on, and we note that the trapped state is essentially unchanged from its initial state, while the output state (more or less) copies the initial state. After some time, the output beam streams downwards under gravity in a well directed uniform beam that resembles the experimental results of the Munich group[63].

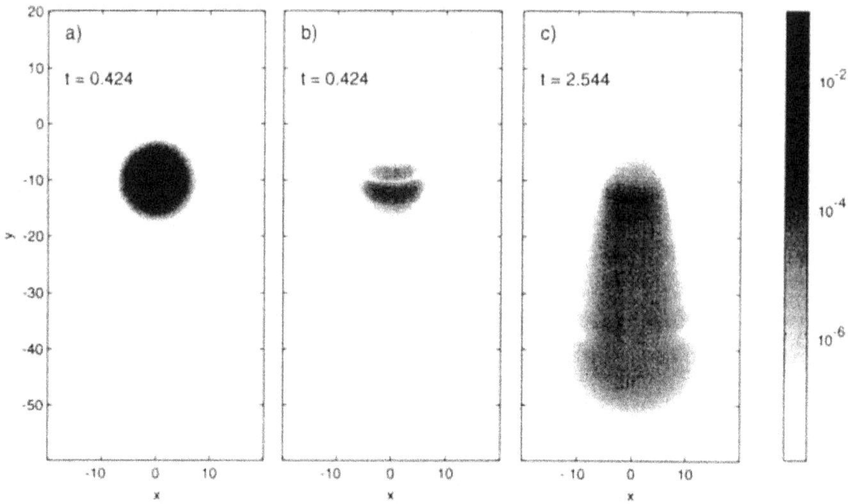

Figure 7. Population densities in states $|1\rangle$ and $|2\rangle$ under cw output coupling in a gravitational field. (a) state $|1\rangle$ at $t = 0.424$; (b) state $|2\rangle$ at $t = 0.424$; (c) state $|2\rangle$ at $t = 2.544$. Densities are plotted logarithmically. RF field is plane wave and parameters are $\Omega = 0.2$, $\Delta_{em} = 40$, $G = 5$, $C = 200$, $V_2 = 0$.

7.2 Analytic treatments

Many of the features of the output coupler behaviour that are found in simulations of Eqs.(44) and (45) can be understood in terms of analytic approximations. The key parameters are the strength of the coupling field (Ω), the net energy match with the output states, and the bandwidth of the coupling. When Ω is large (more precisely[40] $\Omega^2 + \Delta^2 \gg \mu(\mu + 2|\Delta|)$, where μ is the chemical potential of the initial trapped state), it dominates the remaining terms in Eqs.(44) and (45), and the resulting behaviour is synchronous Rabi cycling where each part of the condensate oscillates between being fully in state $|1\rangle$ and fully in state $|2\rangle$, at frequency Ω. The atoms spend such short periods of time in the untrapped state, that they are unable to escape. On the other hand when Ω is small compared to μ, other terms in Eqs.(44) and (45) become more significant, and the difference between the trap potentials becomes important. We can understand this by transforming Eqs.(44) and (45) into a rotating frame using $\tilde{\psi}_2 = \psi_2 \exp(i\Delta_{em}t)$, neglecting kinetic energy (because t is small and the atoms do not have time to move), and then recognising that the equations simply become Rabi cycling with a spatially dependent effective detuning

$$\Delta_{\text{eff}}(\mathbf{r}) = \Delta_{em} - (V_1 - V_2)\,. \tag{47}$$

The mean field contribution to Δ_{eff} cancels (for $w = 1$) because the atom feels the same mean field in either state $|1\rangle$ or $|2\rangle$. Now, because the spectral width Ω of the Rabi cycling is small, significant transfer from $|1\rangle$ to $|2\rangle$ will only occur in a small spatial region near the value of $\mathbf{r}$ for which $\Delta_{\text{eff}}(\mathbf{r}) = 0$. The resulting spatially selective output coupling can be seen in Fig. 7 (b) and (c), where the preferred region of coupling is a narrow crescent near $y = -10$. The location of this region can be calculated by noting that the magnetic traps (i.e. V_1 and V_2) are centered on $x = 0, y = 0$, and thus the condition $\Delta_{\text{eff}}(\mathbf{r}) = 0$ for a given choice of Δ_{em} is equivalent to selecting an equipotential line for the potential difference $V_1 - V_2$, which is a circle centered on $x = 0, y = 0$. Due to gravity, the equilibrium centre of mass position of condensate ψ_1 drops below the centre of the magnetic trap to a position where the spring force is equal to the gravitational force, as shown in Fig. 7(a). Thus the spatially selected region of coupling, where the trap equipotential intersects the trapped condensate, takes a crescent shape. Of course the coupling time τ_c must be long enough that transient broadening does not obscure the spatial selectivity.

7.3 Weak coupling regime

The weak coupling regime is of most interest for cw atom lasers, and perturbative solutions have been derived which describe key aspects of the output field behaviour[66,67]. Following those treatments we make the assumption of an undepleted trapped state and a very dilute output state, which allows us to replace ψ_1 in Eq.(45) by its initial value $\psi_1(t = 0)$ and to neglect the meanfield interaction of the output state with itself. Taking a Thomas-Fermi approximation for $\psi_1(t = 0)$, Eq.(45) becomes a linear Schrodinger equation

$$\frac{\partial \psi_2}{\partial t} = -iH_{oc}\psi_2 + i\frac{\lambda(\mathbf{r}, t)}{2}\Psi_{TF}\exp(-i\mu t) \tag{48}$$

where $H_{oc} = -\nabla^2 + V_{oc}$. The potential for the outcoupled wave is $V_{oc} = V_2 + w(\mu - V_1) + Gy$ and the final term in Eq.(48) is a source term in which the Thomas-Fermi wavefunction is $\Psi_{TF} = [\mu - V_1]^{1/2}$. The eigenfunctions $\phi_E(\mathbf{r})$ of H_{oc} provide a convenient basis for solving Eq.(48). Taking for simplicity the one dimensional case (and neglecting gravity) we obtain the solution

$$\psi_2(x, t) = i\int dE \left\{ e^{i(B-E)t}\left[\frac{\sin Bt}{B}\right]\right\} \Lambda(E)\phi_E(x)\,, \tag{49}$$

where $B = [E - (\mu + \Delta_{em})]/2$ and

$$\Lambda(E) = \frac{1}{2} \int dx \phi_E^*(x) \Omega(x) \exp(ik_{em}x) \Psi_{TF} , \tag{50}$$

is the matrix element of the EM coupling amplitude between the output eigenfunction and the Thomas-Fermi trap state. Eq.(49) provides useful insight into the properties of the output field. The term in square brackets is a familiar factor from perturbation theory, and at small times the term in braces $\{\} \longrightarrow t$ so that $\psi_2(x,t) \approx it[\Omega(x)/2] \exp(ik_{em}x)\Psi_{TF}(x)$. Thus at early times the weak coupler produces an output field that is a copy of the initial trapped state, spatially masked by the EM field amplitude. At long times the term in square brackets $[] \longrightarrow 2\pi\delta(E - \mu - \Delta_{em})$ so that the output state becomes proportional to the eigenfunction $\phi_E(x)$

$$\psi_2(x,t) \approx i2\pi \exp(-iE)t\Lambda(E)\phi_E(x) \tag{51}$$

where E conserves energy according to

$$E = \mu + \Delta_{em} , \tag{52}$$

and the function $\Lambda(E)$ plays the role of a spectral filter. The low energy cutoff of $\Lambda(E)$ can determined[66] by consideration of the classical turning points of the wavefunction $\phi_E(x)$, which shows that $\Lambda(E) = 0$ for $E \leq 0$. Thus (for the case $V_2 = 0$), the lowest possible output energy for an atom is $E = 0$, which gives a corresponding lower limit for the coupling field detuning $\Delta_{em} = -\mu$. [see Eq.(52)] at which value the EM field absorbs all the energy μ of the trapped condensate atom. It is more difficult to find the upper cut-off of $\Lambda(E)$, but the spatial selectivity criterion given by Eq.(47) suggests that output coupling will cease when Δ_{em} exceeds the largest value of $V_1 - V_2$ in the condensate. Setting $V_2 = 0$, neglecting gravity, and using the Thomas-Fermi radius for the edge of the condensate, we can then write the limits for which Δ_{em} produces effective coupling as

$$-\mu \leq \Delta_{em} \leq \mu . \tag{53}$$

The bandwidth of the output coupling $\Delta\omega_B$, which is determined primarily by $\Lambda(E)$, determines the time scale for the output field to reach its steady state ($\sim [\Delta\omega_B]^{-1}$). For the case of a Raman coupler (for which $\exp(ik_{em}x)$ must be retained in Eq.(50)) or a gravitational field, $\Delta\omega_B$ is appreciably larger than for the case of a simple RF coupler in free space. When $\Omega \ll \Delta\omega_B$, and thereby Rabi cycling is suppressed, the faster untrapped atoms escape[68].

7.4 Finite Temperature Output coupling

The Oxford group have extended the theory of output coupling to include the description of finite temperature condensates[67,68] (see the article by Burnett in this volume). Their treatment, which uses operator fields with a number conserving approach is sophisticated, but has an underlying structure that is very similar to that presented in the previous section. The atom has two internal states $|t\rangle$ and $|f\rangle$ (trapped and free) and the atom field operator is divided into $\hat{\psi}_t(\mathbf{r},t)$ for the trapped state and $\hat{\psi}_f(\mathbf{r},t)$ for the output field, (corresponding to ψ_1 and ψ_2). The same initial approximations are made in assuming the output coupling is weak so that the trapped field is undepleted and the output field is very dilute. Inelastic collisions are neglected, and an EM coupling field with amplitude given by Eq.(46) models either RF or Raman coupling. This allows an equation to be written for $\partial\hat{\psi}_f(\mathbf{r})/\partial t$, which is completely analogous to Eq.(48). The output field moves in a potential comprised of the trap potential for state $|f\rangle$ and the meanfield of the (unmodified) initial trapped state, while the self field of $\hat{\psi}_f$ is neglected. The source term involves the equilibrium field operator for the trapped condensate, $\hat{\psi}_t(\mathbf{r},t=0)$ but it is important to note that this is not a meanfield, but includes a thermal component, in which the quasiparticle levels are assumed to be occupied according to a Bose-Einstein distribution of temperature T and chemical potential μ. The operator $\hat{\psi}_f$ is expanded in modes $\phi_\mathbf{k}(\mathbf{r})$ of energy $E_\mathbf{k}$ that are eigenfunctions of a Hamiltonian that is equivalent to H_{oc}, and leads to the solution for the free field

$$\begin{aligned}\hat{\psi}_f(\mathbf{r},t) \approx & -i\Psi_f^0(\mathbf{r},t)\hat{a}_0(0) \\ & -\hat{a}_0(0)\frac{i}{\sqrt{\hat{N}_0}}\sum_j\left[\Psi_f^{j+}(\mathbf{r},t)\hat{\alpha}_j(0)+\Psi_f^{j-}(\mathbf{r},t)\hat{\alpha}_j^\dagger(0)\right], \quad (54)\end{aligned}$$

where $\hat{a}_0$ is the annihilation operator for the condensate mode, $\hat{N}_0$ is the condensate occupation operator $\hat{a}_0^\dagger\hat{a}_0$, and $\hat{\alpha}_j$ is the annihilation operator for the jth quasiparticle level, of energy $\mu+E_j$. The mode functions in Eq.(54) are

$$\Psi_f^\eta(\mathbf{r},t)=\sum_\mathbf{k}\varphi_\mathbf{k}(\mathbf{r})D_{\mathbf{k}\eta}(t)\Lambda_{\mathbf{k}\eta}e^{-iE_\mathbf{k}t} \quad (55)$$

for $\eta = 0, j+, j-$. The quantity $D_{\mathbf{k}\eta}(t) = i(e^{-i(E^\eta_{out}-E_\mathbf{k})t}-1)/(E^\eta_{out}-E_\mathbf{k})$ is an energy selective function in which

$$E^\eta_{out}=\mu+\Delta_{em}+E_\eta\,, \quad (56)$$

with $E_{j\pm} = \pm E_j$, and $E_0 = 0$. The quantity $\Lambda_{\mathbf{k}\eta}$ is a frequency filter analogous to $\Lambda(E)$ in Eq.(50) and is given by

$$\Lambda_{\mathbf{k}\eta} = \frac{1}{2} \int d^3\mathbf{r}\varphi^*_{\mathbf{k}}(\mathbf{r})\Omega(x)\exp(ik_{em}x)\xi_\eta(\mathbf{r}) \tag{57}$$

where ξ_η for $\eta = 0, j+, j-$, represents ψ_0, u_j, v^*_j, the trapped condensate wavefunction, and the usual quasiparticle amplitudes, respectively. We can now interpret the three terms in Eq.(54). The first term is the condensate output and corresponds to the weak coupling mean field calculation given by Eq.(49); the second term (Ψ^{j+}) represents an atom ejected from a trapped quasiparticle mode (stimulated quantum evaporation of the thermal excitations) and the third term (Ψ^{j-}) represents an atom ejected from the condensate and creating a trapped quasiparticle (pair-breaking). Each of these processes obeys energy conservation given by Eq.(56), which for the condensate term ($\eta = 0$) is identical to Eq.(52). For the stimulated evaporation of a quasiparticle ($\eta = j+$) we see that the output atom has energy equal to that of the initial trapped quasiparticle plus the energy provided by the EM field. For the pair breaking term, an atom comes out of the condensate, but leaves behind an extra quasiparticle, so that the output energy is $\mu - E_j$ plus the energy provided by the EM field.

The density of output atoms is given by

$$n_{out}(\mathbf{r}, t) = \langle\hat{\psi}^\dagger_f(\mathbf{r}, t)\hat{\psi}_f(\mathbf{r}, t)\rangle \tag{58}$$

and by assuming that all cross correlations between different operators $\hat{a}_0, \hat{\alpha}_j, \hat{\alpha}^\dagger_j$ vanish and only the populations $\langle\hat{a}^\dagger_0\hat{a}_0\rangle = n_0$, $\langle\hat{\alpha}^\dagger_j\hat{\alpha}_j\rangle = n_j$, and $\langle\hat{\alpha}_j\hat{\alpha}^\dagger_j\rangle = n_j + 1$ remain, we can use Eq.(54) to write the output density

$$n_{\text{out}}(\mathbf{r}, t) = n_0|\Psi^0_f(\mathbf{r}, t)|^2 + \sum_j \left[n_j|\Psi^{j+}_f(\mathbf{r}, t)|^2 + (n_j + 1)|\Psi^{j-}_f(\mathbf{r}, t)|^2\right], \tag{59}$$

where the three separate components of the output field are easily identified. The time evolution of the output field is governed by the output bandwidth $\Delta\omega_B$ as discussed in the previous section, and first discussed extensively by Choi *et al.*[68]. In the weak coupling limit, $\Omega \ll \Delta\omega_B$, we find once again that for short times the output state copies the trapped state $\xi_\eta(\mathbf{r})$

$$\Psi^\eta_f(\mathbf{r}, t) \sim_{t\to 0} [\Omega(\mathbf{r})/2]\xi_\eta(\mathbf{r})t\,. \tag{60}$$

For long times, energy conservation selects only output states for which $E_{\mathbf{k}} = E^\eta_{out}$. Choi *et al.*[68] have given a calculation for RF output coupling in one

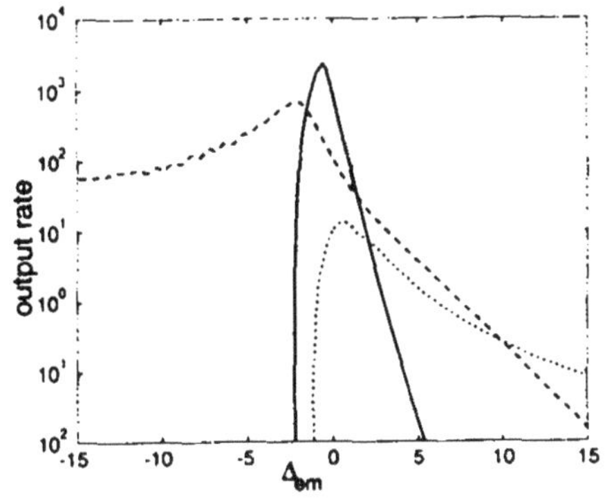

Figure 8. Population output rate from a finite temperature condensate with an RF coupler. Solid line, condensate component. Dashed line, stimulated quantum evaporation component. Dotted line, pair breaking component. Parameters: $T = 0.5T_c$ $(150\hbar\omega/k)$, $\mu = 2.3\hbar\omega/k$, $N_t = 2000$ atoms.

dimension, from a condensate at temperature $T = 0.5T_c$. They calculate the rate dN_{out}/dt in steady state, where

$$N_{\text{out}} = \int n_{\text{out}}(\mathbf{r}, t) d^3r \qquad (61)$$

and the result is presented in Fig. 8, where the three components of the output are plotted (as different line types) as a function of the EM detuning Δ_{em}. A key feature of the output coupling that is readily apparent in this figure is how the detuning Δ_{em} controls the character of the output field. For $\Delta_{em} < -\mu$, only the thermal quasiparticles are ejected from the trap, as we expect from energy conservation (see Eq.(56)): only the quasiparticles have sufficient energy to absorb this amount of energy from the radiation field. For $\Delta_{em} > -\mu$, up to $\Delta_{em} \approx \mu$, fully coherent condensate dominates the output (see also Eq.(53)). Finally for $\Delta_{em} >\approx -\mu$, both pair breaking (where the energy of the initial condensate atom plus the photon can be left in the trap) and thermal evaporation can occur.

As a final comment, we note that the processes of stimulated thermal evaporation and pair breaking reduce the coherence of the output field. Choi *et al.*[68] have calculated the coherence functions $g^{(1)}(x_1, x_2, t)$ and $g^{(2)}(x_1, x_1, t)$ and shown that even in the condensate dominated regime, the first order coherence is degraded, and $g^{(2)}$ exhibits symptoms of atom bunching.

8 Acknowledgements

RJB is grateful for many instructive conversations over the past few years with Professors C. Gardiner and K. Burnett, which have helped shape this article. This research was supported by the Marsden Fund of New Zealand under contract PVT902.

CMS acknowledges collaborations with J. Hope, G.Moy, and N. Robins. Our work is supported by the Australian Research Council.

References

1. H.M. Wiseman, Phys. Rev. A **56**, 2068 (1997); **57**, 674 (1998).
2. C.M. Savage, S. Marksteiner, P. Zoller, in "Fundamentals of quantum optics III", ed. F. Ehlotzky (Springer 1993).
3. Ch.J. Bordé, Phys. Lett. A **204**, 217 (1985); Ann. Phys. Fr. **20**, 477 (1985); in *Proc. of the 12th Int. Conference on Laser Spectroscopy*, eds. M. Inguscio, M. Allegrini, A. Sasso (World Scientific 1995).
4. M. Olshanii, Y. Castin and J. Dalibard, *Proc. of the 12th Int. Conference on Laser Spectroscopy*, edited by M. Inguscio, M. Allegrini and A. Sasso (World Scientific 1995).
5. R.J.C. Spreeuw *et al.*, Europhysics Letters **32**, 469 (1995).
6. M. Holland *et al.*, Phys Rev. A **54**, R1757 (1996).
7. M.-O. Mewes, et al., Phys. Rev. Lett. **78**, 582 (1997).
8. K. Helmerson, D, Hutchinson, K. Burnett, and W.D. Phillips, Physics World, p. 31, August (1999).
9. W. Ketterle, Physics Today **52** no. 12, 30 (1999).
10. A. Peters, Keng Yeow Chung, S. Chu, Nature **400**, 849 (1999).
11. L. Deng *et al.*, Nature **398**, 218 (1999).
12. H.-A. Bachor, *A guide to experiments in quantum optics* (Wiley-VCH, 1998).
13. W. Ketterle and H.-J. Miesner, Phys. Rev. A **56**, 3291 (1997).
14. B.P. Anderson and M.A. Kasevich, Science **282**, 1686 (1998).
15. J.L. Martin, J. Phys. B **32**, 3065 (1999).
16. I. Bloch *et al.*, Phys. Rev. Lett. **82**, 3008 (1999); Nature **403**, 166 (2000).
17. E.W. Hagley, L. Deng, M. Kozuma, J. Wen, K. Helmerson, S.L. Rolston, W.D. Phillips, Science **283**, 1706 (1999).
18. K. Molmer, Phys. Rev. A **55**, 3195 (1997).
19. C.W. Gardiner and P. Zoller, Phys. Rev. A **58**, 536 (1998).
20. E.M. Lifshitz and L.P. Pitaevskii, *Statistical Physics, part 2*, (Pergamon

Press 1980).
21. M. Edwards and K. Burnett, Phys. Rev. A **51**, 1382 (1995).
22. A.E. Siegman, *Lasers* (University Science Books, California 1986).
23. L. Mandel and E. Wolf, *Optical Coherence and Quantum Optics* (1995).
24. A.L. Schawlow and C.H. Townes, Phys. Rev. **112**, 1940 (1958).
25. D.F. Walls and G.J. Milburn, *Quantum Optics* (Springer Verlag: Berlin 1994).
26. R.J. Glauber, Phys. Rev. **130**, 2529 (1963).
27. M.R. Andrews *et al.*, Science **275**, 637 (1997).
28. K. Huang, *Statistical Mechanics*, (Wiley,1987).
29. M.O. Mewes *et al.*, Phys. Rev. Lett. **77**, 416 (1996).
30. M. Holland, *et al.*, Phys. Rev. Lett. **78**, 3801 (1997).
31. E.A. Burt, *et al.*, Phys. Rev. Lett. **79**, 337 (1997).
32. Yu. Kagan, B.V. Svistunov, G.V. Shlyapnikov, JETP Lett. **42**, 209 (1985).
33. E.V. Goldstein and P. Meystre, Phys. Rev. Lett. **80** 5036 (1998); E.V. Goldstein, O. Zobay, P. Meystre, Phys. Rev. A **58**, 2373 (1998).
34. M. Naraschewski and R.J. Glauber, Phys. Rev. A **59**, 4595 (1999).
35. G.M. Moy, J.J. Hope, C.M. Savage, Phys. Rev. A **59**, 667 (1999).
36. G.M. Moy, J.J. Hope, C.M. Savage, Phys. Rev. A **55**, 3631 (1997).
37. A.M. Guzman, M. Moore, P. Meystre, Phys Rev. A **53**, 977 (1996).
38. H.M. Wiseman and M.J. Collett, Physics Lett. A **202**,246 (1995).
39. H.M. Wiseman, A. Martins, D.F. Walls, Quant. Semiclass. Opt. **8** 737 (1996).
40. R.J. Ballagh, K. Burnett, T.F. Scott, Phys. Rev. Lett. **78** 1607 (1997).
41. W. Zhang and D.F. Walls, Phys. Rev. A **57** 1248 (1998).
42. B. Kneer *et al.*, Phys. Rev. A **58** 4841 (1998).
43. N. Robins, C.M. Savage, E. Ostrovskaya, in *Dan Walls Memorial Volume*, eds. H. Carmichael, R. Glauber, M. Scully, in press (Springer, 2000).
44. H.M. Wiseman and M.J. Collett, Phys. Lett. A **202** 246 (1995).
45. J.J. Hope, Phys. Rev. A **55** R2531 (1997).
46. G.M Moy and C.M. Savage, Phys. Rev. A **56** 1087 (1997).
47. J.J. Hope, G.M Moy, M.J. Collett, C.M. Savage, Phys. Rev. A **61** 023603 (2000).
48. D.W. Snoke and J.P. Wolfe, Phys. Rev. B **39**, 4030 (1989); D. Semikoz and I.I. Tkachev, Phys. Rev. Lett. **74**, 3093 (1995).
49. H.T.C. Stoof, Phys. Rev. Lett. **66**, 3148 (1991); Phys. Rev. A **49**, 3824 (1994); Phys. Rev. Lett. **78**, 768 (1997).
50. Yu. Kagan, B.V. Svisunov, G.V. Shlyapnikov, Sov. Phys. JETP **75**, 387 (1992); Yu. Kagan and B.V. Svisunov Phys. Rev. Lett. **79**, 3331 (1997).

51. C.W. Gardiner and P. Zoller, Phys. Rev. A **55**, 2902 (1997).
52. C.W. Gardiner, P. Zoller, R.J. Ballagh, M.J. Davis, Phys. Rev. Lett. **79**, 1793 (1997).
53. C.W. Gardiner and P. Zoller, Phys. Rev. A **58**, 536 (1998).
54. C.W. Gardiner, M.D. Lee, R.J Ballagh, M.J. Davis and P. Zoller, Phys. Rev. Lett. **81**, 5266 (1998).
55. C.W. Gardiner and P. Zoller, Phys. Rev. A **61**, 033601 (2000).
56. M. D. Lee and C. W. Gardiner, cond-mat/9912420, Phys. Rev. A *in press*.
57. H.-J. Miesner, D.M. Stamper-Kurn, M.R. Andrews, D.S. Durfee, S. Inouye, W. Ketterle, Science **279**, 1005 (1998).
58. C.W. Gardiner, Phys. Rev. A **56**, 1414 (1997).
59. M. J. Davis, C. W. Gardiner, R. J. Ballagh, cond-mat/9912439 , Phys. Rev. A *in press*.
60. M. Holland, J. Williams, J. Cooper, Phys. Rev. A **55**, 3670 (1997).
61. M. Bijlsma, E. Zaremba, H. T. C. Stoof, cond-mat/0001323 .
62. M.-O. Mewes, M. R. Andrews, D. M. Kurn, D. S. Durfee, C. G. Townsend, W. Ketterle, Phys. Rev. Lett. **78**, 582 (1997).
63. I. Bloch, T. W. Hänsch and T. Esslinger, Phys. Rev. Lett. **82**, 3008 (1999).
64. H. Steck, M. Naraschewski, H. Wallis, Phys. Rev. Lett. **80**, 1 (1998).
65. B. Jackson, J. F. McCann, C. S. Adams, J. Phys. B **31**, 4489 (1998).
66. M. Edwards, D. A. Griggs, P. L. Holman, C. W. Clark, S. L. Rolston, W. D. Phillips, J. Phys. B **32**, 2935 (1999).
67. Y. Japha, S. Choi, K. Burnett, Y. Band, Phys. Rev. Lett. **82**, 1079 (1999)
68. S. Choi, Y. Japha, K. Burnett, Phys. Rev. A **61**, 063606 (2000).
69. J.M.W. Krüger *Dynamical Behaviour of Radio Frequency Output Couplers for a Bose-Einstein Condensate*, M.Sc Thesis, University of Otago, 2000.

VORTICES AND SOLITONS IN BOSE-CONDENSATES

P. D. DRUMMOND

Physics Department. University of Queensland, St Lucia 4072, Queensland, Australia

E-mail: drummond@physics.uq.edu.au

Web: www.physics.uq.edu.au/people/drummond

The new science of nonlinear atom optics and atom lasers is evolving rapidly. There are similarities between many related areas in modern photonic and atom optics, particularly at the mean-field level. In both cases we can often use classical nonlinear wave equations to describe classical solitons, vortices, and other nonlinear structure. Atom-molecular coupling can be used to play the role of second-harmonic generation. This leads to novel types of soliton. In addition, quantum effects at low densities are likely to be readily observable.

1 Introduction

Atom-optics is the study of matter under conditions where it behaves like a wave. This research field is developing rapidly, due to the development of laser cooling techniques which allow the wave-like properties of atoms to be observed easily. A milestone is reached when low temperature is combined with densities high enough for the atomic wavelength to become as large as the inter-particle spacing. At this point, the observation of a Bose-Einstein condensate is possible, where a large number of atoms all combine in a single, extended quantum state.

The recent discovery of Bose-Einstein condensation (BEC) in ultra-cold alkali-metal vapors[1] at temperatures of 10^{-6}K or lower, has raised the possibility of a coherent atom laser. Recent experimental developments[2] show that this is, indeed, practical. However, theoretical models of coherent atom lasers have often neglected the many-body, many-mode features that are compellingly present[3] in any realistic atom laser or BEC. In simple terms, atom-atom interactions tend to be much stronger and more prevalent than photon-photon interactions are in a conventional laser. This leads to long-lived structural characteristics, including vortices (phase-singularities) and topological and non-topological solitons, which can have both classical and quantum-mechanical features[a].

[a]See WWW pages at: http://www.physics.uq.edu.au/people/drummond

1.1 BEC Quantum Field Theory

The Bose gas quantum Hamiltonian can be expressed in terms of field operators of the atom field, which we assume to be a bosonic field - that is, each atom is composed of an even number of fermions. In the case of massive composite particles, the field operator describes the creation of a composite particle or atom (in a specified electronic state) with center-of-mass position equal to the argument of the field operator. The quantum state will typically be in a continuous superposition of different position eigenstates. When the de Broglie wave-length of each atom is of the order of the inter-atomic spacing, a Bose-Einstein condensation can occur; this involves micro-Kelvin temperatures for low-density atomic vapors.

The basic theory of Bose-Einstein condensation in a non-interacting Bose gas is covered in most texts on statistical mechanics, and by Oitmaa in this volume. The main difference between the usual undergraduate texts and the present treatment, is that interactions and external potentials are explicitly included here. This is known to lead to modifications in the ground-state, but also to more exotic structures like solitons, vortices, and other types of excited state structure.

We will start with the basic Hamiltonian formalism, as covered in many graduate-level texts[3]. This is most easily expressed using quantum fields. Thus, the fundamental fields are $\hat{\Psi}^\dagger(\mathbf{x}), \hat{\Psi}(\mathbf{x})$, where $\hat{\Psi}^\dagger(\mathbf{x})$ creates a bosonic atom at location $\mathbf{x}$, and $\hat{\Psi}(\mathbf{x})$ destroys one; the commutators are

$$[\hat{\Psi}(\mathbf{x}), \hat{\Psi}^\dagger(\mathbf{x}')] = \delta^{(3)}(\mathbf{x}-\mathbf{x}') \ . \tag{1}$$

The Hamiltonian - including both external and inter-particle potentials - is:

$$\begin{aligned} \hat{H} = & \int d^3\mathbf{x} \left\{ \frac{\hbar^2}{2m} \nabla\hat{\Psi}^\dagger(\mathbf{x}) \cdot \nabla\hat{\Psi}(\mathbf{x}) + V(\mathbf{x})\hat{\Psi}^\dagger(\mathbf{x})\hat{\Psi}(\mathbf{x}) \right\} \\ & + \frac{1}{2}\int d^3\mathbf{x} \int d^3\mathbf{x}' U(\mathbf{x}-\mathbf{x}')\hat{\Psi}^\dagger(\mathbf{x})\hat{\Psi}^\dagger(\mathbf{x}')\hat{\Psi}(\mathbf{x}')\hat{\Psi}(\mathbf{x}) \ , \end{aligned} \tag{2}$$

where $\langle\hat{\Psi}^\dagger(\mathbf{x})\hat{\Psi}(\mathbf{x})\rangle$ is the atomic density, m is the atomic mass, $V(\mathbf{x})$ is the external trapping potential, and $U(\mathbf{x}-\mathbf{x}')$ is the atom-atom interaction term. The theory is obviously non-relativistic, since the kinetic energy term is only correct at low velocity, and the interaction is assumed instantaneous (i.e., a-causal). This places a mild restriction on the time-scales; we cannot expect to accurately treat time-scales smaller than the transit-time of a light-field through the condensate.

The resulting Heisenberg equations have the form:

$$i\hbar\frac{\partial}{\partial t}\hat{\Psi}(t,\mathbf{x}) = \left[\frac{-\hbar^2}{2m}\nabla^2 + V(\mathbf{x})\right]\hat{\Psi}(t,\mathbf{x}) + \int d^3\mathbf{x}' U(\mathbf{x}-\mathbf{x}')\hat{\Psi}^\dagger(t,\mathbf{x}')\hat{\Psi}(t,\mathbf{x}')\hat{\Psi}(t,\mathbf{x}) \,. \tag{3}$$

In general, it is a nontrivial problem to solve these equations, which include most of the physics of a trapped single-component BEC. A complication can occur if there is more than one trapped boson field - for example, distinct atomic states, atomic species, or molecules; then the field $\hat{\Psi}$ must be replaced by a vector $\hat{\Psi}_j$, with internal degrees of freedom. In this case, there is an interspecies scattering process, as well as the possibility to have interconversion from one species to another. This is typically[4] induced with an external RF field or Raman coupling, with a corresponding Rabi frequency Ω_{ij}. This results in a new Heisenberg equation:

$$i\hbar\frac{\partial}{\partial t}\hat{\Psi}_j(t,\mathbf{x}) = \left[\frac{-\hbar^2}{2m_j}\nabla^2 + V_j(\mathbf{x})\right]\hat{\Psi}_j(t,\mathbf{x}) + \sum_i \frac{\hbar\Omega_{ij}}{2}\hat{\Psi}_i(t,\mathbf{x}) + \sum_i \int d^3\mathbf{x}' U_{ij}(\mathbf{x}-\mathbf{x}')\hat{\Psi}_i^\dagger(t,\mathbf{x}')\hat{\Psi}_i(t,\mathbf{x}')\hat{\Psi}_j(t,\mathbf{x}) \tag{4}$$

It is useful to make an additional approximation, which is to impose a momentum cut-off on the theory, so only the long wave-length components of the quantum fields are being considered. This is best considered as a renormalization problem, in which the actual interaction term must be modified to take account of the cutoff in a consistent way. As usual, we consider an operational measurement which must remain invariant under the renormalization, and this is usually chosen to be the two-particle S-wave scattering amplitude.

In order to explain this, we might first suppose that the interactions are short-range, so that: $U(\mathbf{x}-\mathbf{x}') \simeq \overline{U}\delta^3(\mathbf{x}-\mathbf{x}')$, where:

$$\overline{U} = \int d^3\mathbf{x} U(\mathbf{x}) \,. \tag{5}$$

In one space dimension, the approach outlined above would be reasonable. In higher space dimensions, it turns out that Nature is more subtle than this. A Dirac delta-function interaction simply has no effect in three dimensions. The relative atom-atom wave-function can 'heal' around a delta-function singularity with no cost in kinetic energy in three dimensions, due to the larger numbers of degrees of freedom available to the quantum fields.

Because this is the case, we must introduce a renormalized effective field theory, which only describes long wave-length behavior. This type of theory implicitly includes the correct short-distance correlations obtained when particles undergo S-wave scattering via the full finite range potential. For this reason, the correlation function (operator product expectation value) doesn't factorize at short distances. Nevertheless, we can still describe long-range behavior of the correlation functions using an equation similar to the one given above, except with the mean potential $\overline{U}$ replaced by an effective potential $\hbar\kappa$, which includes the effects of short-distance correlations. As one might expect, the effective potential equals the mean potential only under conditions where the Born approximation is valid - that is, a very long-range and weak potential.

More generally, κ must be found by means of solving the two-body scattering problem using the full potential, then obtaining or measuring the two-body S-wave scattering cross-section, which defines the low-energy scattering length a through the relation $d\sigma = (2a)^2 d\Omega$. Note that this is four times the differential cross-section for non-identical particles, due to constructive interference effects. Since we cannot tell which of the two particles scattered is which, the total cross-section is not $4\pi d\sigma/d\Omega$ (as it is for non-identical particles), but rather $\sigma = 2\pi d\sigma/d\Omega = 8\pi a^2$. However, in the Born approximation using the effective delta-function potential, we would obtain:

$$d\sigma = \left(\frac{2m\kappa}{4\pi\hbar}\right)^2 d\Omega \tag{6}$$

Hence, the effective potential that would give the correct cross-section with a momentum cut-off in the Born approximation is:

$$\kappa = \frac{4\pi\hbar a}{m}. \tag{7}$$

Thus, provided we assume there is a momentum cut-off for atom wavelengths shorter than a, and we replace the mean potential by the effective potential $\hbar\kappa$, we can still use the simplified equation to describe the long wavelength behavior in the condensate – even in two or three space dimensions. It is also possible to use an effective Hamiltonian with a derivative coupling. This doesn't require a cutoff, but turns out to be non-Hermitian, which creates other problems. For simplicity, we will use a momentum cut-off, together with an ordinary delta-function interaction.

1.2 Mean-field Theory

It is frequently useful to next consider the so-called 'mean-field' approximation, at extremely low temperatures and high densities. In this limit the atoms are wave-like. That is, they behave just like a classical wave, and obey (approximate) classical wave equations. This is similar to the corresponding limiting result with photons at high photon density. If we define $\Psi(t,\mathbf{x}) = \langle\hat{\Psi}(t,\mathbf{x})\rangle$, and also assume (for the moment, at least) that all expectation values of the effective quantum field operator products exactly factorize, so that clearly

$$(\Psi^*(t,\mathbf{x}))^n\,(\Psi(t,\mathbf{x}))^m = \langle\hat{\Psi}^{\dagger n}(t,\mathbf{x})\hat{\Psi}^m(t,\mathbf{x})\rangle\,, \tag{8}$$

then the resulting equations have the form of:

$$i\frac{\partial}{\partial t}\Psi(t,\mathbf{x}) = \left[\frac{-\hbar}{2m}\nabla^2 + V(\mathbf{x})/\hbar + \kappa|\Psi(t,\mathbf{x})|^2\right]\Psi(t,\mathbf{x})\,. \tag{9}$$

The above equation has similarities to the nonlinear Schrödinger equation of fiber optics, but exists in a full three-dimensional space. It describes the defocusing or expanding effect of the inter-particle potential (if it is positive), and a focusing effect if the inter-particle potential is attractive. This equation is called the Gross-Pitaevskii (GP) equation. It is important to remember that this is really an effective field theory, and should be used only to describe long wavelength behavior and correlations over distances larger than a. For a negative potential, the GP equations are unstable, and can lead to a catastrophic self-focusing singularity in free space with $D > 1$ dimensions - although small trapped condensates are meta-stable[5]. For a positive potential, the atoms are further apart, and this delta-function approximation is a reasonable one for atoms with low relative momenta. In the final stages of Bose-Einstein condensation, where all the atoms essentially occupy one macroscopic quantum state, these rather classical equations have a surprising degree of accuracy. This is because atoms, in the extreme low-temperature limit, have such long wave-lengths that they start to behave just like classical waves.

The Hamiltonian of the system that would classically generate the G-P equation, is:

$$H = \int\!\!\int\!\!\int d^3\mathbf{x}\left[\frac{\hbar^2}{2m}|\nabla\Psi|^2 + \frac{\hbar\kappa}{2}(\Psi^*\Psi)^2 + V(\mathbf{x})\Psi^*\Psi\right] \tag{10}$$

The quantized version of this Hamiltonian is the problem of a three-dimensional Bose gas with delta-function interactions - provided one remembers that a momentum cutoff is essential. The ground-state theory can be further improved to include quantum fluctuations, by using perturbation theory[6].

2 Vortices

Vortex formation, or a circulating current, is often regarded as a signature of superfluidity. There is an open debate over the question of superfluidity in the case of low-temperature BEC systems, which are compressible, and behave more like true quantum gases than the relatively incompressible, liquid-like behavior of low-temperature helium. There are some differences between vortices in the two cases. The stability of BEC vortices in a trap environment is still an unresolved question. Generically, we call vortices and vortex rings topological solitons, since they exist by virtue of topological properties of phase singularities - rather than the more common bright solitons, in which the fluid is localized to one region through self-interactions. The simplest case is the toroidal trap vortex, where the current circulates in a ring-shaped trap.

2.1 Gross-Pitaevskii equation in one dimension

In the limit of a deep toroidal trap, we suppose that any mode-structure transverse to the trap circumference can be neglected - that is, the trapped atoms are not sufficiently hot and energetic to excite any transverse mode except the lowest lying mode. This limitation is the principle limitation of the present treatment, and requires a deep trapping potential, and low temperatures.

We therefore consider the condensate wave-function to obey the Gross-Pitaevskii (G-P) equation, which reduces to the nonlinear Schrödinger equation (NLSE), in a one-dimensional ring of radius R and cross-sectional area A. Provided the large radius is much bigger than the small, internal radius, the resulting system has a local cylindrical symmetry. The condensate wave function is factorized into angular and transverse components, so that:

$$\Psi(\mathbf{x},t) = \Psi_\theta(\theta,t)u(\mathbf{r}) \tag{11}$$

where θ is the angle around the ring, and $u(\mathbf{r})$ describes the normalized mode-function transverse to the torus. The cross-sectional area A is defined as:

$$A = \frac{1}{\int\int d^2\mathbf{r}|u(\mathbf{r})|^4} \tag{12}$$

The Hamiltonian in one-dimension, neglecting the transverse kinetic energy terms (as these don't change in a single-mode environment), is:

$$H = \int_0^{2\pi} R d\theta \left[\frac{\hbar^2}{2mR^2} \frac{\partial \Psi_\theta^*}{\partial \theta} \frac{\partial \Psi_\theta}{\partial \theta} + \frac{\hbar \kappa^{(1)}}{2} \left(\Psi_\theta^* \Psi_\theta \right)^2 + V(\theta) \Psi_\theta^* \Psi_\theta \right] \tag{13}$$

Here $\kappa^{(1)} = \kappa/A$ is the effective one-dimensional coupling constant. The quantized version of this Hamiltonian is the problem of a one-dimensional repulsive Bose gas with periodic boundary conditions, and has well-known exact eigenstates at the quantum level, via the Bethe ansatz[7]. In a rotationally uniform trap, the classical equations are:

$$i \frac{\partial \Psi_\theta}{\partial t} = - \frac{\hbar}{2mR^2} \frac{\partial^2 \Psi_\theta}{\partial \theta^2} + \kappa^{(1)} |\Psi_\theta|^2 \Psi_\theta \,, \tag{14}$$

and the number of atoms in the trap N is obtained from noting that the condensate wave function gives the effective particle density:

$$\int_0^{2\pi} R d\theta \Psi_\theta^* \Psi_\theta = N \,. \tag{15}$$

The boundary condition is, from simple periodicity requirements

$$\Psi_\theta(\theta + 2\pi, t) = \Psi_\theta(\theta, t) \,. \tag{16}$$

2.2 *Dimensionless equations*

We define dimensionless variables relative to a characteristic linear density of $\rho_0 = N/(2\pi R)$, where N is the atom number introduced for scaling purposes, by setting:

$$\begin{aligned} \alpha &\equiv \frac{NmRU}{\pi \hbar^2 A} = \frac{4NRa}{A} \,, \\ \tau &\equiv \frac{\hbar}{2mR^2} t \,, \\ \psi(\theta, \tau) &\equiv \Psi(\theta, t) \sqrt{\frac{2\pi R}{N}} \,. \end{aligned} \tag{17}$$

This gives a dimensionless form of the G-P equation:

$$i \frac{\partial \psi}{\partial \tau} = - \frac{\partial^2 \psi}{\partial \theta^2} + \alpha |\psi|^2 \psi . \tag{18}$$

There are three useful dimensionless parameters that describe the solutions, which are also conserved. These are:

- **Particle number:**

$$n = 1 = \frac{1}{2\pi}\int_0^{2\pi} d\theta \psi^* \psi \tag{19}$$

- **Energy:**

$$w \equiv \frac{2mR^2}{\hbar^2 N}H = \frac{1}{2\pi}\int_0^{2\pi} d\theta \left[\frac{\partial\psi^*}{\partial\theta}\frac{\partial\psi}{\partial\theta} + \frac{\alpha}{2}(\psi^*\psi)^2\right] \tag{20}$$

- **Angular momentum:**

$$j = \frac{J}{N} = \frac{1}{4\pi i}\int_0^{2\pi} d\theta \left(\psi^*\frac{\partial\psi}{\partial\theta} - \psi\frac{\partial\psi^*}{\partial\theta}\right) \tag{21}$$

2.3 Ground state and toroidal vortices

For a uniform state, in the case of repulsive interaction ($\alpha > 0$), and zero external potential, we must clearly have (by the normalization condition):

$$\psi^*\psi = 1\,. \tag{22}$$

The lowest energy solution is just the uniform condensate, which evolves in time according to:

$$\psi_0(\theta, \tau) = e^{-i\nu_0\tau}\,, \tag{23}$$

where the dimensionless frequency is given by: $\nu_0 = \alpha$, and the corresponding dimensionless energy is $w_0 = \alpha/2$.

Next, consider solutions whose phase changes linearly around the circumference of the torus. We can call these the uniform vortex states. They correspond to the entire condensate having an angular momentum l, giving rise to the condensate wave function

$$\psi_l(\theta, \tau) = e^{-i\nu_l\tau + il\theta} \tag{24}$$

where l must be an integer, in order to satisfy the boundary condition. The equation of motion gives $\nu_l = \alpha + l$. The energy and momentum per particle of the uniform vortex state are

$$\begin{aligned} w &= \alpha/2 + l^2\,, \\ j &= l \qquad (l = 1, 2, 3, \cdots)\,. \end{aligned} \tag{25}$$

These are the simplest vortices, since they describe a circulation around the circumference of the ring. They have no phase singularity, since the center of the ring is not accessible to the BEC. The circulation is quantized; for these solutions to occur, each particle must have an integer angular momentum, so the whole system has an angular momentum of $Nl\hbar$ in physical units. Another way to think about this, is that each particle is in the same quantum state, with integer angular momentum quantum number.

2.4 *Phonons*

We can obtain the phonon spectrum of small fluctuations around either the ground state or the uniform vortex, as follows. Perturb the vortex state by writing

$$\psi(\theta, \tau) = \psi_l(\theta, \tau) + \delta\psi(\theta, \tau)e^{-i(\nu_l \tau - l\theta)} . \tag{26}$$

and obtain the linearized equation

$$i\frac{\partial}{\partial \tau}\delta\psi = -\frac{\partial^2}{\partial \theta^2}\delta\psi + \alpha\left[\delta\psi + \delta\psi^*\right] - 2il\frac{\partial}{\partial\theta}\delta\psi + (l^2 - l)\delta\psi . \tag{27}$$

This can be solved by putting

$$\begin{aligned} \delta\psi &= Ue^{-i\lambda} + V^* e^{i\lambda} , \\ \lambda &\equiv \omega\tau - n\theta . \end{aligned} \tag{28}$$

We then obtain the phonon dispersion law, which is similar to that of Bogoliubov[6]:

$$\omega = n\left(2l + \sqrt{n^2 + 2\alpha}\right) , \tag{29}$$

where n is an integer, in order to satisfy the periodic boundary conditions. Note that, unlike an infinite, uniform BEC, the phonon spectrum is discrete. In other words, there is no continuum of phonons at low frequency. This is a generic property of finite trap geometries.

2.5 *Higher dimensional vortices*

Vortices in superfluids are usually described by first assuming that the superfluid is incompressible. However, the case of a dilute BEC is somewhat different, as it is described quite well by the Gross-Pitaevskii equation - which definitely allows the system to be compressed. The solutions are obtained as follows, for the case of an ellipsoidal parabolic trap with $V(\mathbf{x}) = m(\omega_x^2 x^2 + \omega_y^2 y^2 + \omega_z^2 z^2)/2$.

First, convert the equations to a simpler form:

$$i\frac{\partial}{\partial t}\Psi(t, \mathbf{x}) = \frac{\hbar}{2m}\left[-\nabla^2 + \frac{1}{\xi^2}[\widetilde{V}(\mathbf{x}) + |\Psi(t, \mathbf{x})|^2/\rho_0]\right]\Psi(t, \mathbf{x}) . \tag{30}$$

Here we have introduced new parameters; a healing length ξ , and a dimensionless potential $\widetilde{V}$:

$$\xi = \hbar/\sqrt{2m\kappa\rho_0} \tag{31}$$

$$\widetilde{V}(\mathbf{x}) = \frac{2m\xi^2}{\hbar^2}V(\mathbf{x}) = \frac{m^2\xi^2}{\hbar^2}(\omega_x^2x^2 + \omega_y^2y^2 + \omega_z^2z^2) \tag{32}$$

Next, look for solutions in the form:

$$\Psi(t, \mathbf{x}) = f\sqrt{\rho_0}e^{il\theta - i\mu t/\hbar} , \tag{33}$$

where $\mu = \hbar\kappa\rho_0$ is the chemical potential for a maximum density ρ_0 . Defining $R = r/\xi$, and $Z = z/\xi$, this gives us the following equations in the steady-state:

$$\frac{\partial^2 f}{\partial R^2} + \frac{1}{R}\frac{\partial f}{\partial R} + \frac{\partial^2 f}{\partial Z^2} - \frac{l^2 f}{R^2} + f - f^3 = -\widetilde{V}(Z, R)f$$

In general, these equations are only soluble numerically. The main feature is that there is a healing length which varies inversely to the mass, density and coupling constant, and this defines the effective radius of the vortex 'core' around which the vortex rotates. To a good approximation, for a first order vortex with $l = 1$, we can find an approximate solution for the atom density of the form (taking $\rho = 0$ when the RHS is negative):

$$\rho = \rho_0\left(\frac{|\mathbf{r} - \mathbf{r}_0|^2}{\xi^2 + |\mathbf{r} - \mathbf{r}_0|^2}\right)\left(1 - \left(\frac{r}{r_m}\right)^2 - \left(\frac{z}{z_m}\right)^2\right) \tag{34}$$

Here $\mathbf{r} = (x, y)$ is the trap radial coordinate, while $r_m^2 = 2\mu/(\hbar m\omega_x^2)$ and $z_m^2 = 2\mu/(\hbar m\omega_z^2)$ are just the maximum radial and longitudinal extent of the BEC in the trap (assuming that the trap has a rotational symmetry, with $\omega_x = \omega_y$). Such vortices have been recently observed[8] by Weiman's group at JILA, using a two-component BEC in order to allow a simple procedure for generating the correct phase structure. Since the vortex center is not constrained to have its phase singularity in one place, it can move, possibly out to the edge of the condensate. In other words, a vortex in a trap is not necessarily dynamically stable, unless pinned to one location in the trap.

3 Soliton solutions

The simplest investigations into a many-body system like a Bose-condensate comprise studies of thermal equilibrium properties, and the physics of small fluctuations about thermal equilibrium. However, it is certainly possible for these systems to display reproducible, even stable behavior in macroscopic states that are very far from thermal equilibrium. The stable, reproducible wave-forms that can result from solutions of nonlinear wave-equations like the Gross-Pitaevskii equation, are called solitons.

The classical view of solitons is that these are stable nonlinear waveforms, which are spatially localized, yet show neither dissipation nor dispersion. In some moving frame with time coordinate τ, this implies that for a complex field Ψ_j, the time-dependent solution must satisfy: $\Psi_j(\tau) = \Psi_j(0)e^{i\omega_j\tau}$. An early example is the nonlinear wave in shallow water [9], described by the KdV equation[10]. In applied mathematics, it is common to use an even more restrictive definition, with conditions on the collisions as well. In this case, the solutions we refer to as solitons, would be called 'solitary waves' or just 'lumps'[11]. A modern example is the short optical pulse in a single-mode fiber, which is rather precisely described by the nonlinear Schrödinger equation[12], and is the subject of many recent experiments[13]. These, however, are restricted to just one space dimension.

Despite these developments, there are many unsolved problems in soliton theory. The following questions seem to be of particular significance:

- When can classical solitons form in more than one dimension?
- When can quantum solitons physically occur?
- Are there essential differences between quantum and classical solitons?
- Do solitons require integrable dynamical equations?
- Are there applications - for example, in communications or computing?
- Is there any novel physics to be found, like different quantum statistics?

We start here by analyzing the one-dimensional solitons of the Gross-Pitaevskii equations. Since an infinite one-dimensional trap is unlikely with current technologies, we focus on the case of a toroidal trap. The relevant solitons have the form of a kink, or phase-singularity in the nonlinear Schrödinger equation. In this state, the atoms have a fractional angular momentum of $\hbar/2$, quite different to the expected integer multiples of $\hbar$ in either the stationary condensate, or in the usually observed vortex states. This particular solution is an interesting one, since it appears likely to be readily accessible within the current BEC techniques. At a more fundamental level, the toroidal soliton must be viewed as a linear superposition of the fundamental Hamiltonian eigenstates or quantum solitons of a repulsive Bose gas in a one-dimensional system with periodic boundary conditions. At the fully quantized level, it is a famous, exactly soluble field theory[7], with unusual properties - including the property that there are excited states with properties that are indistinguishable from an interacting Fermi gas. The fact that there are solitons with an average angular momentum of $\hbar/2$ is further evidence for unexpected behavior in this interacting quantum many-body system.

To set the soliton in motion would require a source of angular momentum, which could originate either in a time-varying magnetic trap potential, or else in a slowly-moving optical standing wave, induced by an external laser. Another possibility is to use a localized, moving potential (obtained, for example, from a blue-detuned 'stirring' laser), to set the condensate in motion. This last technique has already been investigated theoretically as a method for generating vortices, and similar techniques have been used recently for vortex creation in experiments. We suppose that initially the atoms are all trapped in a deep, toroidal magnetic trap - with the central repulsive core potential induced (possibly) by a blue-detuned laser.

Thus, we will investigate the steady-state, classical solutions of the one-dimensional Gross-Pitaevskii equation[14], in a one-dimensional torus or ring. Both vortex-type and kink-type (dark soliton) solutions occur. In this geometry the boundary conditions demand that kink type solutions move, with a fractional angular momentum.

3.1 Dark and Bright Solitons

The dimensionless form of the G-P equation in a torus was, Eq. (18):

$$i\frac{\partial\psi}{\partial\tau} = -\frac{\partial^2\psi}{\partial\theta^2} + \alpha|\psi|^2\psi. \tag{35}$$

This is also called the one-dimensional nonlinear Schrödinger equation (NLSE), and is known to be classically integrable, and to have periodic soliton solutions that correspond to localized disturbances traveling around the circumference of the torus. For a soliton solution, put

$$\psi(\theta,\tau) = f(\theta - v\tau)e^{i\eta(\theta,\tau)}, \tag{36}$$

where v is an arbitrary real parameter, corresponding to the soliton velocity in our dimensionless variables. In general, a continuous set of velocities is possible. Complex values of f correspond to grey solitons, having non-integer velocities. Real values of f correspond to dark and bright solitons, having integer values of v.

In the case of a real envelope function, which we treat here, the boundary condition demands

$$f(\bar{\theta}+\pi) = (-1)^{2l} f(\bar{\theta}-\pi)\,,$$
$$\eta(\theta+\pi,\tau) = \eta(\theta-\pi,\tau) + 2\pi l\,, \tag{37}$$

where $\bar{\theta} \equiv \theta - v\tau$, and $l = 0, \pm\frac{1}{2}, \pm 1, \pm\frac{3}{2}, \pm 2, \cdots$. Assume the form

$$\eta(\theta,\tau) = l(\theta - v\tau) - \omega\tau\,, \tag{38}$$

where ω is an arbitrary real parameter. This satisfies the boundary condition. The angular momentum is $j = l$, and the equation of motion requires $v = 2l$, for f to be real. The equation for $f(\bar{\theta})$ then reads

$$f'' = \alpha f^3 - \beta f \tag{39}$$

where we have introduced a new parameter equivalent to a dimensionless chemical potential, defined as:

$$\beta \equiv \omega + l^2\,. \tag{40}$$

We first consider solutions that exist in the limit of a large torus, where detailed boundary conditions can be neglected to a first approximation. We can look for two types of solution.

- **Dark solitons:** Let $\alpha > 0$, and suppose $f = \tanh(\mu\theta)$. Then we can show that:

$$f'' = 2\mu^2(f^3 - f) \tag{41}$$

  It follows that $\beta = 2\mu^2 = \alpha$, and $f = \tanh(\theta\sqrt{\alpha/2})$. Since the solution has changed sign, we must have half-integer values of l; in a toroidal geometry, dark solitons cannot be stationary. The chemical potential implies that this has a negative binding energy, above the vacuum.

- **Bright solitons:** Let $\alpha < 0$, and suppose $f = \text{sech}(\mu\theta)$. Then, we can show that:

$$f'' = -2\mu^2(f^3 - f/2) \tag{42}$$

  It follows that $\beta = -\mu^2 = \alpha/2$, and $f = \text{sech}(\theta\sqrt{\alpha/2})$. Since the solution has not changed sign, we must have integer values of l only. The chemical potential implies that this has a positive binding energy, below the vacuum.

3.2 Exact solutions

The dimensionless classical field energy is given by

$$w = \frac{1}{2\pi}\int_0^{2\pi} d\theta \left[(f')^2 + l^2 f^2 + \frac{\alpha}{2} f^4\right] = \beta + l^2 - \frac{\alpha}{4\pi}\int_0^{2\pi} d\theta f^4 . \quad (43)$$

Here the second expression was obtained by using the properties of the equation of motion and the normalization condition, together with the usual result that:

$$ff'' = \alpha f^4 - \beta f^2 = \frac{\partial}{\partial\theta}(ff') - \left(\frac{\partial f}{\partial\theta}\right)^2 . \quad (44)$$

We can also add these two expressions for the energy, to obtain a third expression, in which the quartic term does not appear:

$$w = \frac{\beta}{2} + l^2 + \frac{1}{4\pi}\int_{-\pi}^{\pi} d\theta \left[(f')^2\right] . \quad (45)$$

From its equation of motion, the quantity f is the coordinate of a Newtonian particle of unit mass moving in a potential $V(f)$ given by

$$\frac{dV(f)}{df} = -\alpha f^3 + \beta f . \quad (46)$$

The conserved "energy" for this equivalent Newtonian motion is $E = (f')^2/2 + V(f)$, where we take $V(f) = -\alpha f^4/4 + \beta f^2/2$. Thus, the solution can be obtained exactly via the energy integral:

$$\int^f \frac{dx}{\sqrt{2[E - V(x)]}} = \int^{\theta} d\theta . \quad (47)$$

For repulsive interaction $\alpha > 0$, there exist bounded motions between turning points $\pm f_1$. This means that f has a zero, and corresponds to a dark soliton. For attractive interaction $\alpha < 0$, the bounded motion occurs between f_2 and f_1, with $f_2 > f_1 > 0$, and gives a bright soliton. A gray soliton is possible only if f is complex.

3.3 Exact Dark Soliton Solutions

We consider the case $\alpha > 0$. This means that the scattering length is positive, and the inter-particle interaction is repulsive. One can also have attractive interactions, which give rise to localized bright solitons. The Newtonian motion in the potential $V(f)$ is bounded only if the energy is less than a barrier height, so that $E < E_0$, where

$$E_0 \equiv V(f_0) = \frac{\beta^2}{4\alpha} \tag{48}$$

corresponds to the top of the barrier, and we have introduced a constant f_0, defined so that: $f_0^2 = \beta/\alpha$.

The turning points are the roots of

$$E - V(f) = 0\,, \tag{49}$$

which yields the possible solutions f_1, f_2, with

$$\begin{aligned} f_1^2 &= f_0^2 - B_0\,, \\ f_2^2 &= f_0^2 + B_0\,, \end{aligned} \tag{50}$$

where

$$B_0 = \frac{\beta}{\alpha}\sqrt{1 - \frac{E}{E_0}}\,. \tag{51}$$

The turning points of interest for dark solitons are actually at $f = \pm f_1$.

The boundary condition requires that, when $\bar{\theta}$ increases by 2π, $f(\bar{\theta})$ changes sign if l is half-integer, and remain the same if l is integer. In the simplest case, f goes from $-f_1$ to f_1 for half-integer l. For the dark soliton, noting that the integrals are all even in x,

$$\bar{\theta} = \int_0^f \frac{dx}{\sqrt{2\,(E - V(x))}} = \sqrt{\frac{2}{\alpha}} \int_0^f \frac{dx}{\sqrt{(f_1^2 - x^2)\,(f_2^2 - x^2)}}\,.$$

The condition that $f(\pi) = f_1$ then gives the relation between E and ω. The integral we have is just the Jacobi elliptic function with modulus f_2/f_1:

$$\mathrm{sn}^{-1}(x/f_2) = f_1 \int_0^x \frac{dt}{\sqrt{(f_1^2 - t^2)\,(f_2^2 - t^2)}}\,. \tag{52}$$

The explicit solution is therefore

$$f(\bar{\theta}) = f_2 \mathrm{sn}\left(f_1 \sqrt{\frac{\alpha}{2}}\bar{\theta}\right)\,. \tag{53}$$

Assuming f_1 is of order unity, the width of each zero of the soliton is of the order of

$$\Delta\theta = \sqrt{\frac{1}{\alpha}}\,. \tag{54}$$

In other words, $\Delta\theta$ is the healing angle, with: $\Delta\theta \propto 1/\sqrt{N}$.

3.4 *Limiting tanh soliton*

While the elliptic function solutions are the correct solutions to the dark soliton problem, they tend to lack intuition. It is simpler to consider the limiting case of a soliton that is narrow compared to the torus circumference, where an approximate *tanh* solution is obtained. Consider the limiting case $E \to E_0$, in which $B_0 \to 0$ and

$$f_1 = \sqrt{f_0^2 - B_0} \approx f_0 \left(1 - \frac{B_0}{2f_0^2}\right),$$
$$f_2 = \sqrt{f_0^2 + B_0} \approx f_0 \left(1 + \frac{B_0}{2f_0^2}\right). \tag{55}$$

Setting $B_0 = 0$ in the above expressions, we can make the approximation that:

$$\bar{\theta} \approx \sqrt{\frac{2}{\alpha}} \int_0^f \frac{dx}{f_0^2 - x^2} = \frac{1}{f_0}\sqrt{\frac{2}{\alpha}} \tanh^{-1}\left(\frac{f}{f_0}\right), \tag{56}$$

which leads to the explicit solution

$$f(\bar{\theta}) = \sqrt{\frac{\beta}{\alpha}} \tanh\left(\sqrt{\frac{\beta}{2}}\bar{\theta}\right). \tag{57}$$

The normalization condition $\int_{-\pi}^{\pi} d\theta f^2 = 2\pi$ gives

$$\frac{\beta}{\alpha}\int_{-\pi}^{\pi} d\theta \tanh^2\left(\sqrt{\beta/2}\,\theta\right) = 2\pi. \tag{58}$$

Consider now the large N limit. Since $\tanh^2 x < 1$, we have the following inequality:

$$\int_{-\pi}^{\pi} d\theta \tanh^2\left(\sqrt{\beta/2}\,\theta\right) < 2\pi. \tag{59}$$

The normalization requires $\beta > \alpha$, or $\beta > 4NRa/A$. Therefore, $\beta \to \infty$ as $N \to \infty$. In this limit the dark soliton becomes infinitely narrow, and we can carry out the normalization integral to obtain

$$1 = \frac{\beta}{2\pi\alpha}\int_{-\pi}^{\pi} d\theta \tanh^2\left(\sqrt{\beta/2}\,\theta\right) \simeq \frac{\beta}{\alpha}\left(1 - \frac{1}{\pi\sqrt{\beta/2}}\right). \tag{60}$$

This leads to the limiting result that:

$$\beta \simeq \alpha\left(1 + \frac{1}{\pi}\sqrt{\frac{2}{\alpha}}\right). \tag{61}$$

3.5 Conserved Quantities

We can now estimate values of the conserved quantities that characterize the dark soliton with periodic boundaries. The energy can be calculated by using the approximation that $f(\theta) \simeq \tanh\left(\sqrt{\beta/2}\,\theta\right)$. This substitution must be carried out carefully, since the large size of α means that any small error in f can lead to a large error in the nonlinear term contributing to the energy. For this reason, we turn to the third expression for the energy calculated above, Eq. (45), which gives that, to leading order:

$$w = \frac{\beta}{2} + l^2 + \frac{\alpha}{8\pi}\int_{-\pi}^{\pi} d\theta\, \mathrm{sech}^4\left(\sqrt{\alpha/2}\,\theta\right). \tag{62}$$

Next, we must use the approximate expression for β, valid at large N, obtained above. This combination gives us the final result that:

$$w = \frac{\alpha}{2} + \frac{\sqrt{8\alpha}}{3\pi} + l^2 + O(1). \tag{63}$$

The energy and angular momentum for a single tanh soliton and a vortex state are given below for comparison. For narrow solitons:

$$\begin{aligned} w^{soliton} &\simeq \frac{\alpha}{2} + l^2 + \frac{\sqrt{8\alpha}}{3\pi} + O(1), \\ j^{soliton} &= l \qquad (l = \frac{1}{2}, \frac{3}{2}, \frac{5}{2}, \cdots). \end{aligned} \tag{64}$$

For vortices , or uniformly rotating condensates with integer angular momentum:

$$\begin{aligned} w^{vortex} &= \frac{\alpha}{2} + l^2, \\ j^{vortex} &= l \qquad (l = 1, 2, 3, \cdots). \end{aligned} \tag{65}$$

Thus, the dark soliton spectrum might be expected to have a lower energy than that for vortex states, because the lowest soliton state has $j = 1/2$, whereas the lowest vortex state has $j = 1$. However, there is an additional energy due to the compression of the condensate caused by the increase in density in the 'bright' region of the soliton, to allow the low-density hole to form at the phase-singularity. This effect is larger than the increased kinetic energy of the vortex if α is large; although not if α is small. Both effects become vanishingly small relative to the bulk energy of the condensate in the limit of large N, as the singularity of the soliton core occupies a smaller and smaller region. The first observations of dark solitons in cylindrical, cigar-shaped traps in BEC have now been made[15], although these observations were in a non-toroidal environment.

In summary, the soliton can form as a circulating, kink-like solution, which has the unexpected property that it has a fractional (half-integer) angular momentum per particle. This is stable in a rotationally symmetric environment - and in fact, numerical simulations indicate that it is also relatively stable with small (adiabatic) departures from rotational symmetry. However, as it has a higher energy than the non-rotating ground state, one might expect that in an environment that does not conserve angular momentum, it is possible for a decay to occur to the non-rotating state. This would presumably occur via the creation of a grey soliton, which can have a lower angular momentum than the dark, tanh-type soliton.

By using gain and an appropriately synchronized periodic output-coupler, it may be feasible to stabilize the dark soliton, to create an atom laser[16], with a reproducible, coherent waveform. This allows one possible route towards manufacturing a mode-locked, pulsed atom laser. The advantage of mode-locking in lasers, is that all the atom lasers to date have involved a pulsed output field. In a mode-locked laser, the sequence of output pulses are in phase with each other. This allows a wide range of interference experiments and phase-sensitive measurements. In addition, mode-locked lasers can have an enhanced intensity stability relative to their non-mode-locked cousins.

4 Quantum solitons

In the quantum domain, there must be corresponding objects to classical solitons, owing to the correspondence principle. One may even expect these to show more structure than their classical counterparts. A quantum view is that a quantum soliton is an energy eigenstate of a local quantum field theory, which has a spatially localized correlation function, so that: $\hat{H}|\Psi\rangle = E|\Psi\rangle$. It is known that, at least for sufficiently weak couplings, a quantum soliton exists for every classical soliton[17]. Early examples of this were various theoretical models in high-energy physics. However, until recently, the experimental evidence for quantum effects in solitons was extremely scarce. This situation has changed dramatically due to the prediction[18] and observation[19] of quantum effects in optical pulses in fibers, which can be regarded as photonic bound states.

Quantum solitons therefore, are generalizations of the nonlinear soliton solutions of classical wave theory, to include quantum fields. In this sense, there are a wide variety of quantum fields able to be analyzed. Since quantum field theory is generic to many areas of physics, we can expect these entities to be universally significant, in all areas where there are nonlinear interactions involving quantum fields.

Important exactly soluble cases include the many-body bound states of bosons interacting via delta-function interactions in one space dimension. This model (often called the nonlinear Schrödinger model) was solved by Liniger[7], and by McGuire and Yang[20]. Other examples of exactly soluble models like the Hubbard model, are generally restricted to one space dimension - except for Laughlin's highly innovative theory of an idealized model of two-dimensional electron gas in an external magnetic field[21]. This was able to explain the phenomenon of the fractional quantum Hall effect. Analogous behaviour[22] for a BEC may occur as well.

The interesting thing about the latest cold-atom experiments, is that one could hope to combine the temperatures achieved in BEC, with a one-dimensional wave-guide. If this were feasible in the near future, an exactly soluble interacting boson model would be realized in a real laboratory experiment, possibly with as few as two or three bosons.

4.1 Bethe Ansatz

The first exactly soluble model of a Bose gas in one dimension was due to Girardeau[7], who showed that a Bose gas with hard core interactions in one-dimension was exactly soluble. It also turns out that this model is precisely equivalent to a Fermi gas problem as well, which is a remarkable property of this model - thus showing that in the case of interacting Bose gases, it may not be readily possible to distinguish bosonic and fermionic properties - at least, in one dimension. This theory was extended to the case of delta-function interactions later, which is a more realistic case.

The classical equations, in one dimension, are given by:

$$i\frac{\partial\Psi}{\partial t} = -\frac{\hbar}{2m}\frac{\partial^2\Psi}{\partial x^2} + \kappa^{(1)}\Psi^\dagger\Psi^2. \qquad (66)$$

This is the nonlinear Schrödinger (NLS) equation, where $\Psi = \langle\hat{\Psi}\rangle$ is the mean field amplitude. The soliton that occurs for $m/\kappa^{(1)} > 0$ is observed to propagate with solution $|\Psi| \propto \tanh(x/x_0)$, as explained in the previous section.

In quantum theory, the corresponding Hamiltonian[20] is given in a standard notation, by:

$$\hat{H} = \frac{\hbar^2}{2m}\int dx[\partial_x\hat{\Psi}^\dagger\partial_x\hat{\Psi} + C\hat{\Psi}^{\dagger 2}\hat{\Psi}^2]. \qquad (67)$$

Here, $C = m\kappa^{(1)}/\hbar$. The corresponding diphoton eigenstate - as well as bound states with any number of photons - was discovered by McGuire[20] in the case of an attractive interaction, with negative scattering length ($\kappa^{(1)} < 0$). The (stationary) two-particle solution is:

$$|2\rangle = \int\int dx_1 dx_2 e^{C|x_1-x_2|/2} \hat{\Psi}_1^\dagger(x_1)\hat{\Psi}_1^\dagger(x_2)\,|0\rangle\,, \tag{68}$$

Similarly, a localized $N-$ particle solution would be:

$$|N\rangle = \int .. \int dx_1..dx_N e^{C\sum_{i>j}|x_i-x_j|/2} \hat{\Psi}^\dagger(x_1)..\hat{\Psi}^\dagger(x_N)\,|0\rangle\,, \tag{69}$$

where the binding energy is given by:

$$E^b = \frac{N(N^2-1)}{24}\left(\frac{\hbar^2 C^2}{m}\right) = \frac{N(N^2-1)}{24} m(\kappa^{(1)})^2 \tag{70}$$

In principle this type of solution could be found in optical fibers, where the photons are the interacting bosons, and the interaction is mediated through the nonlinear refractive index. However, a quantitative investigation shows that the diphoton solution is far too weakly bound to be readily observable in fibers. To relate these solutions to physical experiments, it is necessary to take into account the fact that practical experiments typically involve 10^9 photons or more. Under these conditions, the bound states exist, but are not easily distinguished from each other, and are subject to disruption by thermal phonons. Thus, the simplest experiments observe classical soliton behavior, which can be regarded as a superposition of quantum solutions[23]. These classical solitons are used (for example) in experimental communications systems. Predicted quantum effects[18] require measurement of quantum interference between the bound states. This leads to the observation of quantum squeezing[19], quantum non-demolition collisions [24], and more recently - photon anti-bunching[25,26].

The interesting thing about BEC experiments, is that this type of localized soliton is potentially observable at relatively low particle number, owing to the existence of trapped atomic species like Li^7, which has a negative scattering length. While this prevents the formation of a stable trapped condensate in three dimensions, there should be no problem in a one-dimensional trap environment, where the equations are stable - provided the particle number is not so large that it causes self-focusing.

The repulsive case exact solutions are very interesting, but too complex to be treated briefly, as it is nontrivial to obtain the eigenvalues - which require the solution of an integral equation[20]. However, we can just write down the form of the general eigenstate in any case, which is:

$$|N\rangle = \int .. \int d^N\mathbf{x}\, e^{\sum_i ik_i x_i} \left[\prod_{i<j}^{N} \left(1 - \frac{iC\epsilon(x_i - x_j)}{k_i - k_j} \right) \right] \hat{\Psi}^\dagger(x_1)..\hat{\Psi}^\dagger(x_N) \left|0\right\rangle \, . \tag{71}$$

Here $\epsilon(x_i - x_j)$ is the sign function, giving the sign of its argument. This more complex case is also exactly soluble, with interesting features including excited states of quasi-particles that have either bosonic or fermionic[7] character!

4.2 Coupled solitons

Next, consider the model of two fundamental Bose fields $\hat{\Psi}_i$, each with mass m. The fields can interact with an effective delta-function interaction (both inter-species and intra-species), while they can also inter-convert from one bosonic species to another. This type of theory with both linear and nonlinear couplings is well-known in fiber optics, where it corresponds precisely to the interactions of photons with opposite helicity in a single-mode, birefringent fiber with anomalous dispersion. This case has some interesting features. It turns out that the soliton is also a Schrödinger 'kitten', or quantum superposition state.

In atomic systems, the model corresponds to the elastic scattering of cold atoms with different internal states in an atomic waveguide. The internal states can be coherently inter-converted via external Raman or microwave fields in a controlled and tunable way, leading to the possibility of investigating fundamental quantum solitons in an environment with a freely adjustable coupling constant. Experiments like this for macroscopic Bose condensates[4] have been carried out. What will happen to the microscopic quantum solitons in this case?

The basic Hamiltonian considered (using Einstein summation conventions for the repeated indices) is:

$$\hat{H} = \frac{\hbar^2}{2m} \int \left[\nabla \hat{\Psi}_i^\dagger \nabla \hat{\Psi}_i + \tilde{\Omega}_{ij} \hat{\Psi}_i^\dagger \hat{\Psi}_j + C_{ij} \hat{\Psi}_i^\dagger \hat{\Psi}_j^\dagger \hat{\Psi}_i \hat{\Psi}_j \right] dx \, . \tag{72}$$

Here $\tilde{\Omega}_{ij}$ is the rate of linear inter-conversion, expressed here in time-units such that $\Omega_{ij} = \frac{\hbar}{m}\tilde{\Omega}_{ij}$. For definiteness, we will take $\Omega_{11} = \Omega_{22} = 0$ and $\Omega_{12} = \Omega_{21} = \Omega$. For definiteness, we choose Ω to be real and positive. Inter-particle elastic scattering is described by $\kappa_{ij}^{(1)} = \frac{\hbar}{m} C_{ij}$, and we will mostly focus here on the common symmetric coupling case, in which $\kappa_{11} = \kappa_{22}$, and $\kappa_{12} = \kappa_{21}$. To obtain bound states, it is necessary to consider negative κ values.

A crucial property of this Hamiltonian is that it conserves the total boson number, $\widehat{N} = \int \widehat{\Psi}_i^\dagger(x)\widehat{\Psi}_j(x)dx$, so that energy eigenstates are also number eigenstates. Combined with translational invariance, this means that the ground state has zero momentum and a fixed particle number.

4.3 Properties of eigenstates

We first notice that the single-particle states have the obvious property that they are dressed by the interaction, with zero momentum eigenstates of:

$$|1_\pm\rangle = \int [\widehat{\Psi}_1^\dagger(x) \pm \widehat{\Psi}_2^\dagger(x)]dx|0\rangle \, , \tag{73}$$

and energy eigenvalues of $E_\pm^{(1)} = \pm\hbar\Omega/2$. These states, which we can call the [+] and [−] bosons, are the dressed free-particle states with linear coupling including. Thus, in the **zero-interaction limit** ($\kappa = 0$), there are three possible two-particle states, which are labeled [++] , [+−] , and [−−] with energies $E^{(2)} = 0, \pm\hbar\Omega$. These are not bound states, of course.

The lowest nontrivial bound state with an interaction, must have the form:

$$|2\rangle = \sum_{ij} \int\int g_{ij}(x-y)\widehat{\Psi}_i^\dagger(x)\widehat{\Psi}_j^\dagger(y)dxdy|0\rangle \, . \tag{74}$$

The energy eigenstates with energy $E = \hbar\omega$ must satisfy an eigenvalue equation involving g_{ij}. This can be written, on transforming to a relative coordinate $r = (x - y)$, as:

$$2[\tilde{\omega} + \frac{\partial^2}{\partial r^2} - C_{ij}\delta(r)]g_{ij}(r) = \tilde{\Omega}_{ik}g_{kj}(r) + \tilde{\Omega}_{jk}g_{ik}(r) \, . \tag{75}$$

Here we have also introduced a scaled eigenfrequency, so $E = \hbar^2\tilde{\omega}/m$.

4.4 Anti-symmetric case

The solutions in the limit of zero linear coupling are straightforward. There are three bound states, each with only a single exponentially decaying wavefunction having the structure:

$$g_{ij}(r) = e^{-\mu_{ij}|r|} \, . \tag{76}$$

The inverse length scales have the form $\mu_{ij} = -C_{ij}/2$, with eigenvalues as in Eq. (70), that is, identical to the previous scalar models:

$$E_{ij}^{(2)} = -\frac{\hbar^2 C_{ij}^2}{4m} \, . \tag{77}$$

As the linear coupling is increased, we can expect the fundamental eigenvalues to be modified. Surprisingly, in the simplest coupling case with $C_{11} = C_{22}$, we find that one of the eigenvalues and its eigenstate is completely unchanged by the coupling. In this case there are degenerate eigenstates for zero linear coupling, which consist of either a pair of $i = 1$ or a pair of $i = 2$ bosons, with energies of $E_{ii}^{(2)}$, as explained above. Any linear combination of these two states is also an eigenstate, including the antisymmetric one.

It is simple to verify that the linear coupling has no effect on the antisymmetric combination, which is therefore invariant under a change of coupling strength. The eigenvalue is $E_0^{(2)} = -\hbar^2 C_{11}^2/4m$, with a length-scale of $x_0 = -2\hbar/(m\kappa_{11}^{(1)})$. This antisymmetric eigenstate has an equal probability of having either pair of bosons:

$$|E_0^{(2)}\rangle = \int\int[\widehat{\Psi}_1^\dagger(x)\widehat{\Psi}_1^\dagger(y) - \{1 \to 2\}]e^{-|x-y|/x_0}dxdy|0\rangle \,. \tag{78}$$

Since the dressed boson energy also changes with the linear coupling, some interesting physics emerges as the linear coupling increases. While up to three localized states exist, these can have an energy above the corresponding continuum energy for two dressed bosons of the type $E_-^{(1)}$, which have a total energy $E = -\hbar\Omega$. This is clearly possible for the antisymmetric case, which by the above arguments has a binding energy of $E^b = m\kappa_{11}^2/4 - \hbar\Omega$, and therefore has an 'auto-dissociation' boundary at $\Omega = m\kappa_{11}^2/4\hbar$.

Thus, while the antisymmetric bound state can exist above the lowest free-particle energy as an energy eigenstate with a discrete eigenvalue, any symmetry-breaking type of perturbation would allow it to decay into the lowest dressed state continuum at large coupling strength Ω. This type of phenomenon is unusual for an effective quantum field theory with delta-function interactions.

4.5 General case

We now consider the more general forms of bound state that occur with $\kappa_{11} = \kappa_{22}$. There should be three possible cases, since there are certainly three solutions in either the zero coupling limit, or the zero interaction limit.

Defining $G(r) = g_{11}(r) + g_{22}(r)$ and $g(r) = g_{12}(r) + g_{21}(r)$, suppose these have the generic expansions:

$$g(r) = \sum_n g_n e^{-\mu_n |r|}$$

$$G(r) = \sum_n G_n e^{-\mu_n |r|} \tag{79}$$

Here the different length-scales μ_n correspond to different structures *within one bound state*! Substituting into the eigenvalue equations for $r \neq 0$, gives (for any eigenstate):

$$\begin{aligned} \tilde{\omega}_n G_n &= \tilde{\Omega} g_n \\ \tilde{\omega}_n g_n &= \tilde{\Omega} G_n \ , \end{aligned}$$

Here we have introduced the notation $\tilde{\omega}_n \equiv \tilde{\omega} + \mu_n^2$, where $\hbar\omega$ is the eigenvalue, and the solution is clearly $\tilde{\omega}_n^2 g_n = \tilde{\Omega}^2 g_n$. There are different possible types of solution, which correspond to different types of eigenstate, with different symmetry properties. The two main alternatives are:

- $g_n = G_n = 0$, so that $g_{11}(r) = -g_{22}(r)$, corresponding to the anti-symmetric solution deduced already. This is a coupling immune state. It has only one length scale.
- $G_n \neq 0$, so that $g_{11}(r) = g_{22}(r)$. These consist - in the zero interaction limit - of two of either the $E_+^{(1)}$ or the $E_-^{(1)}$ dressed bosons, with their corresponding energies. When the interaction is introduced, these two combinations become mixed together, to form new types of localized quantum soliton, which have more than one length scale. Like a Schrödinger cat, we don't know how big they are! Introducing labels of $n = \pm$ for the two length scales involved, the equations are:

$$\begin{aligned} G_\pm &= \pm g_\pm \\ \tilde{\omega}_\pm &= \pm \tilde{\Omega} \ . \end{aligned}$$

In order to calculate the new eigenvalues in detail, we now consider the boundary conditions on the equations at $r = 0$. This imposes the constraints that:

$$\begin{aligned} \sum_n [2\mu_n + C_{12}] g_n &= 0 \\ \sum_n [2\mu_n + C_{11}] G_n &= 0 \ . \end{aligned}$$

Using these equations, together with the results that $G_\pm = \pm g_\pm$, gives the amplitude ratios:

$$\frac{g_+}{g_-} = \frac{-C_{12} - 2\mu_-}{2\mu_+ + C_{12}} = \frac{2\mu_- + C_{11}}{2\mu_+ + C_{11}} \ . \tag{80}$$

Introducing new variables μ, δ defined so that $\mu_\pm = \mu \pm \delta$, the condition $\tilde{\omega}_\pm = \tilde{\omega} + \mu_\pm^2 = \pm\tilde{\Omega}$ implies that $\delta = \tilde{\Omega}/2\mu$. This immediately leads to an exactly soluble quartic for the eigenvalues, with the suggestive form that:

$$f(\mu) = (\mu + C_{12}/2)(\mu + C_{11}/2) = \tilde{\Omega}^2/4\mu^2 \,. \tag{81}$$

I have written this to show the effect of changing the linear coupling $\tilde{\Omega}$.

4.6 Solutions in the general case

- **Equal scattering lengths**
  In order to analyze the eigenvalues in more detail, we consider the case of $\tilde{\kappa}_{12} = \tilde{\kappa}_{11}$ first, which is often (nearly) true in real experiments. In this case, the solution of Eq. (81) can be solved to give:

$$\mu^{(\pm\pm)} = \frac{1}{4}\left(|C| \pm \sqrt{C^2 \pm 8\tilde{\Omega}}\right) , \tag{82}$$

  where each $\pm$ can take either sign independently, giving 4 solutions in total. However, one of them is negative and will be neglected, giving three positive eigenvalues. The smallest of the roots for μ is never physically possible, since the condition of having a normalizable state implies that the inverse length scales are positive. This means that we must have $\mu_- > 0$, which is equivalent to $\mu > \sqrt{\tilde{\Omega}/2}$, thus ruling out the smallest of the three roots as physically acceptable. A critical value of the linear coupling occurs at $\tilde{\Omega}/2_c = (C/4)^2$. This is because these values give $\mu^{(\pm-)} = -C/4 = \sqrt{\tilde{\Omega}/2}$, which is also exactly at the dissociation limit, since this would give an infinite length scale ($\mu_- = 0$). Above this coupling only the largest root is stable, as it occurs at $\mu = -C(1+\sqrt{2})/4 > \sqrt{\tilde{\Omega}/2}$.

- **Unequal scattering lengths**
  More generally, at $\tilde{\Omega} = 0$, the equation is simply a quadratic, and the previous results for quantum solitons in the zero-coupling limit are recovered. As $\tilde{\Omega}$ is increased, the number of roots is initially three, but then decreases from three to one. For this case of two different interaction strengths ($C_{12} \neq C_{11}$), we find that the instability point occurs before the two smaller roots are degenerate. Thus, the smallest root μ_u is always unphysical.

The binding energy (relative to the minimum dissociation energy) for a real, positive root that satisfies the normalizability condition, is:

$$E^b = \hbar^2(\mu - \tilde{\Omega}/2\mu)^2/m \ . \tag{83}$$

It is clear from this equation that the transition point where the smallest length-scale of an eigenstate becomes infinite (i.e., $\mu = \sqrt{\tilde{\Omega}/2}$), also corresponds to the dissociation boundary. In these cases, there can be no auto-dissociation, as the bound states have the same type of symmetry as the lowest continuum manifold. However, the bound state displays an unusual behavior. It is a superposition of states having two dramatically distinct lengths scales just prior to dissociation - with one length scale going to infinity while the other remains of the order of x_0.

In summary, crossing the dissociation boundary is not possible, so there is no auto-dissociation. As $\tilde{\Omega}$ increases, one of the original two bound states generally become more and more destabilized by the linear particle conversion interaction, until only one remains below the dissociation energy. Solving the quartic in the limit of $\tilde{\Omega}/2 \to \infty$, we obtain a larger value of μ, giving the final limiting result that:

$$\mu \approx \sqrt{\tilde{\Omega}/2} - (C_{11} + C_{12})/8$$
$$\delta \approx \sqrt{\tilde{\Omega}/2} + (C_{11} + C_{12})/8 \ .$$

In this limit the two radius scales corresponding to $\mu_\pm$ are completely different to each other, since:

$$\mu_+ = 2\sqrt{\tilde{\Omega}/2} \to \infty$$
$$\mu_- = -(C_{11} + C_{12})/4 \ .$$

Clearly, the exact ground state in the limit of strong linear coupling is the superposition of two completely different and distinguishable quantum structures, one with a relatively large radius comparable to the value x_0, and one with a much smaller value that scales as the inverse linear coupling strength. In more general cases, there can be three different length scales - a Schrödinger 'triplet'!

4.7 *Energy estimates*

Given rapid developments in atom cooling and wave-guiding techniques, we expect that these quantum soliton bound states will become observable in the near future. The spectroscopy of these states is potentially accessible through the technique of Raman photo-association probes, combined with microwave coupling techniques to obtain linear conversion. As an attractive interaction is required, a candidate species is Li^7, which has been trapped and cooled in states with attractive elastic scattering lengths. One can also tune the atomic interaction strength with a Feshbach resonance[27]. The measured Li^7 triplet S-wave scattering length is $a_T = -1.5$nm. This corresponds to length-scales of $x_0 \simeq 100$nm, critical coupling rates in the range $\Omega_c \simeq 10^6$/s, and binding energies of about $E_b \simeq 10^{-28}$J, for a nano-scale atom waveguide with a modal cross-section of $A \simeq 1000\text{nm}^2$. Such waveguides are now under development at Sussex. These energies are greater than thermal with cooled atoms at micro-Kelvin temperatures. We note that to obtain the interesting feature of linear interconversion with two trapped states also requires that the inelastic scattering is small, which may require a different candidate from Lithium.

In summary, the two-particle quantum solitons or bound-states of interacting boson fields with both linear and nonlinear interaction paths, are exactly soluble in one space dimension - and have three types of bound state. The anti-symmetric bound state can auto-dissociate to the continuum of dressed bosons, for large coupling strength. Both the ground state and the first excited symmetric bound state consist of a quantum superposition of two distinct quantum structures. These can have completely different length scales either near the critical coupling, or for a strong linear coupling.

5 Parametric 'super-chemistry' solitons

Each of the one-dimensional soluble cases has led to substantial improvements in our understanding of quantum theory, together with new and interesting physical consequences. However, there are few exact solutions in two or three space dimensions, except for physically inaccessible models like the quantum Davey-Stewartson model[28]. This is especially true if we look for nonlinear quantum field theories which include the most fundamental property that distinguishes quantum mechanics from quantum field theory - that is, the ability to create and destroy particles.

Recent developments in BEC include the possibility of coherent molecule creation. The relevant parametric quantum field theories and their classical analogs[29] are the subject of much current attention[30], due to the existence of stable[31], bright, higher-dimensional solitons[32]. These have now been observed experimentally in photonic systems[33]. The corresponding equations are an example of a classically non-integrable field theory, which generally needs to be treated numerically. The classical equations still can be analyzed for topological structure[34] in a corresponding phase-space, together with analytic results for stability. Thus, there are surprisingly close parallels to nonlinear optics, in the related fields of quantum many-body theory and atom optics. This means we can have nonlinear atom optics – which is now an emerging field of research.

The most elementary treatment of coherent molecule formation is via a cubic coupling between two creation operators and an annihilation operator (and vice-versa, on grounds of hermiticity). Cubic couplings are, of course, basic to QED and QCD, where they involve both fermionic and bosonic fields. These theories do not appear to have exactly known solutions in four dimensional space-time, and are usually treated by various approximations. The most prominent of these is the Feynman diagram method or perturbation theory, which is generally conjectured to be non-convergent.

It would be useful to have a cubic interaction theory which was exactly soluble, to give some guide as to the possible variety of behavior in this class of widely used quantum field theories[36]. In particular, one would like to investigate whether the resulting quantum solitons have any differences resulting from dimensionality, or from the presence of interactions that change particle number. Surprisingly, the simplest cubic interaction involving two boson fields - the parametric interaction of the form $\Psi_2^\dagger \Psi_1^2$ - has not been widely analyzed for bound states in higher dimensions. This problem has become especially interesting since the discovery that the corresponding classical parametric theory has stable higher-dimensional soliton solutions[33], which have now been experimentally observed.

In this section I consider the classical solitons together with the bound states in the parametric quantum field theory, and show the existence of an exactly soluble case in higher dimensions[38]. The model is a traveling wave analog of the quantum theory used to describe squeezed states in quantum optics, and more recently, coherent molecular photo-association in atom optics[39]. The problem is all the more interesting because technical advances in BEC experiments, are now reaching the point that this type of coupling is being experimentally investigated at several laboratories.

The results have a number of entirely novel and unexpected features. These are as follows:

- The parametric theory has quantum bound states in one space dimension.
- It is unstable in higher dimensions.
- This quantum instability shows no trace at the classical level [31]!
- Adding a quartic term gives a stable parametric quantum field theory.

- The resulting theory has an exact solution in three dimensions.
- There is a transition to mean-field 'super-chemistry' at large density.

As a fundamental quantum field theory, the interesting result is that, with a positive quartic S-wave scattering interaction, a rigorous lower bound to the energy does exist, and there are exact two-particle bound-state solutions in higher dimensions. These new types of quantum solitons have a unique character: the solution has a finite binding energy, but the corresponding two-particle wave-function has a zero radius. The point-like structure of these bound states can be termed a 'quantum singularity'. No analogous behavior exists in the corresponding classical theory, which is known[31] to possess stable, finite size classical soliton solutions.

The reason for this unexpected behavior is that the fundamental structure of the new solution is inherently non-classical, being a quantum superposition of two states, with one of them having a single (Ψ_2-type) boson and the other having two (Ψ_1-type) bosons present. This is a bosonic analog of the quark model of mesons, with the Ψ_1-type bosons behaving as 'quarks', and the Ψ_2-type boson taking the role of 'gluon'. However, unlike the usual meson, the system has a finite probability of having no 'quark' present at all. It is somewhat like the 'dressed' states obtained with linear conversion in the last section, except that a molecule is being converted reversibly into two atoms.

To obtain the most physically relevant theory, we should modify the Hamiltonian by imposing a momentum cut-off on the nonlinear couplings. In this case exact two-particle bound states are shown to acquire finite radii in higher dimensions. Moreover, finite-size multidimensional bound states occur even without the stabilizing quartic term, that is, in the simplest version of the theory - pure parametric interaction. It turns out that when this is done, the system can make a transition between the localized, 'meson' type solution - which has the lowest energy at low density and particle number, to a different type of solution with coherently coupled atom and molecular condensates, which has the lowest energy at large particle number. We call this the 'superchemistry' state, as it allows coherent, Bose-enhanced transition between atoms and molecules. These transitions do not follow the standard chemical kinetics equations.

5.1 Hamiltonian

The quantum effective Hamiltonian we consider has the following form:

$$H = H_0 + H_{int} \,, \tag{84}$$

$$H_0 = \hbar \int d^{(D)}\mathbf{x} \left[\frac{\hbar}{2m} |\nabla \Psi_1|^2 + \frac{\hbar}{2M} |\nabla \Psi_2|^2 + \Delta \Psi_2^\dagger \Psi_2 \right] , \tag{85}$$

$$H_{int} = \hbar \int d^{(D)}\mathbf{x} \left[\frac{\chi^{(D)}}{2} \left(\Psi_1^2 \Psi_2^\dagger + \Psi_1^{\dagger 2} \Psi_2 \right) + \frac{\kappa^{(D)}}{2} \Psi_1^{\dagger 2} \Psi_1^2 \right] . \tag{86}$$

Here Ψ_1 and Ψ_2 are two complex Bose fields which we term atomic and molecular fields, respectively, in analogy with the nonlinear optical process of frequency conversion. Their commutation relations are given by:

$$[\Psi_1(\mathbf{x}), \Psi_1^\dagger(\mathbf{x}')] = [\Psi_2(\mathbf{x}), \Psi_2^\dagger(\mathbf{x}')] = \delta(\mathbf{x} - \mathbf{x}') , \tag{87}$$

$$[\Psi_1(\mathbf{x}), \Psi_2^\dagger(\mathbf{x}')] = [\Psi_1(\mathbf{x}), \Psi_2(\mathbf{x}')] = 0 .$$

In addition, m and $M = 2m$ are corresponding effective masses, and $\hbar\Delta = V_2 - 2V_1$ is the energy mismatch. Interactions are provided by χ and κ, which are the coupling constants responsible for the coherent atom-molecule conversion and S-wave scattering processes, respectively, in D ($D = 1, 2, 3$) spatial dimensions. We ignore the trap potential here, for simplicity.

To construct the general candidate for the eigenstate to our Hamiltonian we note that the parametric interaction here transforms pairs of subharmonic quanta into single second harmonic quanta, and vice-versa. That is, the Hamiltonian does not conserve corresponding particle numbers. However, it does conserve a generalized particle number, or Manley-Rowe invariant, equal to:

$$N = N_1 + 2N_2 = \int d^{(D)}\mathbf{x} \left[|\Psi_1|^2 + 2 |\Psi_2|^2 \right] . \tag{88}$$

In addition, since the Hamiltonian is translation-invariant, it must have a momentum conservation law for $\mathbf{P}$, where

$$\mathbf{P} = -\frac{i\hbar}{2} \int d^{(D)}\mathbf{x} \left[\Psi_1^\dagger (\nabla \Psi_1) + \Psi_2^\dagger (\nabla \Psi_2) \right] + \text{h.c.} \tag{89}$$

We therefore search for states that are eigenstates of H, $\mathbf{P}$ and N, with zero momentum and even N. These must have the form of a superposition state:

$$|N\rangle = \sum_{j=0}^{N/2} \int \dots \int d^{(D)}\mathbf{x}_1 \dots d^{(D)}\mathbf{x}_{N-j}\, g_{N-j}^{(N)} (\mathbf{x}_1, \mathbf{x}_2, \dots, \mathbf{x}_{N-j})$$

$$\times \prod_{l=1}^{j} \Psi_2^\dagger(\mathbf{x}_l) \prod_{l=j+1}^{N-j} \Psi_1^\dagger(\mathbf{x}_l) \; |0\rangle \; , \tag{90}$$

where j denotes the number of Ψ_2 field-operators present in each term, and the function $g_{N-j}^{(N)}$ only depends on the relative coordinates. For clarity, we mostly focus on the nontrivial case of a two-particle ($N = 2$) bound-state solution here, which we term a bosonic 'meson', in analogy to the well known quark model, and from now on, drop the dimension superscript on the coupling constants.

5.2 Two-particle eigenstates: bosonic 'mesons'

First consider the two-particle problem. In this case the two-particle eigenstate with $P = 0$ has the following explicit form:

$$|2\rangle = \left[\int d^{(D)}\mathbf{x} \;\; \Psi_2^\dagger(\mathbf{x}) \right.$$

$$\left. + \int\int d^{(D)}\mathbf{x}_1 d^{(D)}\mathbf{x}_2 g\left(\mathbf{x}_1 - \mathbf{x}_2\right) \Psi_1^\dagger(\mathbf{x}_1)\Psi_1^\dagger(\mathbf{x}_2)\right] |0\rangle \; , \tag{91}$$

where $g(\mathbf{x}_1 - \mathbf{x}_2) \equiv g_2^{(2)}(\mathbf{x}_1, \mathbf{x}_2)$ is the two-particle wave function and $g_1^{(2)}(\mathbf{x}_1) = 1$.

To prove a lower bound on the Hamiltonian energy we apply Eq.(86) to the ansatz $|2\rangle$ and use the symmetry property of the two-particle wave-function: $g(\mathbf{x}) = g(-\mathbf{x})$. Neglecting the positive term $\sim (2\hbar^2/m) \int d^{(D)}\mathbf{x} |\nabla g(\mathbf{x})|^2$, appearing by integration by parts, we reduce the calculations to a chain of algebraic inequalities, and finally obtain that if $\kappa > 0$ and

$$\chi^2 > 2\Delta\kappa \; , \tag{92}$$

then the lower bound E_l can be defined by:

$$E_l = \hbar\Delta - \frac{\hbar\chi^2}{2\kappa} \leq \frac{\langle\varphi^{(2)}|\, H \,|\varphi^{(2)}\rangle}{\langle\varphi^{(2)}|\, \varphi^{(2)}\rangle} \; . \tag{93}$$

We can also obtain a variational upper bound to the ground-state energy, by using a trial wave-function. The remarkable result is that in two and three dimensions, these results agree exactly - meaning that we have found the ground state energy! To add spice to the meal, it turns out that the ground state energy for the N particle case is exactly $N/2$ times the energy for the case given above. In other words, there is an exact N particle solution for this interacting many-body theory in $D = 2, 3$ dimensions - and it is a Bose condensate of the 'mesons' [39].

5.3 Exact solutions

Next, we wish to understand the exact nature of these solutions in a more traditional approach; we will solve for the kernel function. Applying our Hamiltonian to the ansatz $|2\rangle$, one can obtain that the eigenvalue problem is equivalent to the following simultaneous equations:

$$\nabla^2 g(\mathbf{r}) - \mu^2 g(\mathbf{r}) = q\delta(\mathbf{r}) \ , \tag{94}$$

$$E = \hbar\Delta + \hbar\chi g(0) \ = -\hbar^2\mu^2/m \ , \tag{95}$$

where $\mathbf{r} = \mathbf{x}_1 - \mathbf{x}_2$, $K = |\mathbf{K}|$, and we have defined

$$\mu^2 = -\frac{Em}{\hbar^2} \tag{96}$$

and

$$q = \frac{m}{\hbar}\left[\frac{\chi}{2} + \kappa g(0)\right] \ . \tag{97}$$

Here μ is an inverse scale length; the solution is bound (against two particle decay) if μ is real and positive.

Eqs.(94),(95) can be easily analyzed using the Fourier transform method. In this approach we seek for a solution to Eq.(94) in the form

$$g(r) = \frac{1}{(2\pi)^D}\int d^{(D)}\mathbf{k}\ G(\mathbf{k})\exp(i\mathbf{k}\mathbf{r}) \ , \tag{98}$$

where $r = |\mathbf{r}|$.

Expanding the δ-function into a Fourier integral, we then find

$$g(r) = -\frac{q}{(2\pi)^D}\int d^{(D)}\mathbf{k}\ \frac{\exp(i\mathbf{k}\mathbf{r})}{\mu^2 + k^2} \ , \tag{99}$$

where $k = |\mathbf{k}|$.

In the one-dimensional case ($D = 1$) the integration gives:

$$g(r) = -\frac{q}{2\pi}\int_{-\infty}^{+\infty} dk\ \frac{\exp(ikr)}{\mu^2 + k^2} = -\frac{q}{2\mu}\exp(-\mu|r|) \ . \tag{100}$$

Using this result at $r = 0$ and the definition of q, we solve for $g(0)$ and find:

$$g(0) = -\frac{\chi}{2}\left[\kappa + \frac{2\hbar\mu}{m}\right]^{-1} \ . \tag{101}$$

Correspondingly, the energy eigenvalue then becomes:

$$E = \hbar\Delta - \frac{\hbar\chi^2}{2}\left[\kappa + \frac{2\hbar\mu}{m}\right]^{-1} = -\hbar^2\mu^2/m \,. \tag{102}$$

Here μ must be positive for a localized bound state. We note that this equation can be rewritten in the form of a cubic equation. Hence, we can show that there always exists one positive solution for μ. This proves[37] the existence of a one-dimensional bosonic 'meson' of a finite size.

The two- and three-dimensional results are qualitatively different. In these cases we obtain, from Eq.(99):

$$g(0) = -\frac{q}{2\pi^{D-1}}\int_0^\infty dk\,\frac{k^{D-1}}{\mu^2+k^2}\,, \quad (D = 2,3)\,, \tag{103}$$

where we have transformed to polar (for $D = 2$) and spherical (for $D = 3$) coordinates. Using the definition of q we next solve for $g(0)$ and obtain:

$$g(0) = -\frac{\chi}{2}\left[\kappa + \frac{\hbar/m}{f_D(\mu)}\right]^{-1}\,, \tag{104}$$

where we have defined

$$f_D(\mu) = \frac{1}{2\pi^{D-1}}\int_0^\infty dk\,\frac{k^{D-1}}{\mu^2+k^2}\,. \tag{105}$$

This integral diverges for $D = 2, 3$. Hence we find that $g(0) = -\chi/(2\kappa)$, and the energy E is given by:

$$E = \hbar\Delta - \frac{\hbar\chi^2}{2\kappa}\,. \tag{106}$$

With this result for $g(0)$ it follows also that $q = 0$, and hence (see Eq.(99)) $g(r) = 0$ if $r \neq 0$. That is, the exact solution agrees with the variational result. The results of this section show that our model Hamiltonian provides quantum solitons or two-particle eigenstates in one and more spatial dimensions. They are superpositions of either a molecule or two atoms.

An important difference between the one dimensional and multidimensional solutions is in their structure and dependence on the additional quartic interaction. In one dimension the bound state has finite characteristic size and is available even without a quartic term in the Hamiltonian. In two and three dimensions the bound states have a singular point-like structure. The corresponding binding energy is finite if and only if $\kappa > 0$. If, however, $\kappa = 0$ we obtain (see Eq.(106)) an energy collapse: $E \to -\infty$. Thus, while the additional quartic interaction prevents an energy collapse and makes the multidimensional quantum solitons available in this simple model, it does not prevent singularities in space.

5.4 Cut-off dependent results

These singularities represent a rather unusual situation, since the classical counterpart of the theory has well-behaved and widely used multidimensional nonlinear-optical soliton solutions. Since we expect the quantum theory to be correct, this leads to the 'paradox' of how a singular quantum field theory can describe real physical processes.

A possible way is to incorporate the fact that parametric couplings are usually restricted to a finite range of relative momenta or wave-numbers. To represent this we can introduce a cut-off at $|\mathbf{k}| = k_{\max}$ in the relative momenta of the interacting fields. We will choose $k_{\max} \sim 2\pi/a$, where a is the S-wave scattering length of the atomic field Ψ_1, as a natural cut-off scale. The interaction part of the Hamiltonian (86) can then be expressed in terms of $a(\mathbf{k})$ and $b(\mathbf{k})$, the Fourier components of Ψ_1 and Ψ_2, so that its cut-off dependence is implemented through the limits of the corresponding integrals:

$$H_{int} = \frac{(2\pi)^D \hbar}{2} \int d^{(D)}\mathbf{K} \left\{ \chi \int_{|\mathbf{k}|=0}^{k_{\max}} d^{(D)}\mathbf{k} \left[a^\dagger(\mathbf{k}_+) a^\dagger(\mathbf{k}_-) b(\mathbf{K}) + \text{h.c.} \right] \right.$$

$$\left. + \kappa \int \int_{|\mathbf{k}_{1,2}|=0}^{k_{\max}} d^{(D)}\mathbf{k}_1 d^{(D)}\mathbf{k}_2 \; a^\dagger(\mathbf{k}_{1+}) a^\dagger(\mathbf{k}_{1-}) a(\mathbf{k}_{2+}) a(\mathbf{k}_{2-}) \right\} . \quad (107)$$

Here $\mathbf{k}_{(i)\pm} \equiv \mathbf{K}/2 \pm \mathbf{k}_{(i)}$, and the commutation relations for $a(\mathbf{k})$ and $b(\mathbf{k})$ are given by $[a(\mathbf{k}), a^\dagger(\mathbf{k}')] = [b(\mathbf{k}), b^\dagger(\mathbf{k}')] = \delta(\mathbf{k} - \mathbf{k}')/(2\pi)^D$.

We can now analyze the eigenvalue problem $H|2\rangle = E|2\rangle$ directly, in Fourier space, by introducing a cut-off dependent Fourier transform of $g(\mathbf{r})$, so that

$$g(\mathbf{r}) = \frac{1}{(2\pi)^D} \int_{|\mathbf{k}|=0}^{k_{\max}} d^{(D)}\mathbf{k} \; G(\mathbf{k}) \exp(i\mathbf{k}\mathbf{r}) , \quad (108)$$

This implies that, due to the cut-off in the nonlinearities, we need only investigate eigenstates for which $G(\mathbf{k})$ vanishes if $|\mathbf{k}| > k_{\max}$. This leads to the following simultaneous equations for an eigenstate:

$$(k^2 + \mu^2)\, G(\mathbf{k}) = -q , \quad (109)$$

$$E = \hbar\Delta + \hbar\chi g(0) = -\hbar^2\mu^2/m , \quad (110)$$

where $k = |\mathbf{k}|$ and we have used the same notations for μ and q as before. The above equations are valid for $|\mathbf{k}| < k_{\max}$.

In order to evaluate the binding energy and the effective radius, we must next solve for $g(0)$. After a little algebra, we find:

$$g(0) = -\frac{\chi}{2}\left[\kappa + \frac{\hbar/m}{f_D(\mu, k_{\max})}\right]^{-1} , \tag{111}$$

where the cut-off structure functions

$$f_D(\mu, k_{\max}) = \frac{1}{2^{[D/2]}\pi^{[(D+1)/2]}}\int_0^{k_{\max}} dk\, \frac{k^{D-1}}{\mu^2 + k^2} , \quad (D = 1, 2, 3) ,$$

(with $[x/2]$ being the integer part of $x/2$) are given by:

$$f_1(\mu, k_{\max}) = \frac{1}{\pi\mu}\tan^{-1}\left(\frac{k_{\max}}{\mu}\right) ,$$

$$f_2(\mu, k_{\max}) = \frac{1}{4\pi}\ln\left(1 + \frac{k_{\max}^2}{\mu^2}\right) , \tag{112}$$

$$f_3(\mu, k_{\max}) = \frac{1}{2\pi^2}\left[k_{\max} - \mu\tan^{-1}\left(\frac{k_{\max}}{\mu}\right)\right] .$$

This result clearly shows the difference caused by the dimensionality of the space. In one dimension, $f_1(\mu, k_{\max})$ approaches a constant value if $k_{\max} \to \infty$, while in two and three dimensions, $f_D(\mu, k_{\max})$ has a logarithmic or linear divergence, respectively.

The effect of this divergence depends on whether or not the additional quartic interaction is present. If it is present (with $\kappa > 0$), there are exact solutions without cut-off, and $g(0) = -\chi/(2\kappa)$, so that the energy eigenvalue E takes the form of Eq.(106), and $g(\mathbf{r}) = 0$ if $|\mathbf{r}| > 0$. In other words, the solutions in two and three dimensions have a finite energy but zero radius in the limit of $k_{\max} \to \infty$. If $\kappa = 0$, then we must impose a finite cut-off on the couplings to prevent an energy divergence.

The general result for the energy eigenvalue E is given by:

$$E_D = \hbar\Delta - \frac{\hbar\chi^2}{2}\left[\kappa + \frac{\hbar/m}{f_D(\mu, k_{\max})}\right]^{-1} = -\hbar^2\mu^2/m , \tag{113}$$

where μ must be positive for a localized bound state. Analysis of this equation with respect to μ shows that under certain condition a positive solution for μ is always available. This condition, in the cases of one and two dimensions can be written in the form of Eq.(92), while in the three-dimensional case it is modified to

$$\chi^2 > 2\Delta_2 \left(\kappa + \frac{2\pi^2 \hbar}{m k_{\text{max}}} \right) . \tag{114}$$

In the limit $k_{\text{max}} \gg \mu$ one can write down a simple approximate result for the eigenvalue in three dimensions ($D = 3$), including a cut-off at the inverse scattering length:

$$E_3 \simeq \hbar\Delta - \frac{2\hbar\chi^2}{5\kappa} . \tag{115}$$

However, while this is reasonable as a purely mathematical theory, there is a slight complication here when we turn to the physical origin of the cutoff. We have taken $k_{\text{max}} \simeq 2\pi/a$, where a is the S-wave scattering length. This means that the result is rather sensitive to the cut-off in three-dimensions, due presumably to the strongly localized nature of the solution. The calculation is therefore to be regarded as a model calculation, which needs to be improved by taking into account the exact atom-atom potential at short range, in a realistic Bose condensate with real atoms rather than idealized fields.

In summary, exact two-particle solutions to an idealized field-theory model are obtained in one, two and three space dimensions. The results have the remarkable character that the theory, while having a well-behaved (and widely used) classical counterpart, has quantum singularities in the eigenstates in more than one space dimensions, corresponding to zero-radius structures. The reason for this behavior is the inherently non-classical structure of the bound state, which is a quantum superposition state.

The physical interpretation of these bound states is that they are a superposition of a molecule and two atoms, which can be inter-converted coherently through a process like Raman photo-association. This is currently under experimental investigation at the University of Texas. It is not impossible that this parametric quantum theory, as well as being theoretically interesting, could result in the first experimental test of multidimensional quantum soliton theory for Bose fields. Thus, complementary to high-energy physics and particle accelerators, investigation of particle-like quantum structures may become available in a larger variety of physical systems, including nonlinear optics and laser physics.

5.5 Super-chemistry limit

Physical mechanisms that can realize coherent atomic dimerization and produce ultra-cold molecules include Feshbach resonance and Raman photo-association[40,41]. Feshbach resonances have already been observed[27], while experiments of this type with the Raman photo-association are underway in the case of homogeneous (degenerate) BEC [42].

In such coupled atomic/molecular BEC systems the two-particle (di-boson) quantum soliton solutions are 'dressed' molecules each of which exists in a superposition with a pair of atoms. With a characteristic χ_3-value estimate of about $\chi_3 \sim 10^{-6}\,\mathrm{m}^{3/2}/\mathrm{sec}$ [41,42], the atomic masses $m_1 \simeq m_2 \sim 10^{-25}\,\mathrm{kg}$, and the longest S-wave scattering length $a \sim 5\,\mathrm{nm}$, these results give a quantum soliton binding energy which is comparable to achievable temperatures in current BEC experiments.

Of more importance, from the point of view of BEC experiments, are N-particle eigenstates (where we assume N is even) and the ground-state energy of this quantum many-body system. While this is a difficult problem, some important results can be obtained using a variational approach. Here we consider two important examples:

- a variational ansatz that corresponds to $N/2$ independent di-bosons [39]
- a coherent or mean-field theory ansatz.

A remarkable result that emerges with the treatment of the first type of ansatz is that, in the limit $k_{\max} \to \infty$, it provides the exact ground-state energy for N atoms, of:

$$E = \hbar(N/2)\left[\Delta - \chi^2/2\kappa\right] .$$

The ground state energy has no lower bound as $\kappa \to 0$; this is in contrast to the mean-field behavior corresponding to the classical Hamiltonian energy. The classical Hamiltonian is known to have a rigorous lower bound and to support classical solitons[29].

With a finite cutoff, the ansatz corresponding to $N/2$ independent di-bosons, is no longer the exact eigenstate, and therefore does not necessarily result in the lowest possible energy. The corresponding variational estimate of the energy, in three dimensions, is:

$$E = \hbar(N/2)\left[\Delta - 2\chi^2/5\kappa\right] .$$

This shows that the theory can be consistent with the classical result, which does not diverge, provided there is a cut-off. Slightly different estimates are obtained if we assume κ is kept constant.

The second type of ansatz that we employ here – the coherent or mean-field theory ansatz:

$$|N_C\rangle = \exp\left[\int d^{(3)}\mathbf{x} \sum_{i=1}^{2} \Psi_i(\mathbf{x})\hat{\Psi}_i^\dagger(\mathbf{x})\right]|0\rangle \qquad (116)$$

can provide (at large N) a lower energy. To show this we use trial functions $\Psi_i(\mathbf{x})$ in the form of Gaussians. The final result for the corresponding minimum energy shows evidence that at sufficiently large particle number (around 10^5 atoms at a first estimate), the coupled condensate approach - where atoms interact coherently - is favored. If this transition does exist, it has properties similar to a quantum liquid-gas transition.

Thus, at low particle density, the formation of individual 'dressed' molecules is favored, as atoms couple to molecules in a *particle-like* way, in a kind of Rabi limit. These dressed states have interesting properties, reminiscent of Cooper pairs, but cannot be described by the classical parametric soliton equations. At large density (but not too large so that S-wave scattering is dominant) the coherent coupling of two entire condensates is dominant, in a kind of Josephson limit. With large enough parametric coupling, and provided other recombination processes are negligible, there are coherent nonlinear *wave-like* interactions between the atomic and the molecular Bose condensates (just as in nonlinear optics), which make it possible to form stable, three-dimensional, BEC solitons.

More than this, the coherent nonlinear dynamics of these atom-molecular BEC interactions may result in a completely novel type of coherent chemical kinetics at ultra-low temperatures or a **'superchemistry'** behavior. The parametric interaction here could provide enormous (Bose-enhanced) chemical reaction rates ($\propto N_1\sqrt{N_2}$), due to the effect of bosonic stimulated emission, similar to lasing. This is in sharp contrast to the predictions of the usual chemical kinetics, where the rates do not depend on the number of product particles, and go to zero at low temperatures, according to the Arrhenius law.

As this is a novel area of BEC theory, it is under rapid development. The results given here should be regarded as preliminary. They are not yet tested in experiment, and are more in the category of model calculations. Real atomic systems certainly have a great deal more complexity than I have included here. The results do show, however, that a simple cubic quantum field theory has a great potential for interesting new physics, including unexpected behaviour in both classical and quantum regimes.

6 Summary

Examples treated here were as follows:

- **Vortices and vortex rings** have a uniform circulating bosonic current.
- **Classical solitons** generally have a point of maximum or minimum density.
- **Quantum solitons** are the exact solutions that give rise to classical solitons.
- **Atom-molecular solitons** can potentially exist as localized objects, with distinct quantum and classical regimes.

The theory of the evolution of BEC atoms in a trap differs from nonlinear optics chiefly in that there is an external potential, which behaves analogously to the confining dielectric mode structure of a fiber mode. In the atom laser case, with relatively short wavelengths, there are typically many more modes that can contribute to the dynamics than in photonic systems. Thus, a multi-mode, three-dimensional theory is generally appropriate for a BEC. Finite temperature effects are also important, but these are treated elsewhere. I have assumed here that the experiments described are in the limit of low enough temperatures to largely ignore any thermal effects.

There are also significant quantitative differences between the photonic and atom optic cases. In particular, nonlinear effects tend to be much more significant, even with relatively low particle number. This raises the possibility that distinctive quantum 'granularity' effects may be observed, perhaps leading to new physics not contained in the classical mean-field wave equations. Similar effects were observed in the 1980's in the case of the two-dimensional electron gas, leading to the discovery of the fractional quantum Hall effect. So far, the nonlinear *quantum* structures of a BEC have only just started to be investigated!

Acknowledgments: These lectures were prepared from research notes and publications involving co-workers in Australia and the USA. Particular credit and acknowledgments are especially due to K. Huang, K. Kheruntsyan and H. He. Any resulting errors or misinterpretations are my own responsibility entirely.

References

1. M.H. Anderson, J.R. Ensher, C.E. Wieman, and E.A. Cornell, Science **269**, 198 (1995); C.C. Bradley, C.A. Sackett, J.J. Tollett, and R.G. Hulet,

Phys. Rev. Lett. **75**, 1687 (1995); K. B. Davis, M.-O. Mewes, M. R. Andrews, N. J. van Druten, D. S. Durfee, D.M. Kurn, and W. Ketterle, Phys. Rev. Lett. **75**, 3969 (1995); K. Burnett, Contemporary Physics **37**, 1 (1996); K.B. Davis, M.O. Mewes and W. Ketterle, Appl. Phys. B **60**, 155 (1995).

2. M.-O. Mewes, M. R. Andrews, D. M. Kurn, D. S. Durfee, C. G. Townsend, and W. Ketterle, Phys. Rev. Lett. **78**, 1997 (582); K. Burnett, P. S. Julienne, and K.-A. Suominen, Phys. Rev. Lett. **56**, 1416 (1996); H. Wiseman and M. J. Collet, Phys. Lett. A **202**, 246 (1995); A. M. Guzman, M. Moore, and P. Meystre, Phys. Rev. A **53**, 977 (1996); M. Holland, K. Burnett, C. Gardiner, J. I. Cirac, and P. Zoller, Phys. Rev. A **54**, R1757 (1996).
3. A. L. Fetter and J. D. Walecka, *Quantum theory of many-particle systems* (McGraw-Hill, New York, 1991); A. A. Abrikosov, L. P. Gorkov, and I. E. Dzyaloshinski, *Methods of Quantum Field Theory in Statistical Physics* (Dover, New York, 1963); N. N. Bogoliubov, J. Phys. (USSR) **11**, 23 (1947); K. A. Brueckner and K. Sawada, Phys. Rev. **106**, 1117 (1957); K. Huang, C. N. Yang, and J. M. Luttinger, Phys. Rev. **105**, 776 (1957); N. P. Proukakis, K. Burnett, and H. T. C. Stoof, Phys. Rev. A **57**, 1230 (1998).
4. D. S. Hall, M.R. Matthews, C. E. Wieman, E. A. Cornell, Phys. Rev. Lett. **81**, 1543 (1998).
5. C. C. Bradley, C. A. Sackett, J. J. Tollett and R. G. Hulet, Phys. Rev. Lett. **75**, 1687 (1995); P. A. Ruprecht, M. J. Holland, K. Burnett and M. Edwards, Phys. Rev A **51**, **4704** (1995).
6. N. Bogoliubov, J. Phys. USSR **11**, 23 (1947).
7. H. A. Bethe, Z. Phys. **71**, 205 (1931), M. Girardeau, J. Math. Phys. **1**, 516 (1960), E. H. Lieb and W. Liniger, Phys. Rev. **130**, 1605, 1616 (1963).
8. M.R. Matthews, B.P. Anderson, P.C. Haljan, D. S. Hall, C. E. Wieman, E. A. Cornell, Phys. Rev. Lett. **83**, 2498-2501 (1999).
9. J. Scott-Russell, Proc. Roy. Soc. Edinburgh, **1**, 433-434 (1844).
10. D. J. Korteweg and G. de Vries, Phil. Mag. **39**, 422 (1895).
11. S. Coleman, *Aspects of Symmetry*, (Cambridge University Press, Cambridge, 1985).
12. A. Hasegawa and F. Tappert, Appl. Phys. Lett. **23**, 142-144 (1973).
13. L. F. Mollenauer, R. H. Stolen and J. P. Gordon, Phys. Rev. Lett. **45**, 1095-1098 (1980).
14. S. A. Morgan, R. J. Ballagh, and K. Burnett, Phys. Rev. A **55**, 4338 (1997); W. P. Reinhard and C. W. Clark, J. Phys. B: At. Mol. Opt.

Phys. **30**, L785 (1997); T. F. Scott, R. J. Ballagh, and K. Burnett, *ibid.* **31**, L329 (1998); O. Zobay, S. Pötting, P. Meystre, and E. M. Wright, Phys. Rev. A **59**, 643 (1999).

15. S. Burger, K. Bongs, S. Dettmer, W. Ertmer, K. Sengstock, A. Sanpera, G. V. Shlyapnikov and M. Lewenstein, Phys. Rev. Lett. **83**, 5198-5201 (1999).
16. P. D. Drummond, A. Eleftheriou, K. Huang and K. V. Kheruntsyan, to appear.
17. N. H. Christ and T. D. Lee, Phys. Rev. D **12**, 1606 (1975); T. D. Lee, *Particle Physics and Introduction to Field Theory* (Harwood Academic Publishers, London, 1988).
18. S. J. Carter *et al.*, Phys. Rev. Lett. **58**, 1841 (1987); P. D. Drummond and S. J. Carter, J. Opt. Soc. Am. B **4**, 1565 (1987).
19. M. Rosenbluh and R. M. Shelby, Phys. Rev. Lett. **66**, 153 (1991); P. D. Drummond, R. M. Shelby, S. R. Friberg, and Y. Yamamoto, Nature **365**, 307 (1993).
20. I. B. McGuire, J. Math. Phys. **5**, 622 (1964); H. B. Thacker, Rev. Mod. Phys. **53**, 253 (1981).
21. R. B. Laughlin, Phys. Rev. Lett. **50**, 1385 (1983); D. C. Tsui, H. L. Störmer, and H. C. Gossard, Phys. Rev. Lett. **48**, 1559 (1982).
22. N. K. Wilkin, J. M. F. Gunn, and R. A. Smith, Phys. Rev. Lett. **80**, 2265 (1998).
23. K. Bergman and H. A. Haus, Opt. Lett. **16**, 663 (1991).
24. S. R. Friberg, S. Machida, and Y. Yamamoto, Phys. Rev. Lett. **69**, 3165 (1992).
25. M. Werner, Phys. Rev. **A** 54, R2567 (1996); M. J. Werner and S. R. Friberg, Phys. Rev. Lett. **79**, 4143 (1997).
26. S. Spalter, M. Burk, and G. Leuchs, Europhysics Lett. **38**, 335 (1997).
27. S. Inouye *et al.*, Nature **392**, 151 (1998); Ph. Courteille, R. S. Freeland, and D. J. Heinzen, Phys. Rev. Lett. **81**, 69 (1998).
28. C. L. Schultz, M. J. Ablowitz, and O. BarYaacov, Phys. Rev. Lett. **59**, 2825 (1987); G. D. Pang, F.-C. Pu, and B.-H. Zhao, *ibid.* **65**, 3227 (1990); G. D. Pang, Physics Lett. A **184**, 163 (1994).
29. Y. N. Karamzin and A. P. Sukhorukov, Zh. Eksp. Teor. Fiz. **68**, 834–847 (1975); Y. N. Karamzin and A. P. Sukhorukov, Moscow University Physics Bulletin **20**, 339–342 (1974).
30. M. J. Werner and P. D. Drummond, J. Opt. Soc. Am. B **10**, 2390–2393 (1993).
31. A. A. Kanashov and A. M. Rubenchik, Physica D **4**, 122 (1981).
32. K. Hayata and M. Koshiba, Phys. Rev. Lett. **71**, 3275–3278 (1993).

33. B. A. Malomed *et al.*, Phys. Rev. E **56**, 4725–4735 (1997); X. Liu, L. J. Qian, and F. W. Wise, Phys. Rev. Lett. **82**, 4631-4634 (1999).
34. H. He, M. J. Werner and P. D. Drummond, Phys. Rev. E **54**, 896-911 (1996).
35. R. Friedberg and T. D. Lee, Phys. Rev. B **40**, 6745 (1989); Y. Wang, Phys. Lett. A **176**, 19 (1993).
36. T. D. Lee, Phys. Rev. **54**, 1329 (1954); L. Van Hove, Physica **21**, 901 (1955); see also S. S. Schweber, *An Introduction to Relativistic Quantum Field Theory* (Row, Peterson, New York, 1961); R. Friedberg and T. D. Lee, Phys. Rev. B **40**, 6745 (1989); Y. Wang, Phys. Lett. A **176**, 19 (1993).
37. H. He and P. D. Drummond, Phys. Rev. **A56**, R1107-R1109 (1997).
38. K. V. Kheruntsyan and P. D. Drummond, Phys. Rev. A **58**, 2488 (1998).
39. P. D. Drummond, K. V. Kheruntsyan, and H. He, Phys. Rev. Lett. **81**, 3055 (1998).
40. J. M. Vogels *et al.*, Phys. Rev. A **56**, R1067 (1997); P. S. Julienne *et al.*, Phys. Rev. A **58**, R797 (1998).
41. P. Tommasini, E. Timmermans, M. Hussein, and A. Kerman, cond-mat/9804015.
42. R. H. Wynar, R. S. Freeland, D. J. Han, C. Ryu and D. J. Heinzen, Science **287**, 1016 (2000); D. J. Heinzen, R. H. Wynar, P. D. Drummond and K. V. Kheruntsyan, Phys. Rev. Lett **84** 5029 (2000).

COOLING, TRAPPING AND MANIPULATION OF NEUTRAL ATOMS AND BOSE-EINSTEIN CONDENSATES BY ELECTROMAGNETIC FIELDS

KRISTIAN HELMERSON AND WILLIAM D. PHILLIPS
Atomic Physics Division, Physics Laboratory
National Institute of Standards and Technology - Gaithersburg, Maryland 20899
E-mail: kristian.helmerson@nist.gov
Web: physics.nist.gov/StaffOrg/div842/group4.html

We give a general discussion of the mechanical effects of light, and of laser cooling and trapping techniques. This is followed by a description of experiments in the manipulation of Bose-Einstein condensates with optical laser pulses.

1 Introduction

Bose-Einstein condensation (BEC) in a dilute atomic vapor of bosons occurs when the thermal de Broglie wavelength of the atoms is comparable to the interatomic spacing. For a dilute vapor (one at a low enough density such that the atoms can be considered as weakly interacting) extremely low temperatures are required to achieve sufficiently long de Broglie wavelengths. To obtain and maintain such extraordinarily low temperatures required the development of cooling and trapping techniques for neutral atoms. Certainly, the techniques developed by the spin polarized hydrogen community, especially evaporative cooling (see the lecture notes of Andrew Wilson in these proceedings), have been essential in producing Bose-Einstein condensates; however, laser cooling and trapping of neutral atoms has allowed BEC in dilute gases to be observed first with alkalis, and has contributed greatly to many of the beautiful experiments with condensates.

These notes are based on a series of lectures given at the ANU Physics Summer School on Bose-Einstein Condensation, held in Canberra in January of 2000. The first two lectures were a general discussion of the mechanical effects of light, and laser cooling and trapping techniques for neutral atoms. The third and fourth lectures describe experiments with Bose-Einstein condensed sodium atoms at the National Institute of Standards and Technology (NIST) in Gaithersburg, Maryland. These experiments represent a direct application of many of the wonderful techniques developed for manipulating neutral atoms.

The topics to be covered in the general discussion of the mechanical effects of light, and laser cooling and trapping techniques include: radiative forces, both spontaneous and dipole; diffraction of atoms by optical standing waves; Doppler cooling and the Doppler cooling limit; deceleration and cooling of atomic beams; optical molasses and the discovery of sub-Doppler laser cooling; new laser cooling mechanisms, including polarization gradient and Sisyphus cooling; various topics related to trapping neutral atoms, including magnetic traps, both static and time-averaged orbiting potentials (TOP); laser dipole traps and far-off resonant traps (FORTs); radiation pressure traps and the optical Earnshaw theorem; magneto-optical traps (MOTs); mechanisms limiting the loading of MOTs and techniques to increase the densities, including dark spot MOTs.

These topics have been selected as representing some of the important developments in the interaction of electromagnetic fields and atoms which have enabled the creation and manipulation of Bose-Einstein condensates. More details on these topics, especially as they are applied to BEC research, can be found in the other lectures at this summer school as well as in other sets of lecture notes, special journal articles and review articles [1-10]. In addition, there are other related topics in the manipulation of matter by light; atom interferometry, atom holography, and atom optics in general, which are not covered in these notes but may be of great significance in BEC research. Again, an introduction to these topics as well as detailed discussions can be found in various lecture notes and review articles [11,12,13].

The last two sections of these notes describe experiments in the manipulation of Bose-Einstein condensates with optical laser pulses. After a brief description of the strategy employed to obtain condensates of sodium atoms in a TOP trap, we discuss coherent atom optics with condensates: specifically, normal and Bragg diffraction of BEC, output coupling of atoms from the condensate using optical Raman pulses to realize a highly directional, well-collimated atom laser, and matter-wave interferometry to measure the spatial phase variations across the condensate. Finally, we end with a discussion of nonlinear atom optics experiments with weakly interacting Bose-Einstein condensates. This includes the first observation of multi-wave mixing of matter-waves, and the generation of solitons in a condensate by optical phase imprinting.

Before beginning the discussion of the mechanical effect of atom-light interactions, we define for convenience some frequently used notation and symbols, with frequencies expressed in radians/second. The reader is cautioned that other authors may have used other conventions:

$\delta = \omega_{laser} - \omega_{atom}$ is the detuning of the laser frequency from the natural resonant frequency of the atom.

$\Gamma = \tau^{-1}$ is the decay rate of the population in the excited state, the inverse of the natural lifetime, the linewidth of the transition.

$k = 2\pi/\lambda$ is the laser photon wave vector.

k_B is Boltzmann's constant.

Ω is the on-resonance Rabi frequency for a laser field, the precession frequency of the Bloch vector representing the 2-level atom when $\delta = 0$.

$I/I_0 = 2\Omega^2/\Gamma^2$ is the normalized intensity of the laser.

M is the mass of the atom.

$v_{rec} = \hbar k/M$ is the recoil velocity of an atom upon emission or absorption of a single photon.

$E_{rec} = Mv_{rec}^2/2 = \hbar^2 k^2/2M$ is the kinetic energy of an atom having velocity v_{rec}.

2 Radiative forces

Light comes in quantized packets of energy and momentum called photons. The transfer of energy and momentum between the photon and an atom, through either coherent or incoherent scattering, results in a force exerted on the atom. This is the basis for the mechanical effects of atom-light interactions. Radiative forces or the mechanical effects of light are generally divided into two categories - the scattering force and the dipole forces.

2.1 The scattering force

The scattering force, also called the spontaneous force or radiation pressure, is the force exerted on an atom by the incoherent scattering of photons. More precisely it is the force on an atom corresponding to absorption of a photon followed by spontaneous emission. The photon absorbed transfers $\hbar k$ of momentum to the atom. The spontaneous emission of the photon is symmetrically distributed and so the momentum transfer due to spontaneous emission, averaged over many absorption-emission events, is zero. The average force on a two-level atom moving with velocity $\mathbf{v}$ in a plane wave of wave vector $\mathbf{k}$ and detuning δ is

$$\mathbf{F}(\mathbf{v}) = \hbar\mathbf{k}\frac{\Gamma}{2}\frac{I/I_0}{1 + I/I_0 + [2(\delta - \mathbf{k}\cdot\mathbf{v})/\Gamma]^2}. \quad (1)$$

This force is the maximum scattering rate $\Gamma/2$, times the resonance Lorentzian times the photon momentum, i.e. the rate of absorbing photon momentum. The detuning $\delta - \mathbf{k} \cdot \mathbf{v}$ accounts for the Doppler shift, and the force is large when $|\delta - \mathbf{k} \cdot \mathbf{v}| \leq \Gamma$. The spontaneous force is limited by the rate at which spontaneous emissions can occur. These occur at a rate Γ for excited atoms, whose maximum fractional population is 1/2.

For real, multilevel atoms, the situation can be more complicated. A common occurrence, typical of alkali atoms, is that the ground state is split by the hyperfine interaction into two states separated in frequency by many times the optical linewidth Γ. An atom excited by a laser from one of these hyperfine levels to an optically excited state may decay by spontaneous emission to the other hyperfine level. Transitions from this level are then so far out of resonance that effectively no further absorption occurs and no force is applied to the atom. While various schemes involving selection rules and polarization of the light may be used to avoid this problem of optical pumping, the most straightforward method is to apply a second laser frequency, tuned to resonance between the "wrong" hyperfine state and the optically excited state. This "repumper" keeps the atom out of the wrong ground state and allows the atom to effectively feel the force of the laser acting on the main transition.

Equation 1 is only valid if the force can be meaningfully averaged over many absorption-emission events. If a single event changes the atomic velocity so much that the resonance condition is not satisfied, such an average is not possible. This imposes a validity condition $\hbar k^2/M \ll \Gamma$ on Eq.(1). This condition is well satisfied for most atoms of interest in laser cooling. For example, $M\Gamma/\hbar k^2$ is 1200 for cesium laser cooled on its resonance transition at 852 nm and is 200 for sodium cooled at 589 nm.

2.2 The dipole force

The dipole force, also called the gradient force or stimulated force, is the force exerted on an atom due to coherent redistribution of photons. The dipole force can be considered as arising from stimulated Raman events - the absorption and stimulated emission of photons. (Note that the absorption and stimulated emission cannot be thought of as successive and independent events; their correlation is central to the proper understanding of the force.)

We can also understand the dipole force in analogy to a driven, classical oscillator. A harmonically bound charge driven by an oscillating electric field $\mathbf{E}$ has an oscillating dipole moment $\boldsymbol{\mu}$ which is in phase with the driving field when driven below resonance, and out of phase when driven above resonance.

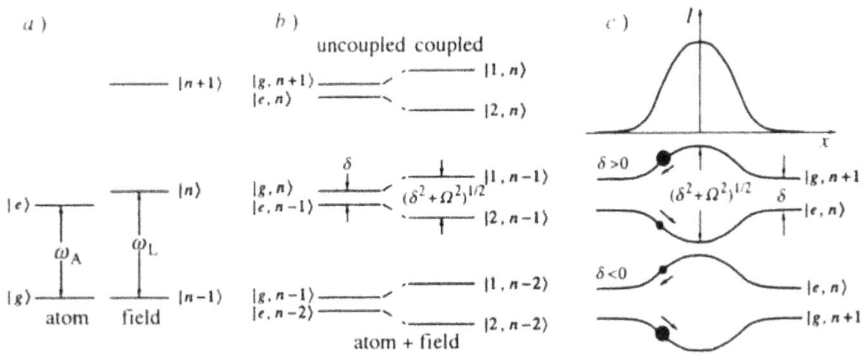

Figure 1. *a*) Energy levels of a 2-level atom and a laser field in the bare basis, *b*) dressed basis with the atom uncoupled and coupled to the field. *c*) Top: the intensity profile of the laser field. Bottom: the dressed energy levels as a function of position for this intensity profile, for laser detuning above and below resonance. The sizes of the black dots indicate the relative population of the dressed states.

The energy of interaction between the dipole and field is $W = -\boldsymbol{\mu} \cdot \mathbf{E}$. Below resonance the energy is negative and the oscillator will be drawn toward a more intense field, while above resonance it will be drawn to the weaker part of the driving field.

A particularly powerful and intuitive way of describing the dipole force is in the dressed-atom picture. This has been treated in detail by Dalibard and Cohen-Tannoudji [14] and we present the basic idea here. Consider a 2-level atom with ground state $|g\rangle$ and excited state $|e\rangle$. Separately consider a single mode of the radiation field, close to resonance with the atomic transition, having energy levels labeled $...|n-1\rangle$, $|n\rangle$, $|n+1\rangle$,... according to the number of photons in the mode. The energy levels of these two distinct systems are shown in Fig.1*a* for the case where the photon energy ω_L is greater than the difference in energy between the atomic states ω_A. This is the "bare" basis in which we consider the atom's energy levels and those of the photon field separately. If we now consider the atom and laser field together as a single system we have the dressed basis (atom dressed by laser photons) of Fig.1*b*. If there is no interaction between the atom and the laser field (as, for example, when the laser field does not spatially overlap the atom's position) the dressed level energies are simply the sum of the atom and field energies. This gives a ladder where each rung is a pair of nearly degenerate energy levels, since an atom in the ground state with n photons in the laser field has nearly the same energy as an atom in the excited state with $n-1$ photons in the field (with the energy difference being equal to the laser detuning). When the atom interacts with the field, the dressed levels (the eigenstates of the full Hamiltonian) become superpositions of ground and excited states and of numbers of photons in the field. By convention, the higher of the two levels in a rung is called $|1\rangle$, and the rungs are labeled by the photon number n associated with the excited state in the uncoupled basis [15]. According to the general rule that interacting energy levels repel each other by an amount depending on the coupling, the spacing between the dressed levels increases from the detuning δ to the "effective Rabi frequency" $\Omega_{eff} = \sqrt{\delta^2 + \Omega^2}$ as the interaction turns on. Because each of the dressed levels has both ground and excited state character, an atom can make spontaneous emission transitions between the rungs of the ladder. These transitions establish an equilibrium population between the two types of levels [14].

Now consider a light field whose intensity varies in space, such as a focused laser beam with a Gaussian intensity profile, or a plane standing wave. The Rabi frequency seen by an atom now varies in space, so the energy of the dressed levels also varies. The dipole force arises from this variation in energy and the relative populations of the dressed levels. Figure 1c illustrates the idea. For laser detuning above resonance ($\delta > 0$) the upper of the two dressed levels is the one that connects to the ground state in the limit of small interaction. This upper level always has the largest population, and its potential tends to repel the atom from the region of most intense light. For $\delta < 0$ it is the lower level that connects to the ground state and that has the higher population; its potential attracts the atom to the higher intensity. The actual dipole force is the average force, weighted by population, for the two potentials. The dipole force is derivable from a potential [14,16–19] which can be written as

$$U = \frac{\hbar\delta}{2} \ln\left[1 + \frac{I/I_0}{1 + (2\delta/\Gamma)^2}\right]. \tag{2}$$

Where the spatial dependence of the potential comes in through the dependence of the intensity I on position.

2.3 Doppler cooling

Doppler cooling results from near resonant radiation pressure acting on an atom and thus it is a particularly instructive example of an application of the scattering force. To see how cooling works we shall consider the one-dimensional motion of an atom in counter-propagating plane waves. If the laser beams are tuned below the atomic resonance, an atom moving in the direction opposite to one of the beams will, because of the Doppler shift, see that beam shifted closer to its resonance frequency. At the same time it sees the other laser beam, propagating in the same direction as its velocity, shifted further away from resonance. The atom absorbs more photons from the beam propagating opposite to its velocity and thus slows down. From Eq.(1), the forces due to the positive and negative-going waves, along the direction of propagation are

$$F_{\pm}(v) = \pm\hbar k \frac{\Gamma}{2} \frac{I/I_0}{1 + I/I_0 + [2(\delta \mp kv)/\Gamma]^2}. \tag{3}$$

If we assume that $I/I_0 \ll 1$, so that the intensity is low enough for us to add the forces from the two waves independently, and that $kv \ll \Gamma$, the total force is given by

$$F_{tot} = F_+ + F_- = \frac{4\hbar k^2 I/I_0}{\left[1 + (2\delta/\Gamma)^2\right]^2} \frac{2\delta}{\Gamma} v = -\alpha v. \tag{4}$$

For $\delta < 0$, α is positive and the force damps the velocity at a rate $\dot{v}/v = -\alpha/M$. The largest damping force is obtained for $\delta \cong -\Gamma/2$.

Atomic motion in a strong standing wave [14,16,17,20] is beyond the scope of this treatment, however a simple approximation for small velocity, ignoring interference of the two beams and stimulated redistribution of photons between the beams [21] shows that the friction coefficient α maximizes at about $\alpha = \hbar k^2/4$ for $I/I_0 = 1$ in each beam and $2\delta/\Gamma = -1$. Under these conditions the velocity damping time $v/\dot{v}$ for sodium, cooled on the 589 nm resonance line, would be 13 μs.

So far we have only considered the average force on an atom in a light field. We should remember that the force arises from discrete photon scatterings, and so must fluctuate about its average. The fluctuations can be thought of as arising from two sources: fluctuations in the number of photons absorbed in a given time and fluctuations in the direction of the spontaneously emitted photons. Both of these effects arise because of the randomness of spontaneous emission.

The fluctuations represent a random walk of the atomic momentum, with each random walk step being of magnitude $\hbar k$. For simplicity we will assume a fictitious one-dimensional situation where photons are emitted as well as absorbed along a single axis. Each scattering event represents two random walk steps, one from the absorption, which could be from either of the counter-propagating beams, and one from the spontaneous emission, which can be in either direction along the axis. The mean square momentum of the atom increases linearly with the number of scattering events (random walk steps), with a rate

$$\frac{d}{dt}\left\langle p^2\right\rangle = 2R\hbar^2k^2 \tag{5}$$

where R is the scattering rate; the factor of 2 comes from the two steps per scattering. We define a momentum diffusion coefficient

$$2D_p = \frac{d}{dt}\left\langle p^2\right\rangle \tag{6}$$

so that D_p/M is the rate of increase in kinetic energy, the heating rate. The damping force decreases the kinetic energy as $\mathbf{F}\cdot\mathbf{v} = -\alpha v^2$, the cooling rate. At equilibrium we set the sum of the heating and cooling rates to zero, finding

$$D_p/\alpha = M\left\langle v^2\right\rangle = k_BT. \tag{7}$$

Here we have replaced v^2 with its mean value and used the equipartition theorem to identify $k_BT/2$ as the mean kinetic energy $M\left\langle v^2\right\rangle/2$ in the single degree of freedom. Using Eq.(4) to get α, Eqs.(5) and (6) to get D_p, and remembering that the total scattering rate from the two laser beams, each of intensity, I, is

$$R = \Gamma\,\frac{I/Io}{1+(2\delta/\Gamma)^2}, \tag{8}$$

we find, for low intensity and small velocity:

$$k_BT = \frac{\hbar\Gamma}{4}\,\frac{1+(2\delta/\Gamma)^2}{2\delta/\Gamma}. \tag{9}$$

This is the Doppler temperature, and we emphasize that it applies to the fictional one-dimensional case we have constructed. A true 1D experiment, such as cooling an atomic beam along one axis, would produce a lower temperature depending on the distribution of scattered photons in 3D. However, it can be shown [21] that in a symmetrical three dimensional case, the temperature is also given by Eq.(9). The temperature minimizes at a detuning of $\delta = -\Gamma/2$ where

$$k_BT_{Dopp} = \frac{\hbar\Gamma}{2}, \tag{10}$$

defines the Doppler cooling limit. To derive the Doppler temperature we assumed that the velocity was small enough that $kv \ll \Gamma$. At the Doppler limit, sodium atoms would have $v_{r.m.s.} = 30$ cm/s, corresponding to a temperature of 240μK and $kv_{r.m.s.} = \Gamma/20$, so the assumption is justified in this case (and most others). We may also ask whether the velocity distribution corresponds to a temperature. That is, is it Maxwell-Boltzmann? It can be shown that for sodium, and similar atoms where the recoil energy $E_{rec} = \hbar^2k^2/2M \ll \hbar\Gamma$, the velocity distribution is indeed very close to being thermal [21]. What happens if the recoil energy is not small? Then we violate the validity condition on Eq.(1), and the detuning due to the Doppler shift changes significantly with each emission or absorption. In the limit where the linewidth Γ is small compared to the recoil energy, we feel intuitively that the cooling limit, rather than being related to Γ, as in Eq.(10), is related to the recoil energy. Indeed, it can be shown [21,22] that in this case the lowest temperature attainable (the recoil temperature) is given by

$$\frac{1}{2}k_BT = \frac{\hbar^2k^2}{2M} = E_{rec}, \tag{11}$$

that is, one recoil energy per degree of freedom. While it might seem that this is the ultimate limit of laser cooling, in fact it is possible to break the recoil limit under certain circumstances where the interaction with the light is turned off as the atomic velocity becomes small. One way this may be achieved is by velocity selective coherent population trapping in multilevel atoms [23,24]. Atoms are optically pumped into a coherent superposition of both internal states and center-of-mass velocities, a superposition that cannot absorb the laser light. Another way is velocity-space optical pumping [25,26]. Atoms are cooled on a transition with a narrow linewidth, but with a distribution of laser frequencies such that the excitation rate for zero velocity atoms is small or vanishing. This has been accomplished by use the use of pulsed two-photon Raman transitions [27].

The damping force of Eq.(4) is similar to the viscous force on an object moving in a fluid. Because of this, the configuration of pairs of counter-propagating laser beams is often called "optical molasses". The viscosity is so high that a velocity corresponding to the Doppler cooling limit is damped out and randomized while the atom travels only a few tens of micrometers, much smaller than the size of the molasses, which is typically a centimeter. Thus the atoms executes a Brownian-like motion with a short mean free path, moving diffusively rather than ballistically [28,29]. The evidence for this is the long residence time of atoms in optical molasses. Atoms require several seconds to diffuse out of a typical optical molasses [29,30], whereas, moving ballistically, atoms cooled to the Doppler cooling limit would traverse a region the size of a typical molasses in a few tens of milliseconds.

3 Deceleration and cooling of an atomic beam

Experimentally, atoms must first be at reasonably slow velocities ($kv \leq \Gamma$) before Doppler cooling can be effective. The two general ways this has been accomplished is by laser deceleration of an atomic beam [28,31] and by collection of slow atoms from a thermal gas [32,33]. Loading from a thermal gas has the advantage of allowing a more compact and simpler apparatus, while the beam deceleration technique usually allows lower background pressure and faster production of slow atoms.

Deceleration of an atomic beam is usually accomplished by directing a near resonant laser beam so as to oppose the atomic beam. The atoms absorb photons at a rate determined by the intensity of the laser beam, the detuning from resonance and the atoms' velocity. For each photon absorbed, the atomic velocity changes by v_{rec} in the direction of the laser propagation. The spontaneously emitted photons are emitted randomly in a pattern that is symmetric on reflection through the atom, so there is no net average change in the atomic velocity due to these emissions. If the absorption is followed by stimulated emission into the same direction as the incident laser beam (we assume the laser beam is a plane wave), there is no net momentum transfer from the absorption-emission process. Only absorption followed by spontaneous emission contributes to the average force, which is given by the rate of scattering photons times the momentum of a photon. For a two level atom this force is given by Eq.(1). At high intensity this force saturates to the value $\hbar k\Gamma/2$.

The acceleration of an atom due to the saturated radiation pressure force is $a_{max} = \hbar k\Gamma/2M = v_{rec}\Gamma/2$, which can be quite large. For sodium with $\lambda = 2\pi/k = 589$ nm, $1/\Gamma = 16$ ns and $M = 23$ a.m.u., $v_{rec} \approx 3$ cm/s and $a_{max} \approx 10^6$ m/s^2. For cesium, with $\lambda = 2\pi/k = 852$ nm, $1/\Gamma = 30$ ns and $M = 133$ a.m.u., $v_{rec} \approx 3.5$ mm/s and $a_{max} \approx 6\times10^4$ m/s^2. This acceleration would stop in 50 cm, a thermal, 1000 m/s Na atom scattering ≈33,000 photons in 1 ms; and in 80 cm, a thermal, 300 m/s Cs atom scattering ≈84,000 photons in 5 ms.

Implicit in Eq.(1) is one of the major impediments to effective deceleration of an atomic beam using a counter-propagating laser beam. The force acting on the atom is large if $|2(\delta - \mathbf{k}\cdot\mathbf{v})| \leq \Gamma\sqrt{1 + I/I_0}$. Atoms much outside this resonant-velocity range will experience little deceleration, and atoms initially within this range will be decelerated out of it. This process results in a cooling or velocity compression of a portion of the atomic beam's velocity distribution, and was first observed by Andreev et *al.* [34]. Atoms initially at the resonant velocity decelerate out of resonance. Other atoms with nearby velocities will also decelerate, those with larger velocities first decelerating into resonance, then to slower velocities out of resonance, while initially slower atoms decelerate to still lower velocities. The atoms will "pile up" at a velocity somewhat lower than the resonant velocity. Both deceleration and cooling occur because a range of velocities around the resonant velocity are compressed into a narrower range at lower velocity. However, only a small portion of the total velocity distribution has been decelerated by only a small amount.

There are a number of possible solutions to this problem, some of which have been discussed in ref. 28. These include Zeeman tuning [28] where a spatially varying magnetic field compensates the changing Doppler shift as the atoms decelerate, so as to keep the atoms near resonance; white-light deceleration [35] where a range of laser frequencies ensures that some light is resonant with the atoms, regardless of their velocity (within the range to be decelerated); diffuse-light deceleration [36] where light impinges on the atoms from all angles so that, with the Doppler shift, some of the light is resonant with each velocity; Stark cooling [37] where a spatially varying electric field is used to Stark shift atoms and keep them near resonance as they Doppler shift due to deceleration; intense standing wave deceleration [38] where the linewidth is sufficiently power broadened to capture a large velocity distribution for laser deceleration; and "chirp cooling" [39] in which the frequency of the laser is swept up, or chirped, in time such that the laser stays in resonance with atoms that have been decelerated, and they continue to absorb photons and decelerate. Zeeman tuning and chirp cooling were among the earliest atomic beam slowing techniques demonstrated and they continue to be the most widely used techniques to date.

3.1 Chirp cooling

In chirp cooling, the frequency of the laser is swept up, or chirped, in time [39]. Because of the chirp, atoms that have been decelerated by the laser stay in resonance, continue to absorb photons, and continue to decelerate. Furthermore, the chirp brings the laser into resonance with additional atoms having lower velocities than the original group around the velocity initially resonant with the laser.

In order to analyze this process, let us consider atoms having positive velocities near some velocity V opposed by a laser beam propagating in the negative direction. We express any atomic velocity as $v = V + v'$. The acceleration of atoms having velocity V ($v' = 0$) is $a = F(V)/M$, where $F(V)$ is given by Eq.(1). Therefore, we write $V(t) = V(0) + at$. Also we let the detuning vary as $\delta(t) = \delta' - kV(t)$. That is, we chirp the laser frequency so as to stay a constant detuning δ' from resonance with atoms having the decelerating velocity $V(t)$. Now we transform to a frame decelerating with $V(t)$. In this frame the atomic velocity is v' and the laser detuning is Doppler shifted to δ'. The force on an atom in this frame, for small v', is

$$F(v') = 2\hbar k^2 \frac{I}{I_0} \frac{(2\delta'/\Gamma)\, v'}{\left[1 + I/I_0 + (2\delta'/\Gamma)^2\right]^2} \tag{12}$$

The term multiplying v' is minus the friction coefficient α. When $\delta' < 0$, the force opposes the velocity v' and tends to damp all velocities to zero in the decelerating frame, which is $V(t)$ in the laboratory frame. Maximum damping occurs for $I/I_0 = 2$ and $2\delta'/\Gamma = -1$. The final velocity to which the atoms are decelerated is determined in practice by the final frequency to which the laser is chirped. The result of chirp cooling an atomic beam is that all of the atoms in the initial distribution below the velocity resonant with the laser at the beginning of its chirp are decelerated.

The first definitive experiment showing such chirp cooling was in ref. 40, with deceleration to zero velocity first achieved in ref. 41. The analysis given above is similar to that given in ref. 42. The robust character of this sort of cooling is evident. Atoms within a range of velocities around $V(t)$ are damped (in velocity) toward $V(t)$. Lower velocities, not initially close to $V(t)$, come within range as the laser chirp brings $V(t)$ into coincidence with them. If the laser intensity changes during the time an atom is being decelerated (because, for example, the laser beam is not collimated), the atoms will continue to decelerate according to the chosen chirp rate, but with a different effective detuning δ'. The chosen chirp rate, however, must be consistent with the achievable deceleration for the given I/I_0. That is, the chirp rate must satisfy

$$\dot{\delta} = ka = \frac{\hbar k^2 \Gamma}{2M} \frac{I/I_0}{1 + I/I_0 + [2\delta'/\Gamma]^2} \tag{13}$$

This means that $\dot{\delta}$ has an allowable upper limit of ka_{max}. We have noted that for the velocities to be damped in the decelerating frame we must have $\delta < 0$ and it is easy to show that the conditions for best damping lead to a deceleration half as large as the maximum.

3.2 Zeeman tuning

In the Zeeman tuning technique, a spatially varying magnetic field is used to keep the frequency of an atom resonant with a counter-propagating laser beam, as the atom slows down by scattering photons from this laser. Because

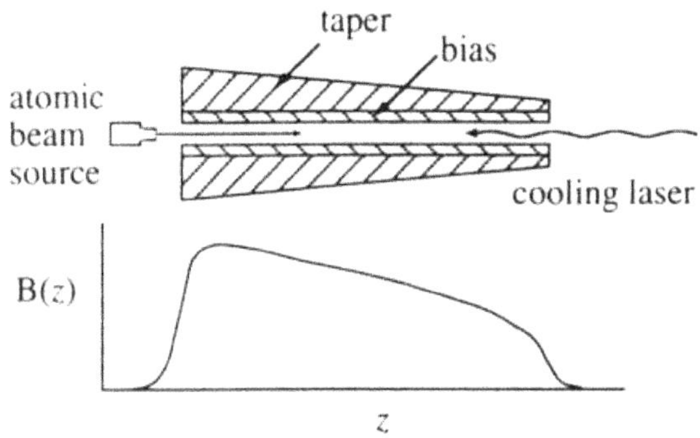

Figure 2. Upper: Schematic representation of a Zeeman slower. Lower: Variation of the axial field with position.

the atoms are slowed down and the spread of the velocity distribution is narrowed, this process is often referred to as "Zeeman cooling". Figure 2 illustrates the general idea of this scheme. The atomic beam source directs atoms, which have a range of velocities, along the axis (z direction) of a tapered solenoid. This magnet has more windings at its entrance end, near the source, so the field is higher at that end. The laser is tuned so that, given the field-induced Zeeman shift and the velocity-induced Doppler shift of the atomic transition frequency, atoms with velocity v_0 are resonant with the laser when they reach the point where the field is maximum. Those atoms then absorb light and begin to slow down. As their velocity changes, their Doppler shift changes, but is compensated by the change in the Zeeman shift as the atoms move to a point where the field is weaker. At this point, atoms with initial velocities slightly lower than v_0 come into resonance and begin to slow down. The process continues with the initially fast atoms decelerating and staying in resonance while initially slower atoms come into resonance and begin to be slowed as they move further down the solenoid. Eventually all the atoms with velocities lower than v_0 are brought to a final velocity that depends on the details of the magnetic field and laser detuning. The magnetic field profile of the tapered solenoid is

$$B(z) = B_0\sqrt{1 - 2az/{v_0}^2} \tag{14}$$

with $0 \leq 2az \leq v_0^2$. B_0 is the magnetic field producing a Zeeman shift equal to the Doppler shift for atoms with velocity v_0 and $a \leq a_{max}$ is the deceleration rate.

The first experiment on the deceleration of atoms using the Zeeman technique is in ref. 43. Subsequently, neutral sodium atoms were stopped in ref.

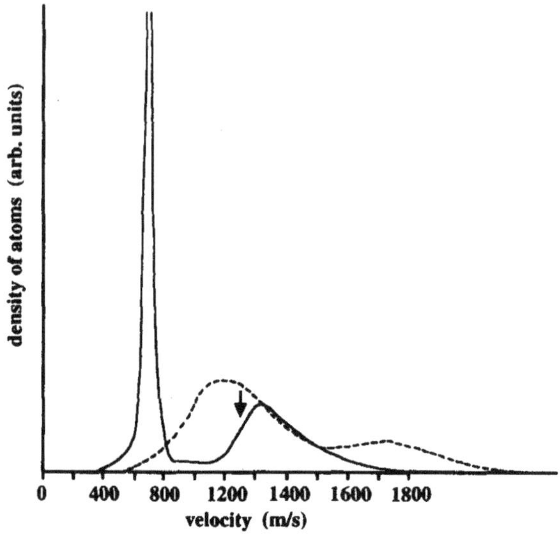

Figure 3. Velocity distribution before (dashed) and after (solid) Zeeman cooling. The arrow indicates the highest velocity resonant with the slowing laser. (The extra bump at 1700 m/s is from $F = 1$ atoms, which are optically pumped into $F = 2$ during the cooling process.)

44. Figure 3 shows the velocity distribution resulting from Zeeman cooling: a large fraction of the initial distribution had been swept down into a narrow final velocity group. The velocity distribution after deceleration was measured in a detection region some distance from the exit end of the solenoid using a separate detection laser. We were able to determine the velocity distribution in the atomic beam by scanning the frequency of the detection laser and observing the fluorescence from atoms having the correct velocity to be resonant.

4 Traps for neutral atoms

Trapping of atoms usually refers to their confinement by the application of external fields rather than by the use of a material container. In contrast with the trapping of ions by electric and magnetic fields, the trapping forces that can be applied to neutral atoms are relatively weak. Ions have a charge on which an electromagnetic field can exert a large Coulomb or Lorentz force. Neutral atoms, however, may be acted upon through their permanent magnetic dipole moments or induced electric dipole moments, allowing generally smaller forces to be applied. The strongest traps for neutral atoms have energy depths of only a few kelvin, while ion traps can typically hold room temperature ions, and have trapped particles with energies of a few thousand electron volts.

Another possible difficulty with traps for neutral atoms is that the trapping potentials represent changes in the internal energy of the atoms. That is, the positions of and, in general, the spacings between energy levels are changed by the trapping fields. This means, in particular, that high accuracy spectroscopy of trapped atoms is problematic.

Neutral atoms have the advantage that the lack of space charge effects means that one can generally trap larger numbers and densities of neutral atoms than of ions. Furthermore, some applications demand that one work with neutral atoms (as for example in Bose condensation of an atomic gas). Fortunately, even rather weak forces are capable of trapping atoms that have been laser cooled, and many different kinds of neutral atom traps have been demonstrated. Among these are:

i) Magneto-static traps, first demonstrated in 1985 [45] rely on the force exerted by a gradient magnetic field on the permanent magnetic dipole moment of an atom such as a ground-state alkali.

ii) Laser dipole traps, first demonstrated in 1986 [46] use the dipole force that results from the gradient of the energy of the oscillating dipole moment induced on an atom in an inhomogeneous laser field.

iii) Radiation pressure traps use the scattering force of Eq.(1), but are not stable in 3D for two-level atoms. The magneto-optical trap (MOT), using multilevel atoms and an inhomogeneous magnetic field, was the first radiation pressure trap to be demonstrated [47].

iv) Magneto-dynamic traps use the micromotion driven by oscillating magnetic field gradients to allow trapping of high-field-seeking states not stably trapped in static magnetic fields. Such a trap, first demonstrated in 1991 [48], is analogous to the radio frequency Paul trap for ions.

v) Microwave traps are low-frequency, spontaneous-emission-free analogs of laser dipole traps. Such a trap was first demonstrated near a magnetic resonance transition [49].

vi) Electrostatic traps, while never demonstrated in 3D, have been proposed [50] for excited atoms.

vii) Gravito-optical traps, which combine optical dipole forces with gravity to produce stable trapping [51], are only one example of hybrid traps that combine different types of forces to achieve trapping of atoms.

viii) TOP traps (for Time-averaged Orbiting Potential) [52], are an important modification of one type of magnetostatic trap. While time dependent, they are not dynamic in the sense of atomic micromotion being essential.

These notes will not treat each of these kinds of traps in detail, nor attempt to give more complete references about them. We shall, however, discuss the radiation pressure, laser dipole and magnetic traps in some more detail below. Since these are the traps that are currently in extensive use in BEC research.

4.1 *Dipole force traps*

The dipole force discussed in section 2.2 can be used to trap atoms. A single, focused laser beam, tuned below resonance is the simplest dipole trap and was first proposed by Ashkin in 1978 [53]. As an example, consider sodium atoms interacting on their strongest transition ($I_0 = 6$ mW/cm^2) with a modest power, 10 mW Gaussian laser beam focused to a $1/e^2$ radius of 10 μm. This gives $I/I_0 \cong 10^6$ at the focus. For the detuning maximizing U, we find $U_{max}/k_B \cong 100$ mK. For a 1W beam we find $U_{max}/k_B \cong 1$ K. Such traps can easily confine laser cooled atoms.

The dipole potential is a conservative potential, so it does not have any dissipative mechanism associated with it. It was not until the demonstration of laser cooling in optical molasses [29] that such a trap could be loaded and finally realized [46]. One of the difficulties involved in such trapping is that although the trap is tuned below resonance (the proper sign of the detuning to achieve cooling) the cooling provided by the trapping light does not reduce the thermal energy below the trap depth. Auxiliary cooling [54] is required, but even this is difficult because the inhomogeneous light shifts induced by the trapping laser interfere with the cooling process. Dalibard *et al.* [55,56] proposed a solution in which the trapping and cooling are alternated in time, and this procedure was used for the first dipole trap [46].

Another difficulty with a single focus dipole trap is that the radiation pressure force pushes the atoms away from the focus, while the dipole force is attracting them to it. While more complicated, counter-propagating beam geometries can avoid this difficulty [53,57,58], one can, at the expense of reduced trap depth, solve the problem by detuning the laser [46]. According to Eq.(1), the destabilizing radiation pressure force varies as $1/\delta^2$ (for sufficiently large δ), while the dipole trapping force obtained from the dipole potential given by Eq.(2) varies as $1/\delta$. Thus, for large enough detuning, the radiation pressure will be negligible compared to the dipole trapping force.

Large detuning has other advantages, particularly when coupled with multi-level laser cooling. When the detuning is large enough that the trap depth is comparable to the natural linewidth, the thermal energy of laser cooled multilevel atoms can still be considerably less than this depth, which is comparable to the Doppler cooling limit of Eq.(10). Furthermore, the optimum detuning for good multilevel laser cooling is at least several times the linewidth, so the inhomogeneous light shifts should not much affect the cooling. Such a far-off-resonance-trap (FORT) has been demonstrated to work without the need to alternate cooling and trapping phases [59,60].

Another advantage [61] of such a FORT is that for sufficiently large detuning the population of the trapped atoms is almost entirely in the ground state. The trap is then nearly free of heating due to spontaneous emission [62] and of many of the collisional perturbations involving excited atoms. FORTs have been used to hold atoms for collision experiments [63], to trap atoms for evaporative cooling [64], and to confine Bose-Einstein condensates [65].

Variations of the dipole force trap include using the evanescent wave created by total internal reflection of light (detuned blue of resonance) to act as a mirror to reflect atoms [51,66], crossing two red-detuned laser beams at their foci to achieve a strong gradient in all directions [64], and crossing sheets of blue-detuned light to form a box bounded by light (with confinement in the vertical direction sometimes provided by gravity) [67,68].

4.2 Radiation pressure force traps

Radiation force traps use the spontaneous or scattering force to confine atoms. Unlike the dipole force, the force generated by absorption of a photon followed by spontaneous emission does not depend on the gradient of the intensity. Hence larger volume traps can be created using the scattering force compared to the dipole force for a given flux of incident photons. A simple radiation force trap, shown in Fig.4, can be made in 1-D using two counter-propagating focused laser beams with separated foci [53]. Midway between the foci the radiation pressure from the two beams is balanced and the net force on an atom is zero. As the atom moves away from this equilibrium point toward one of the foci it is pushed back by the higher intensity near that focus. In contrast to the dipole force trap, radiation pressure trap depths on the order of a kelvin can be achieved with a modest, near saturation intensity, i.e., a few milliwatts per square centimeter [69]. While the trap of Fig.4 produces radiation pressure force trapping along only one axis, the dipole force can provide trapping along the two orthogonal axes [53], a configuration first demonstrated in ref. 58.

It is tempting to extend the trap idea of Fig.4 to three dimensions, but it can be shown that such as extension is not possible when the radiation pressure force is proportional to the photon flux from the laser beams. The impossibility of such trapping is related to the fact that the divergence of the

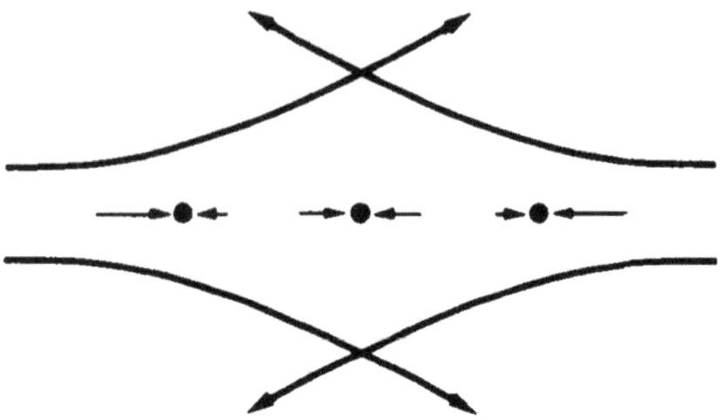

Figure 4. A one dimensional radiation pressure force trap formed from two counter-propagating, focussed laser beams with separated foci. The length of each of the pairs of arrows indicates the magnitude of the force from the respective laser beams on an atom at the indicated positions.

Poynting vector is zero, and has been called the optical Earnshaw theorem [70] by analogy with the theorem from electrostatics forbidding stable trapping of a test charge in a charge-free region.

In spite of the optical Earnshaw theorem, which applies to 2-level atoms in a weak, static laser field, it is possible to create a 3-D radiation force trap by making use of such features as saturation, multiple levels, optical pumping and Zeeman shifts. The most successful radiation pressure trap that circumvents the optical Earnshaw theorem is the magneto-optical trap or MOT. Conceived by Dalibard [71] and demonstrated in an MIT-Bell Labs collaboration [47] its principle is illustrated in Fig.5 for 1-D operation and a simple $J = 0 \rightarrow J = 1$ transition.

A pair of current-carrying coils with opposing currents creates a quadrupole magnetic field that is zero at the origin and whose vector value is proportional to the displacement from the origin. The simple $J = 0 \rightarrow J = 1$ transition gives us a Zeeman shifted transition frequency with a non-degenerate ground state. (This non degeneracy leads to Doppler rather than sub-Doppler cooling.) Two circularly polarized laser beams with opposite helicity counter-propagate along the coils' axis. The $\sigma^{\pm}$ beam excites atoms to the $m = \pm 1$ excited states, respectively. Thus, for a red-detuned laser frequency ($\delta < 0$) atoms displaced in the positive direction will experience a Zeeman shift that brings the $m = -1$ state into resonance with the laser frequency, and the σ^{-} laser beam that excites this state is the one that pushes it back toward the origin. Similarly, an atom displaced in the negative direction is pushed back by the σ^{+} beam. In addition to the restoring force, there is

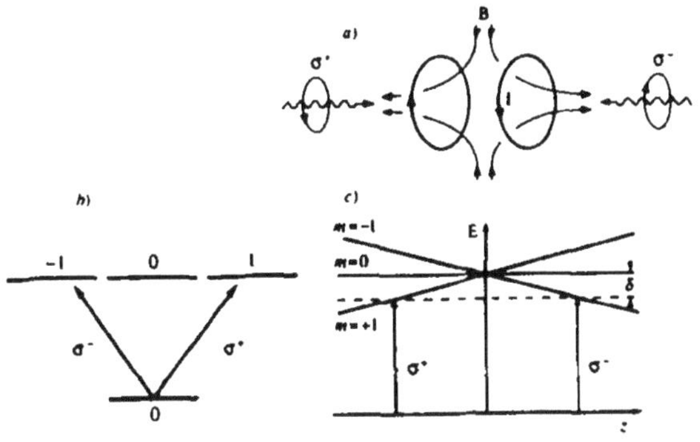

Figure 5. *a*) Magnetic field and laser configuration for a 1-D MOT. *b*) Transition scheme. *c*) Energy levels and transitions in the spatially varying magnetic field. The designation of the *m*-state is with respect to a space-fixed axis, as is the laser polarization.

also the usual Doppler-cooling damping force.

The force on an atom with velocity v and position z can be obtained from Eqs.(3) and (4) by replacing the effective detuning $\delta \mp kv$ with $\delta \mp (kv + \beta z)$, where βz is the magnitude of the Zeeman frequency shift at position z:

$$F(v, z) \cong \frac{4\hbar k I/I_0}{\left[1 + (2\delta/\Gamma)^2\right]^2} \left(\frac{2\delta}{\Gamma}\right) (kv + \beta z) \qquad (15)$$

where in the second expression we are in the limit of low velocity, magnetic field and laser intensity. For negative detuning the above force represents a damped harmonic oscillator. Typical operating conditions for a MOT might be $I/I_0 = 1$, $2\delta/\Gamma = -1$ and $\beta = 10$ MHz/cm. This would lead to an oscillation frequency of about 1 kHz for Na, but with strong overdamping.

While we have considered only the 1D case for a simple transition, the MOT was first demonstrated in 3D, on an atom with a degenerate ground state (Na). The theory in 3D has been worked out in detail for the $J = 0 \to J = 1$ transition [72] and for transitions allowing sub-Doppler cooling, some insights have been gained by studying the forces in 1D on a moving atom in a magnetic field[73–76]. Experiments have shown that, with a degenerate ground state, sub-Doppler temperatures are achieved in a MOT [75] along with larger trapping and damping than predicted by the $J = 0 \to J = 1$ theory.

The MOT has become an important tool in the study of cold atoms. A particularly useful feature is that it can capture atoms from an uncooled, thermal atomic vapor [32,33]. This often allows considerable simplification of the apparatus compared to one where an atomic beam is first decelerated.

A MOT can concentrate atoms to the point that collisions [77,78] and radiation pressure exerted by the atoms' fluorescence [79] are factors limiting the density. A major advance in reducing such problems is the "dark spot" MOT [80] in which atoms are, for the most part, optically pumped into a state from which the excitation rate is considerably smaller than in a conventional MOT. In practice this is accomplished, for atoms with ground state hyperfine structure, by pumping the atoms into one of the hyperfine states. Normally laser cooling and trapping of such atoms is performed by applying a separate laser frequency to excite each of the ground hyperfine states to the electronically excited state, ensuring atoms are not pumped into a state from which they cannot be excited. In a dark spot MOT, one of the laser frequencies (the re-pumper) is eliminated from the central region of the trap. Atoms then accumulate in the hyperfine state that would have been excited by the missing frequency. These atoms are rarely excited (only by off-resonant light or indirectly scattered resonant light) so problems due to excited atoms are greatly reduced, even though the atoms are still cooled and trapped. Such techniques have achieved atomic densities near 10^{12} cm^{-3}.

4.3 *Magneto-static traps*

Both laser dipole and radiation pressure traps involve optical excitation of the atoms being trapped. Although in principle a sufficiently intense laser dipole trap can be tuned far enough off resonance for the excitation rate to be small, it is difficult in practice to make it truly negligible. As a result, optical trapping generally results in some heating of the atoms due to the random nature of spontaneous emission, and cooling is required to keep the atoms in equilibrium. Magnetic traps do not suffer this problem and so can be used to store atoms for long periods of time without the need for additional cooling. On the other hand, such traps only work for atoms having a magnetic dipole moment and only for those states of such atoms whose Zeeman energy increases in increasing magnetic field.

The idea of magnetic trapping was an extension of the magnetic focusing of atomic beams [81] and was discussed by Paul during the 1950s [82]. The first published proposals for magnetic trapping of neutral atoms were in the 1960s [83,84,85]. It was not until after the demonstration of laser cooling of neutral atoms that the first magnetic trapping of atoms was achieved [45].

The principle of the magnetic trap can be understood by considering an atom with a Zeeman sub-structure such as that shown in the excited state of Fig.5c. (Note that magnetic trapping is generally done on ground state atoms with such Zeeman structure.) Since the energy varies with magnetic field, an atom feels a force in a magnetic field gradient. For states whose energy increases with magnetic field (low-field-seeking states), a trap is formed by the field of the coils in Fig.5*a*. This quadrupole field is zero on the axis midway between the coils, and its magnitude increases linearly along any line away from this central point. Low-field-seeking atoms experience a restoring force towards, and are trapped around, this point. (It can be shown [86,87] that in a current-free region no magnetic field can have a relative maximum in its magnitude, so that high-field-seekers cannot be trapped.) The depth of a magnetic trap is given by the maximum Zeeman shift along the easiest escape path. A magnetic field of 1 mT (10 G) gives a typical shift of 14 MHz, equivalent to 670 μK, so laser cooled atoms are easily trapped by modest fields.

The quadrupole magnetic field was used in the first magnetic trap [45] and is the simplest of magnetic traps. It works for low-field-seekers as long as they do not change their spin orientation with respect to the local field. This means than they must adiabatically follow the changes in the direction of the field as they orbit in the trap. The condition for adiabatic following is that

$$\frac{d\theta}{dt} \ll \omega_{Zeeman}, \tag{16}$$

where θ is the angle of the magnetic field at the atom's position and ω_{Zeeman} is the Zeeman frequency separating states of different spin orientation. Since laser cooled atoms move so slowly, they generally experience small field rotation rates, and their motion is usually adiabatic. The quadrupole trap, however, has a point where the magnetic field is zero and the adiabatic condition is impossible to satisfy. Atoms passing sufficiently close to the trap center will fail to follow adiabatically, change their spin orientation (undergo Majorana transitions) and be ejected from the trap [45]. This difficulty can be avoided by a variety of traps that do not have a point of zero magnetic field [88], but such traps are generally not as "stiff" as a quadrupole trap. That is, the restoring force near the center of the trap is not as large as for a quadrupole trap. In practice, non-adiabaticity is usually a problem only for very cold atoms confined very near the center of a quadrupole trap, which is exactly what occurs on the way to BEC.

In order to achieve Bose-Einstein condensation through runaway evaporative cooling, it is desirable to have a "stiff" trap to maintain a high collision or thermalization rate. For atoms cold enough to Bose condense, the problem of non-adiabatic spin flip transitions at the zero-field point of a quadrupole trap are quite severe [89,90]. The TOP trap solved this problem by superposing onto the quadrupole field a magnetic field rotating in the plane of symmetry.

The rotating field guarantees that the field is not zero at the trap center, and produces a rounding of the sharp-cusp, quadrupole, trapping potential. This time-averaged, orbiting potential (TOP) trap [52], although time-dependent, is not a magneto-dynamic trap in the sense that the TOP trap does not rely on micromotion of the atoms. The atoms respond to the time averaged potential at a given point in the trap. A more detailed discussion of the TOP trap can be found in the discussion of the BEC experiment at NIST-Gaithersburg, in section 6.1 of these lecture notes.

5 Sub-Doppler laser cooling

The early experiments [29,91] on optical molasses produced a satisfying agreement of the observed temperature and spatial diffusion with the predictions of the theory of Doppler cooling as outlined in the section 2.3. Some experiments, however, were in disagreement with expectations [92]. Finally, in 1988, careful temperature measurements [93] showed conclusively that atoms in optical molasses were much colder than the Doppler cooling limit. The time-of flight (TOF) method used to measure the temperature in those experiments has become a standard technique. Atoms are collected and cooled in the optical molasses at the intersection of the molasses laser beams. The molasses beams are suddenly extinguished and the released atoms fall toward a probe. As the atoms pass through the probe laser beam they absorb and fluoresce light. This fluorescence is measured, with time resolution, by a photodetector. The distribution of detected fluorescence in time gives the distribution of times of flight from the molasses to the probe, and the temperature can be determined from that distribution.

5.1 Observation of sub-Doppler temperatures

Figure 6*a* shows an example of an experimental TOF signal for sodium atoms cooled in optical molasses [21]. The experimental points correspond reasonably well with the predicted signal for a temperature of 25 μK, while 250 μK, about the Doppler cooling limit for Na, is completely inconsistent with the experimental points. Furthermore, the dependence of the temperature on detuning was found to be inconsistent with the theory of Doppler cooling. Figure 6*b* (points) shows the measured temperature as a function of laser detuning for Na, where the linewidth is 10 MHz. The temperature decreases for detunings larger than $\Gamma/2$ (until the laser frequency approaches resonance with another hyperfine state, about 60 MHz to the red of the chosen resonance). This is in sharp contrast to the prediction of Doppler cooling theory (Fig.6*b*, solid

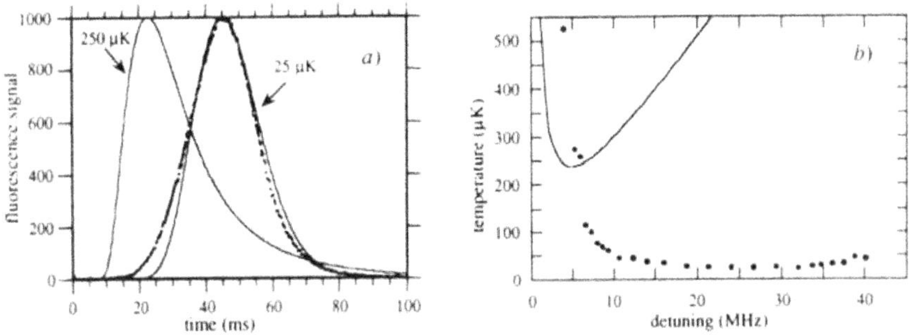

Figure 6. *a*) Experimental time-of-flight distribution (points) for Na atoms released from an optical molasses. Expected TOF distributions for temperatures of 25μK and for 250μK (approximately the Doppler cooling limit) are shown. *b*) Measured temperature as a function of laser detuning for sodium atoms cooled by a 3D optical molasses. The temperature predicted by Doppler cooling theory is shown by the solid line.

line), which has the temperature increasing for detunings greater than $\Gamma/2$. In addition, the temperature was found to be linearly dependent on laser intensity [94], again in contrast to the predictions of Doppler cooling theory, and the temperature was found to depend on magnetic field and laser polarization [93,94]. These latter facts, particularly, suggest that the magnetic sublevels of the atom play an important role.

The observation of sub-Doppler-limit temperatures was quite surprising. Doppler cooling theory at low intensity was simple and compelling. Furthermore, at least for 1D, there was a complete theory, taking into account the effects of high intensity and interference between laser beams that were ignored in the treatment presented in the previous section. The theory had been restricted to 2-level atoms, but it was widely believed that this restriction was not particularly important. At low intensity the Doppler temperature depended on the transition linewidth and the detuning, and these were the same for any of the degenerate Zeeman sublevels in a given alkali hyperfine level. Nevertheless, the magnetic field and polarization dependence of the sub-Doppler-limit temperatures pointed to the importance of the Zeeman sublevels.

5.2 New cooling mechanisms

The explanation for the sub-Doppler temperatures soon came in the form of a new theory of multi-level laser cooling. The key elements of the new

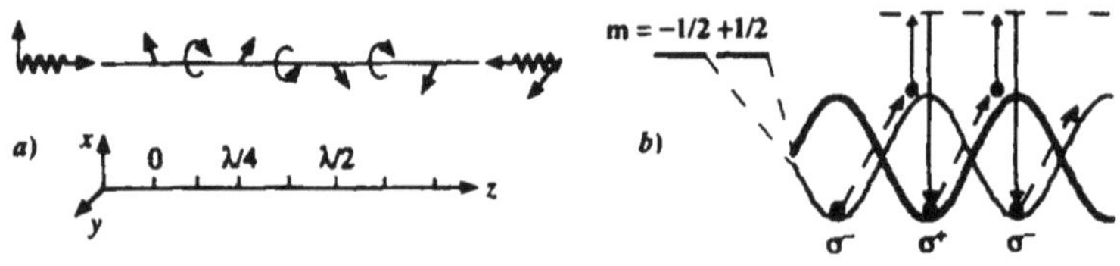

Figure 7. *a*) Interfering, counter-propagating beams having orthogonal, linear polarizations create a polarization gradient. *b*) The different Zeeman sublevels are shifted differently in light fields with different polarizations; optical pumping tends to put atomic population on the lowest energy level, but non-adiabatic motion results in "sisyphus" cooling.

theory [73,95,96] were optical pumping among the magnetic sublevels of the electronic ground level and differential light shifts of the sublevels. While the original theories treated cases where a spatial gradient of the polarization of the optical field was important, it was later demonstrated that such multilevel laser cooling was possible even without polarization gradients[96–102].

Since the theory of multilevel laser cooling is treated extensively in the literature a detailed description will not be given here. The most important of the multilevel cooling mechanisms is Sisyphus cooling [19,73,103,104]. Semi-classically (when the atom can be considered to be well-localized on the scale of an optical wavelength) the simplified physical picture for Sisyphus cooling can be understood by considering the atom and laser field situation illustrated in Fig.7.

Figure 7*a* shows a 1-D set of counter-propagating beams with equal intensity and orthogonal, linear polarizations. The interference of these beams produces a standing wave whose polarization varies on a sub-wavelength distance scale. At points in space where the linear polarizations of the two beams are in phase with each other, the resultant polarization is linear, with an axis that bisects the polarization axes of the two individual beams. Where the phases are in quadrature, the resultant polarization is circular and at other places the polarization is elliptical. An atom in such a standing wave experiences a fortunate combination of light shifts and optical pumping processes.

Because of the differing Clebsch-Gordon coefficients governing the strength of coupling between the various ground and excited sublevels of the atom, the light shifts of the different sublevels are different, and they change with polarization (and therefore with position). Figure 7*b* shows the sinusoidal variation of the ground-state energy levels (reflecting the varying light shifts or dipole forces) of a hypothetical $J_g = 1/2 \rightarrow J_e = 3/2$ atomic system. Now imagine an atom to be at rest at a place where the polarization is σ^-, e.g. at $z = \lambda/8$ in Fig.7*a*. As the atom absorbs light with negative angular momentum and radiates back to the ground states, it will eventually be optically pumped into the $m_g = -1/2$ ground state, and simply cycle between this state and the excited $m_e = -3/2$ state. For low enough intensity and large enough detuning we can ignore the time the atom spends in the excited state and consider only the motion of the atom on the ground state potential. In the $m_g = -1/2$ state, the atom is in the lower energy level at $z = \lambda/8$, as shown in Fig.7*b*. As the atom moves, it climbs the potential hill of the $m_g = -1/2$ state, but as it nears the top of the hill at $z = 3\lambda/8$, the polarization of the light becomes σ^+ and the optical pumping process tends to excite the atom in such a way that it decays to the $m_g = +1/2$ state. In the $m_g = +1/2$ state, the atom is now again at the bottom of a hill, and it again must climb, losing kinetic energy, as it moves. The continual climbing of hills recalls the Greek myth of Sisyphus, so this process, by which the atom rapidly slows down while passing through the polarization gradient, is called Sisyphus cooling. In Sisyphus cooling, the radiated photons, in comparison with the absorbed photons, have an excess energy equal to the light shift. While in Doppler cooling, the energy excess comes from the Doppler shift.

In contrast to the case for Doppler cooling (see Eq.(4)) the friction force is independent of laser intensity, and proportional to detuning at low (but not too low) intensity and low velocity, while the momentum diffusion coefficient is proportional to intensity and independent of detuning. This leads, according to Eq.(7), to a temperature that depends linearly on intensity and inversely on detuning. That is, the temperature is proportional to the light shift

$$k_B T \propto \hbar\Delta_{lightshift} = \frac{\hbar\Omega^2}{4\delta} \tag{17}$$

where the expression for the light shift is valid in the limit of low intensity and large detuning.

A less restricted treatment [105] shows that the friction force is not linear in the atomic velocity, nor is the momentum diffusion constant independent of velocity. Nevertheless, the temperature remains approximately linear in the light shift as long as the intensity is sufficiently above a critical intensity. The lower limit to the temperature obtainable by Sisyphus cooling is set by this lowest intensity for which the cooling process works [105,106]. This is the intensity at which the light shift is comparable to the recoil energy, and it leads to a lower limit for the thermal velocity on the order of a few times the recoil velocity.

These qualitative features of multi-level laser cooling have been confirmed by experiments on atoms cooled in 3D optical molasses [21,107]. The experimental results showed the linear dependence of the temperature on intensity and light shift for all but the largest intensity at the smallest detuning, outside the limits of validity of the simple results listed above. The constancy of the lowest temperature, once the detuning is large enough, is consistent with its depending only on the recoil energy. The temperature of 2.5 μK obtained for Cs represents an *r.m.s.* velocity of about three recoil velocities, similar to the case for the lowest temperatures observed for Na [21,98] and Rb [108].

6 Coherent manipulation of Bose-Einstein condensates with light

The creation of Bose-Einstein condensates (BECs) in dilute atomic vapors of Rb, Na and Li [109,110,111] is one of the major triumphs of laser cooling and trapping of neutral atoms. Alternatively, the creation of BECs has renewed interest in the applications of laser cooling and trapping techniques for atom optics; the manipulation of atoms with mirrors, beamsplitters and lenses analogous to the manipulation of light.

Until recently, atom optic experiments have used thermal sources of atoms much as early experiments in optics used lamps. What was lacking was a coherent source of matter-waves similar to the laser for light. The creation of a Bose-Einstein condensate (BEC) of a dilute atomic gas has opened up the possibility of realizing a matter-wave source analogous to the optical laser. The macroscopic occupation of the ground state of a trap by a BEC is similar to the occupation of a single mode of an optical cavity by photons. The atoms forming the condensate all occupy the same wavefunction - both in terms of their internal and external degrees of freedom.

Atoms from a BEC are nearly the ideal, monochromatic source for atom optics. Many atom optical elements involve the interaction of the atoms with an optical field and the associated transfer of the photon momentum to the atoms. Because of the repulsive atom-atom interaction, which can be described by a mean field, the BEC swells to a size significantly larger than the ground state wavefunction of the harmonic trap confining the atoms [112]. The spatial extent of the resulting wavefunction can be several orders of magnitude larger than the optical wavelength. Hence the momentum width, given by the Heisenberg uncertainty principle, can be much less than the photons momentum. Not all experiments will realize this reduced, intrinsic momentum width. The interaction energy may be converted to kinetic energy when the atoms are released from the trap. Nonetheless, the resulting additional momentum spread, due to the atom-atom interaction, can still be significantly less than the momentum of a single photon.

The principal focus of research on Bose-Einstein condensates in the Laser Cooling and Trapping Group of NIST is in atom or matter-wave optics. This includes optically induced diffraction as beamsplitters and mirrors for the condensate atoms, as well as applications of diffraction in atom interferometry and the realization of an "atom laser", the atomic analogue of the optical laser. These experiments are basically demonstrations of "single atom" phenomena applied to a collection of atoms with a high degree of first-order coherence. When the interactions between the atoms is considered, collective matter-wave phenomena can be observed which are analogous to effects in nonlinear optics with light. We have observed such effects with matter-waves and they are described in the subsequent section titled, "Nonlinear atom optics with Bose-Einstein condensates."

Our atom optics experiments are performed on a BEC of sodium atoms. We begin with a brief description of our experimental apparatus and our approach for creating a Bose-Einstein condensate. We do this essentially for two reasons: our strategy for achieving BEC is somewhat different than other approaches, and it illustrates an application of many of the techniques of laser cooling and trapping developed over the last 15 years.

6.1 The tri-axial TOP trap for sodium

Our Bose-Einstein condensates of sodium atoms are produced in a time-averaged orbiting potential or TOP trap [52]. Our TOP trap differs from those in other BEC experiments in two respects. First, all other experiments with TOP traps that have resulted in BEC use rubidium. We are currently the only group making condensates of sodium in a TOP trap. The lighter mass of sodium poses a greater technical challenge in the design of the TOP trap compared to rubidium. Since the oscillation frequency of an atom in a trap is higher for the lighter atom, the frequency of the rotating bias field of the TOP trap must be correspondingly higher so that the atom experiences the time-averaged potential. Typically, sub-Doppler laser cooling can cool a sample of atoms to an energy which is a few times the recoil energy. Since the recoil energy is inversely proportional to the mass of the atom, the TOP trap must also be deeper or stronger to contain the lower mass, laser cooled atoms. Second, the geometry of our TOP trap is different from other TOP traps resulting in a totally anisotropic or tri-axial, time-averaged potential.

A time orbiting potential or TOP trap is a magnetic trap consisting of a quadrupole magnetic field and a constant magnitude rotating bias field. The potential resulting from a superposition of these two magnetic fields, averaged over a rotation period, is harmonic for small displacements. The time-averaged value of the minimum magnetic field in the TOP trap is just the magnitude of the rotating bias field. That is, the rotating bias field has effectively "plugged" the zero field region of the quadrupole field.

In the NIST-Gaithersburg TOP trap, the bias field rotates in a plane containing the symmetry axis of the quadrupole field. For small displacements, the time-averaged magnetic field is,

$$\langle \mathbf{B} \rangle_t = B_0 + \frac{b_q^2}{4B_0}\left(x^2 + 2y^2 + 4z^2\right) \tag{18}$$

where B_0 is the magnitude of the rotating bias field and $2b_q$ is the gradient of the quadrupole field along the symmetry (z) axis. The spring constants for this TOP magnetic trap are in the ratio of 1:2:4 in the x, y and z direction, respectively. In the standard configuration for the TOP trap fields, such as in the original trap of JILA [52] used to create the first BEC of Rb, the bias field rotates in the symmetry plane of the quadrupole field, and The spring constant in the radial (x, y) direction is a factor of eight less than the axial direction (z) producing a disk shaped time-averaged potential. The magnetic trap given by Eq.(18) is tri-axial; that is, unlike the JILA TOP trap, it has no rotational symmetry. In addition, it is closer to spherical than the JILA TOP trap and better matched for loading from the nearly spherical clouds of laser cooled atoms from the MOT.

Although the TOP trap has a number of desirable properties, including some independent adjustments of the spring constants in the three principle directions, there are certain limitations associated with trapping a mixed state such the $F = 1$, $m_F = -1$ state of sodium. The most important is the quadratic Zeeman effect which reduces the "effective" magnetic moment of that state.

Equation 18 shows that in order to achieve the stiffest TOP trap for a given radius of the "circle of death" (a radius typically chosen to be larger than the radius of the sample of trapped atoms), the largest possible quadrupole field should be used. If the strength of the quadrupole field is increased then the magnitude of the rotating field must be increased to keep the radius of the "circle of death" constant. The energy of the $F = 1$, $m_F = -1$ state of sodium as a function of magnetic field initially increases linearly (the linear Zeeman effect), but the slope decreases with magnetic field due to the quadratic Zeeman effect. Eventually the energy of this state reaches a maximum around 31.5 mT (315 Gauss) above which it becomes an anti-trapped state. Thus for a fixed radius of the "circle of death", the stiffness of the TOP trap no longer increases linearly with the strength of the quadrupole field, becoming extremely weak for sufficiently large values of the bias field.

6.2 BEC of sodium in a TOP trap

Similar to other BEC experiments, our Bose-Einstein condensate of alkali atoms is produced by evaporative cooling in a magnetic trap loaded with laser cooled and trapped atoms, however, the details of our technique differ from other groups. More specifically, we load a dark spot MOT [80] from an effusive source of sodium at 625 K using Zeeman slowing. The slowing laser beam passes through the trap so that the capture area of the trap subtends a large solid angle of the flux of slow atoms. In order to minimize the effect of the slowing laser on the trapped atoms, we use a hybrid slower geometry [113]. A conventional Zeeman slower is used to slow atoms from about 800 m/s down to about 160 m/s, followed by a short section of reverse slower [114] to slow atoms from about 160 m/s down to a few m/s, which is within the capture velocity of the dark spot MOT. To avoid the loss of atoms in the slowing process due to optical pumping as the atoms pass through the zero field region between the conventional and reverse Zeeman slowing magnets, a second laser frequency is in the slowing beam to pump atoms from the $F = 1$ hyperfine level to the $F = 2$ level where they can continue to be slowed. In addition, a dark spot is placed in the slowing laser beam, creating a shadow in the region of the cloud of atoms in the dark spot MOT [115]. This further reduces the effect of the slowing laser beam on the trapped atoms. Typically, we load more than 10^{10} atoms into the MOT in less than 0.5 seconds.

After loading into the dark spot MOT, the magnetic fields are rapidly switched off and the atoms are further cooled to $\approx 200\mu$K by a brief period of dark molasses (there is a dark spot in the repumper light in the molasses beams), followed by optical pumping of the entire population into the $F = 1$ hyperfine level. The magnetic trap is then rapidly switched on, trapping atoms in the $m_F = -1$ sublevel of the $F = 1$ ground state. Since all three sublevels are present in equal populations at the time when the magnetic trap is turned on, two-thirds of the sample of laser cooled atoms are necessarily lost in the transfer. Experimentally, we find that we are able to trap 4 to 5 $\times 10^9$ sodium atoms in the magnetic trap.

The atoms are confined in the benign environment of a magnetic trap in order to be evaporatively cooled [116,117]. We have developed two strategies for evaporatively cooling atoms to Bose-Einstein condensation. The first strategy evolved in response to the quadratic Zeeman effect problem in the $m_F = -1$ state of sodium, discussed in section 6.1. That is, for the large cloud of atoms initially confined in the TOP trap, the stiffness of the trap could not be increased, while keeping the "circle of death" well outside the cloud, to sufficiently compress the sample for runaway evaporation. Instead, we initially compress and evaporatively cool the sample of atoms trapped in a quadrupole magnetic field. When the sample is sufficiently cold and dense, we transfer them to the TOP trap by rapidly switching on the rotating bias field while the quadrupole field is on, after which further evaporative cooling can proceed in the TOP trap. Evaporative cooling in the TOP trap can be achieved by removing the higher energy atoms to untrapped states with either rf-induced transitions or the circle of death. For most experiments we use rf-induced transitions because it allows us greater control and flexibility. By appropriate choice of frequency and power, we can control the final energy of the atoms we are removing as well as the width of cut made into the sample of atoms. In addition, the parameters of the rf can be changed rapidly compared to changing values of magnetic fields for the circle of death. This technique works for a large range of initial numbers of trapped atoms.

When the initial number of trapped atoms in the MOT is $\geq 10^9$, we can achieve BEC with atoms in the TOP trap, directly. We initially lose a large number of atoms after transfer into the TOP trap because the circle of death cannot be placed sufficiently outside the cloud of atoms. After this initial loss and the cloud of atoms has rethermalized to a lower temperature and higher density, runaway evaporation can be achieve by compressing the sample and removing high energy atoms with the circle of death. Both strategies produce approximately the same number of final condensate atoms, about 3×10^6, at a BEC transition temperature of 1.2 μK.

We probe our samples of atoms using the technique of absorption imaging [109]. The TOP trap is rapidly shut off and after a variable delay, a short laser pulse optically pumps the atoms from $F = 1$ to $F = 2$, after which another short laser pulse resonant with the $3S_{1/2}$, $F = 2 \rightarrow 3P_{3/2}$, $F' = 3$ transition is applied to the atoms along the direction of gravity (the x direction). The light absorbed from this laser beam is imaged onto a CCD camera. From this image we extract the transverse spatial dependence of the optical depth along the direction of the probe beam.

6.3 Diffraction of atoms by a standing wave

When an atomic beam passes through a periodic optical potential formed by a standing light wave, it diffracts similar to the diffraction of light by a periodic grating. The diffraction can be divided into two regimes, normal and Bragg diffraction. Both diffraction processes can be thought of as arising from the simultaneous absorption of a photon from one laser beam of the optical standing wave, and stimulated emission of a photon due to the counter-propagating laser beam. (This is a similar picture for the origin of the optical dipole force

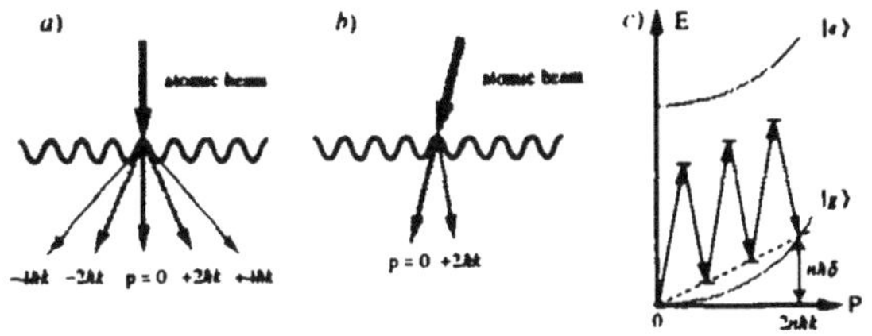

Figure 8. *a*) Normal incidence diffraction *b*) Bragg diffraction *c*) Bragg diffraction as a $2n$-photon Raman transition.

and the momentum transfer resulting from diffraction can be thought of as arising from this force.) This necessarily means that the momentum transfer to the atomic beam by the optical standing wave is quantized in units of $2\hbar k$, twice the momentum associated with a single photon.

In normal diffraction, illustrated in Fig.8*a*, the incident atomic beam non-adiabatically enters the light field at normal incidence. As there is no difference in frequency between the two laser beams comprising the standing wave, the exiting atomic beam is symmetrically diffracted with respect to the incident atomic beam. Energy conservation is satisfied by the spread in energies associated with the non-adiabatic "turn-on" and "turn-off" of the standing wave. For short interaction times such that the atoms do not move appreciably along the direction of the standing wave, the standing wave potential can be considered a thin phase grating that modifies the atomic de Broglie wave with a phase modulation, which for a square profile laser beam is given by $\phi(x) = (U_0\tau/\hbar)\cos^2(kx)$, where U_0 is the maximum depth of the optical potential given by Eq.(2) and τ is the interaction time of the atomic beam with the standing wave. An atom with zero momentum is therefore split by the standing wave into multiple components with momenta $p_n = 2n\hbar k$, $(n = 0, \pm 1, \pm 2, \ldots)$, with populations $P_n = J_n^2(U_0\tau/2\hbar)$, where $J_n(x)$ are Bessel functions of the first kind. Normal incidence diffraction of atoms by a near resonant optical standing wave was first demonstrated in Pritchard's group at M.I.T. [118] in 1983.

In the early experiments demonstrating diffraction of atoms by optical standing waves [118,119], a beam of atoms passed through an optical standing wave and the diffracted beam was detected downstream. In our experiments [120], we start with BEC at a temperature sufficiently below the transition temperature such that no discernible thermal fraction is present. We then adiabatically reduce the strength of the confining potential, which low-

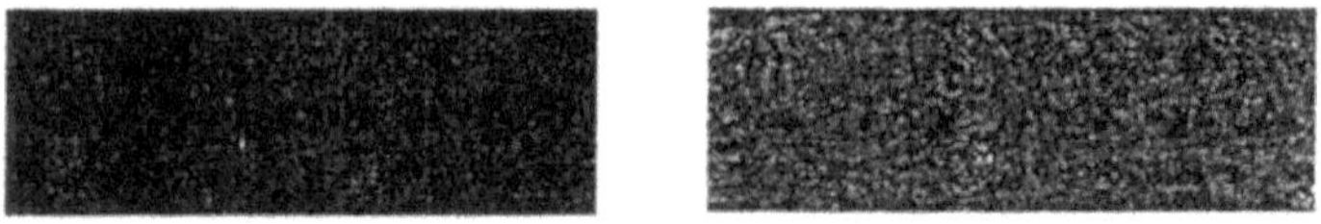

Figure 9. Diffraction of a BEC by a short pulse, optical standing wave. *a*) For low intensities only first order diffraction into $\pm 2\hbar k$ momentum states is visible. *b*) At higher intensities, higher order diffraction ($\pm 4\hbar k$ and $\pm 6\hbar k$) is observed.

ers the energy of the condensate by both reducing the mean-field interactions and increasing the size of the condensate wavefunction. We then expose the atoms to a short pulse of the optical standing wave, while they are either still in the TOP trap, or shortly after releasing them from the trap by rapidly shutting off the magnetic fields. Hence the condensates are essentially at rest and we expose them to the optical standing wave temporally. We detect the momentum transferred to the atoms from the diffraction process by taking an absorption image after a sufficient time delay, such that the various atomic wave-packets with different momenta have spatially separated.

Figure 9 is an example of normal diffraction of BEC by a short pulse of a weak and strong optical standing wave. The optical standing wave is applied along the z-axis, 2 ms after the condensate atoms have been released from the adiabatically expanded trap. For a weak pulse there is only a small phase modulation imposed on the cloud of atoms by the optical standing wave and only the first diffraction orders with momentum $\pm 2\hbar k$ are observed (Fig.9*a*). When the pulse intensity is increased by a factor of 5 there is a substantial phase modulation imposed on the released condensate atoms and higher diffraction orders are observed. In Fig.9*b*, both second and third order diffraction, corresponding to momentum transfer of $\pm 4\hbar k$ and $\pm 6\hbar k$, are clearly evident. The images in Fig.9 were taken 10 ms after the application of the diffraction pulse. The momentum spread of the undiffracted atoms is approximately 0.06 $\hbar k$ and so the diffracted components are clearly resolved.

6.4 *Bragg diffraction of atoms*

When the atoms enter and exit the monochromatic standing wave adiabatically, energy conservation must be explicitly satisfied in the interaction between the atoms and the light field. In this regime, also known as the Bragg regime, the energy difference of the atom before and after the change of momentum of $2\hbar k$ must come from the photon field. This is typically accomplished by having the atomic beam incident on the standing wave at an angle such that the atoms see a differential Doppler shift between the two counter-propagating laser beams comprising the standing wave. This geometry is shown schematically in Fig.8*b*. Under these conditions, Bragg diffraction can be understood as a stimulated Raman transition between two momentum states. Figure 8*c* shows nth order Bragg diffraction as a $2n$-photon, stimulated Raman process in which photons are absorbed from one beam and stimulated into the other. Conservation of energy and momentum requires $(n2\hbar k)^2/2M = 2n\hbar kv\sin\theta$, where M is the mass of the atom, v is the longitudinal velocity of the atomic beam and θ is the angle of incidence of the atomic beam on the standing wave. Bragg diffraction of atoms by a near resonant optical standing wave was also first demonstrated in Pritchard's group at M.I.T. [119] in 1988.

In the case where we start with a BEC essentially at rest, two laser frequencies are needed to satisfy energy conservation in the Bragg diffraction process. The difference in frequencies results in a moving standing wave, that would correspond to the differential Doppler shift observed by the atom in the stationary frame of the standing wave. In our experiments, we create our moving standing wave by having a frequency difference δ between the two counter-propagating waves that make up the standing wave. In the presence of this moving standing wave, an atom initially at rest will simultaneously absorb photons from the higher frequency laser beam and be stimulated to emit photons into the lower frequency beam acquiring momentum $\pm 2n\hbar k$ in the process. In order to satisfy energy conservation, the detuning δ must be chosen such that $n\delta = n^2 4E_{rec}/\hbar$, where E_{rec} is the recoil energy.

Figure 10 shows first, second and third order Bragg diffraction of Bose condensed atoms released from the magnetic trap. When the frequency difference between the two lasers is 100 kHz, atoms initially at rest can resonantly absorb a photon from the higher frequency laser beam and be stimulated to emit a photon into the lower frequency beam. The result of this process is a transfer of two units of photon momentum to the atoms, which then travel ballistically with a velocity of 6 cm/s. Similarly, by setting the frequency difference of the lasers to -100 kHz, the momentum transfer to the atoms from the Raman process will be in the opposite direction. Since Bragg diffraction of the atoms can be thought of as a two level system (the initial and final momentum states) coupled by the Raman process, it is possible to transfer all of the atoms to the final momentum state. We have observed first order Bragg diffraction of 100% of the condensate atoms. The amount of transfer was reduced in the images of Fig.10 so that the location of the condensate

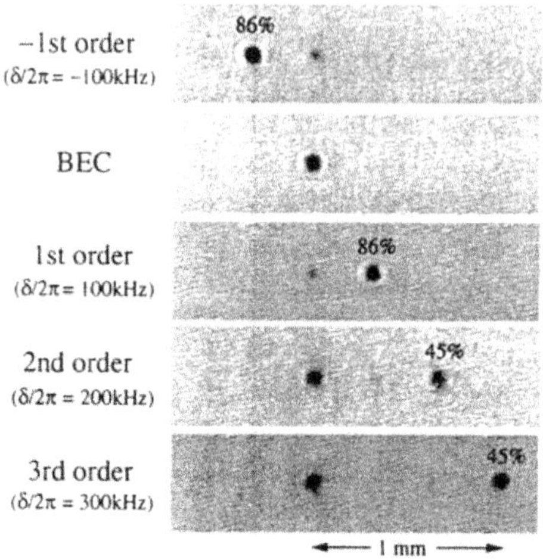

Figure 10. Bragg diffraction of a BEC: By applying a moving standing wave (whose velocity is determined by the frequency difference of the two waves comprising the standing wave) we can Bragg diffract a portion of the condensate into a well-defined momentum state.

atoms, initially at rest, could be easily identified. Second and third order Bragg diffraction was observed when the laser detuning was increased to 200 kHz and 300 kHz, respectively. We have observed up to sixth order Bragg diffraction with a momentum transfer of $\approx 12\hbar k$ (corresponding to a velocity of 0.35 m/s).

6.5 *Raman output coupling: Demonstration of a CW atom laser*

In order to realize an atom laser from BEC, it is necessary to coherently extract the condensed atoms: that is, an atom output coupler is needed. The first demonstration of an output coupler for BEC was reported in 1997 [121], where coherent, rf-induced transitions were used to change the internal state of the atoms from a trapped state to an untrapped one. This method, however, did not allow the direction of the output coupled atoms to be chosen. The extracted atoms fell under the influence of gravity and expanded because of the intrinsic repulsion of the atoms. We have developed a highly directional method to optically couple out a variable fraction of a condensate. We use stimulated Raman transitions between magnetic sublevels to coherently transfer trapped condensate atoms to an untrapped state while giving them a momentum kick [122].

The Bragg diffraction of atoms [120] discussed earlier involves a stimulated Raman transition between different momentum states while keeping the atoms in the same magnetic sublevel. If the frequency difference between the lasers includes the additional Zeeman energy between two magnetic sublevels, a simultaneous change in the momentum and internal state of the condensate atoms can be achieved. This is illustrated in Fig.11, where BEC trapped in the $F = 1$, $m_F = -1$ state is transferred to the $F = 1$, $m_F = 0$ state. Two units of photon momentum are transferred in the Raman process, so the cloud of atoms in the $m_F = 0$ state has a velocity of 6 cm/s with respect to those atoms in the $m_F = -1$ state (Fig.11*a*).

We can repeatedly apply the Raman pulses to achieve multiple output coupling of atoms from a BEC. This is shown in Fig.11$a - d$. In order to avoid changes in the Raman resonance frequency between different magnetic sublevels we synchronized the application of the Raman pulses to our rotating TOP field. (Our condensate atoms were displaced by gravity away from the zero of the quadrupole field, so that the local magnetic field was modulated by the rotating TOP bias field.) Figures 11*a-c* are optical absorption images of the condensate after one, three and seven Raman pulses respectively. For these images, the TOP trap was held on for a 9 ms window during which time 6 μs Raman pulses were applied at a subharmonic of the rotating TOP bias frequency. The magnetic fields were then extinguished and the atoms were imaged 1.6 ms later. In Fig.11*d*, the TOP trap was held on for a 7 ms window during which time 140 Raman pulses were applied at the 20 kHz frequency of the rotating bias field and the distribution of atoms was imaged 1.6 ms later. The Raman pulse duration was reduced to 1 μs in order to couple less atoms out of the condensate during each Raman pulse. In the time between two Raman pulses each output coupled wavepacket moves only 2.9 μm. These pulses strongly overlap because this spatial separation of 2.9 μm is much smaller than the $\approx 50\mu$m size of the condensate, therefore the output coupled atoms form a continuous coherent matter-wave.

Our Raman output coupling scheme dramatically reduces the transverse momentum width of the extracted atoms compared to other methods such as rf output coupling [121]. This dramatic reduction occurs because the output coupled atoms have received a substantial momentum kick from the Raman process. If the atoms were simply released from the trap with no momentum transfer, they would undergo a burst of expansion due to the repulsive interactions with the other condensate atoms. In our output coupling scheme, however, this additional expansion energy is primarily channeled into the forward direction. The increase in the transverse momentum width due to the interaction between the atoms is reduced by roughly the ratio of the timescale

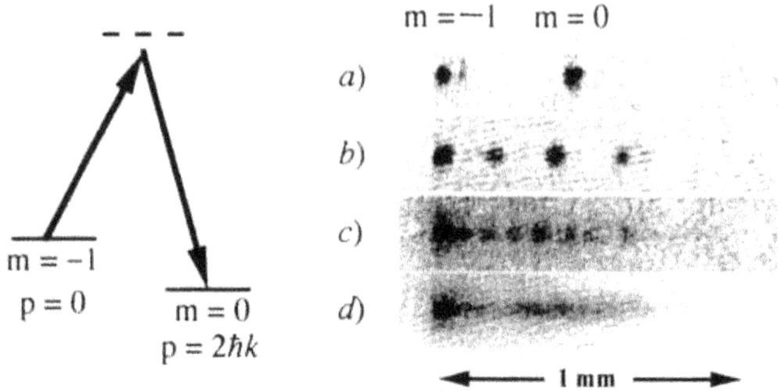

Figure 11. Raman output coupler: (left) A stimulated Raman transition is used to transfer $2\hbar k$ of momenta, and change the magnetic sublevel from the trapped $m = -1$ to the untrapped $m = 0$ state. (right) Series of images demonstrating multiple Raman output coupling of atoms from a BEC en route to demonstrating a quasi-continuous stream of coherent atoms. In *a*)-*c*), one, three and seven 6 μs Raman pulses were applied to the condensate respectively. *d*) is the result of the application of 1 μs Raman pulses at the full repetition rate of ≈20 kHz imposed by the frequency of the rotating TOP bias field (140 pulses in 7 ms). The pictures are absorption images taken after a time-of-flight period.

over which the mean field repulsion acts on the freely expanding condensate, divided by the characteristic time it takes the output coupled atoms to leave the still trapped condensate. In our case, the reduction ratio is about a factor of ≈ 20 which results in a well collimated beam of atoms.

One of the important properties of an optical laser is that the coherence length of the emitted beam of photons is much longer than the size of the cavity. A similar property for an atom laser would be highly desirable. The output coupled beam of our atom laser beam is much longer than the characteristic size of the condensate. Since stimulated Raman transitions are coherent processes, we expect the coherence length of this beam to be much longer than the size of the condensate, but this has yet to measured. We have, however, shown in an interference experiment [123], that successive pulses are fully coherent.

7 Nonlinear atom optics with Bose-Einstein condensates

The advent of the laser as an intense, coherent, light source enabled the field of nonlinear optics to flourish. The interaction of light in materials, whose index of refraction depends on the intensity, has led to effects such as multi-wave mixing of optical fields to produce coherent light of a new frequency, and optical solitons, pulses of light that propagate without dispersion. Nonlinear optics now plays an important role in many areas of science and technology. With the experimental realization of Bose-Einstein condensation (many atoms in a single quantum state) and the matter-wave or atom "laser" (atoms coherently extracted from a condensate), we now have an intense source of matter-waves analogous to the source of light from an optical laser. This has led us to the threshold of a new field of physics: nonlinear atom optics [124].

The analogy between nonlinear optics with lasers and nonlinear atom optics with Bose-Einstein condensates (BECs) can be seen in the similarities between the equations that govern each system. For a condensate of interacting bosons, in a trapping potential V, the macroscopic wave function Ψ satisfies a nonlinear Schrödinger equation [112],

$$i\hbar\frac{\partial\Psi}{\partial t} = \left(-\frac{\hbar^2}{2M}\nabla^2 + V + g\left|\Psi\right|^2\right)\Psi \tag{19}$$

where M is the atomic mass, g describes the strength of the atom-atom interaction ($g > 0$ for sodium atoms), and $|\Psi|^2$ is proportional to atomic number density.

7.1 *Four-wave mixing with matter-waves*

The nonlinear term in Eq.(19) is similar to the third-order susceptibility term, $\chi^{(3)}$, in the wave equation for the electric field describing optical four-wave mixing. We therefore expect that if three coherent matter-waves are spatially overlapped with the appropriate momenta, a fourth matter-wave will be produced due to the nonlinear interaction, analogous to optical four-wave mixing. In contrast to optical four-wave mixing, the nonlinearity in matter-wave four-wave mixing comes from atom-atom interactions, described by a mean-field: there is no need for an external nonlinear medium.

Using the atoms from a BEC, we have observed such four-wave mixing of matter-waves. This work is described in detail in ref. 125. In our four-wave mixing experiment, we used optically induced Bragg diffraction [120] to create three overlapping wavepackets with appropriately chosen momenta. When the three wavepackets spatially separated, a fourth wavepacket, due to the wave-mixing process, was observed (see Fig.(12)).

The process of four-wave mixing of matter-waves (and also optical waves), can be thought of as Bragg diffraction off of a matter grating. In this picture, two of the matter-waves interfere to form a standing wave grating. The third wave can Bragg diffract off of this grating, giving rise to the fourth wave. An

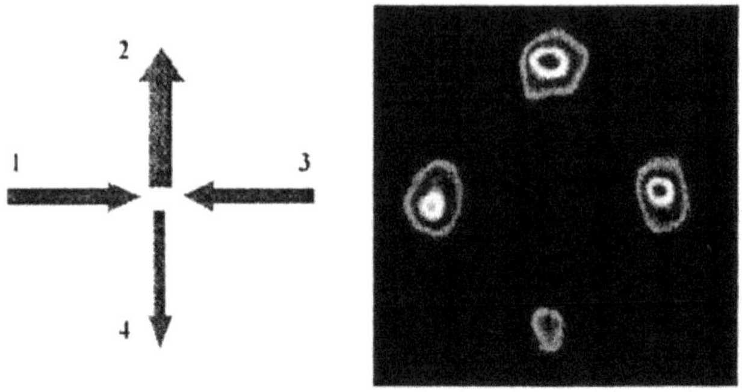

Figure 12. The process of four-wave mixing of matter-waves can always be transformed to a reference frame where the mixing process is degenerate (all of the waves have the same energy; left). The nonlinear term describing the mean-field, s-wave interaction of the atoms is responsible for the four-wave mixing. Atoms from waves 1 and 3 scatter off each other and go off back-to-back. The scattering process can be stimulated by wave 2, so that it is more likely that one of the scattering pairs goes into this wave. By momentum conservation, wave 4 is created. The small cloud of atoms in the image on the right is the fourth wave generated by four-wave mixing of matter-waves.

alternative picture of four-wave mixing is in terms of stimulated emission. In this picture it is helpful to view the four-wave mixing process in a reference frame where the process looks like degenerate four-wave mixing; that is, all of the waves have the same energy.

In four-wave mixing, both energy and momentum (corresponding to phase matching) must be conserved. Since atoms, unlike photons, cannot be created out of the vacuum we have the additional requirement, for matter-waves, of particle number conservation. (If we included the rest mass of the atom, particle number conservation is contained in energy conservation.) Given these three conditions, one can show that the only four-wave mixing configurations possible with matter-waves are those that can be viewed in some frame of reference as degenerate four-wave mixing. This is also the geometry of phase conjugation. Figure 12 shows the four-wave mixing geometry for matter-waves viewed in the degenerate or phase conjugation frame.

In the picture of four-wave mixing as arising from stimulated emission, atoms in waves 1 and 3 can be considered as undergoing an elastic collision. The scattering process results in atoms going off back-to-back in order to conserve momentum, but at some arbitrary angle with respect to the incident direction. (The scattering process is typically s-wave and the outgoing waves can be considered spherical.) In the presence of wave 2, however, this scattering process can be stimulated. There is an enhanced probability that one of the atoms from the collision of waves 1 and 3 will scatter into wave 2. (This probability is enhanced by the atoms in wave 2.) Because of momentum conservation, the enhanced scattering of atoms into wave 2 results in an enhanced number of atoms in wave 4. In this picture, it is obvious that the four-wave mixing process removes atoms from waves 1 and 3 and puts them into waves 2 and 4. This may have some interesting consequences in terms of quantum correlations between the waves.

7.2 Quantum phase engineering

A three-dimensional image of an arbitrarily complex object can be constructed by sending light, with sufficient spatial coherence, through the appropriate phase and/or amplitude mask. This is the basic principle behind physical optics, which includes wave phenomena like diffraction and holography. Diffraction can be achieved with a periodic phase and/or amplitude mask; while a more complicated mask is needed to construct a complex holographic image. In each case, the mask modifies the incoming wave and subsequent propagation produces the desired pattern of light. This idea can be readily adapted to atom optics, especially when the "incoming" matter-wave is from a highly coherent source such as a Bose-Einstein condensate.

We have developed a technique to optically imprint complex phase patterns onto a Bose-Einstein condensate in order to create interesting topological states. This technique is analogous to sending a wave through a thin phase mask. The basic idea is to expose the condensate atoms to a short pulse of laser light with a spatially varying intensity pattern. The laser detuning is chosen such that spontaneous emission is negligible. (The phase mask can also serve as an amplitude mask by tuning closer to resonance, so that spontaneous emission is significant.) The pulse duration is sufficiently short that the atoms do not move an appreciable distance (i.e. the wavelength of light) during the pulse. This is sometimes referred to as the Raman-Nath regime. During the laser pulse, the AC Stark effect shifts the energy of the atoms by $U(r,t)$. Hence the effect on the atomic wavefunction is to "instantaneously" change its phase. This effect can be represented by multiplying the wavefunction by the phase factor $\exp(i\phi(r))$, where $\phi(r) = -\int U(r,t)dt/\hbar$. Since the AC Stark or light shift is proportional to the intensity of the light, any spatial intensity variation in the light field will be written onto the BEC wavefunction as a spatial variation in its phase.

Optically induced phase imprinting is a tool for "quantum phase engineering" of the wavefunction to create a wide variety of states. For example, as discussed in the section on diffraction of the condensate, the application of a short pulse of standing wave light will imprint a sinusoidal phase onto the condensate. The imprinted wavefunction subsequently evolves in momentum states differing by $2\hbar k$. It should be possible to use quantum phase engineering to produce collective states of excitation of the interacting BEC, such as solitons and vortices. The application of a uniform intensity light field to half of the BEC imprints a relative phase difference between the two halves. This phase step is expected to give rise to dark solitons (see following section). Such solitons will propagate with a speed related to the phase difference [126], which can be adjusted by the intensity of the laser pulse.

It should also be possible to produce one or more vortices by applying a laser pulse which has a linearly-varying, azimuthal, intensity dependence [127]. This will produce a topological winding of the BEC phase, which if large enough (i.e. 2π) should produce a vortex. Numerical solutions to a 3-D Gross-Pitaevskii equation [128] show that this is the case; and also show that such a vortex, although unstable because it is created in a non-rotating trap, will live for a sufficient time to be observable. Increasing the phase winding will generate multiple vortices (vortices with more than $\hbar$ of angular momentum are not stable and will immediately split into multiple vortices each with angular momentum $\hbar$). Quantum phase engineering can generate arbitrary phase patterns, and perhaps other interesting quantum states. In this sense, it is a form of atom holography [129]. The technological challenge is mostly one of imaging. Any complicated pattern must be imaged to the size of the BEC, typically of order 50 μm.

7.3 Solitons in a BEC

Solitons are stable, localized waves that propagate in a nonlinear medium without spreading. They may be either bright or dark, depending on the details of the governing nonlinear wave equation. A bright soliton is a peak in the amplitude while a dark soliton is a notch with a characteristic phase step across it. Equation 19, which describes the weakly interacting, zero-temperature BEC also supports solitons. The solitons propagate without spreading (dispersing) because the nonlinearity balances the dispersion; for Eq.(19) the corresponding terms are the nonlinear interaction $g|\Psi|^2$, and the kinetic energy $-(\hbar^2/2M)\nabla^2$, respectively. Our sodium condensate only supports dark solitons because the atom-atom interactions are repulsive [126,130] ($g > 0$).

A distinguishing characteristic of a dark soliton is that its velocity is less than the Bogoliubov speed of sound [126,130] $v_0 = \sqrt{gn/M}$, where n is the unperturbed condensate density. The soliton speed v_s can be expressed either in terms of the phase step δ $(0 < \delta \leq \pi)$, or the soliton "depth" n_d, which is the difference between n and the density at the bottom of the notch [126,130]:

$$v_s/v_0 = \cos(\delta/2) = \sqrt{1 - n_d/n} \tag{20}$$

For $\delta = \pi$ the soliton has zero velocity, zero density at its center, a width on the order of the healing length [130], and a discontinuous phase step. As δ decreases the velocity increases, approaching the speed of sound. The solitons are shallower and wider, with a more gradual phase step. They travel opposite to the direction of the phase gradient. Because a soliton has a characteristic phase step, optically imprinting a phase step on the BEC wavefunction should be a way to create a soliton.

7.4 Observation of solitons in a BEC

We modified the phase distribution of the BEC by employing the technique of quantum phase engineering discussed in an earlier section. The condensate atoms were exposed to a pulsed, off-resonant laser beam, coaxial with the absorption probe beam, with a spatial intensity profile such that only half of the BEC was illuminated. This was accomplished by blocking half of the laser beam with a razor blade and imaging this razor blade onto the condensate. The intensity pattern at the condensate, as observed by our absorption imaging system, had a light to dark (90% to 0%) transition region of 7 μm. The intensity required to imprint a phase of π was checked with a Mach-Zehnder atom interferometer based on optically induced Bragg diffraction [131,132]. Our Bragg interferometer [133] differs from previous ones in that we can independently manipulate atoms in the two arms (because of their large separation) and can resolve the output ports to reveal the spatial distribution of the condensate phase. When a phase of π was imprinted on one half of the condensate relative to the other half, the two output ports of the interferometer displayed the complementary halves of the condensate.

To observe the creation and propagation of solitons, we do not use interferometry, but instead measure BEC density distributions with absorption imaging after imprinting a phase step. Figure 13 shows the evolution of the condensate after the top half was phase imprinted with $\phi = 1.5\pi$, a phase for which we observe a single deep soliton (the reason for imprinting a phase step larger than π is discussed below). Immediately after the phase imprint, there is a steep phase gradient across the middle of the condensate such that this portion has a large velocity in the $+x$ direction. This velocity can be

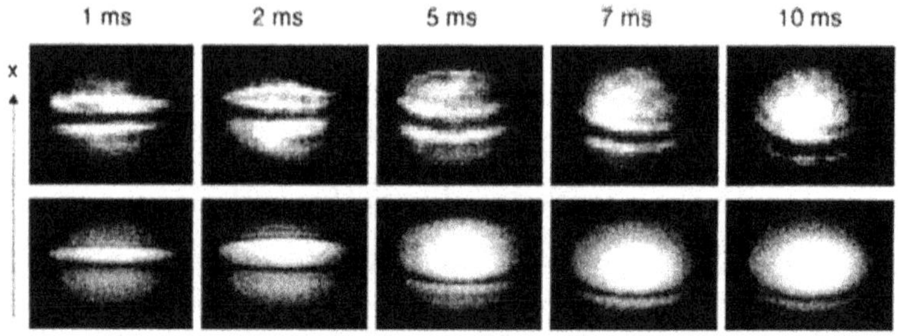

Figure 13. Experimental (upper) and theoretical (lower) images of the integrated BEC density for various times after we imprint a phase of about 1.5π on the top half of the condensate with a $1\mu s$ pulse. The measured number of atoms in the condensate was $1.7(3) \times 10^6$, and this value was used in the calculations. A positive density disturbance moved rapidly in the $+x$ direction and a dark soliton moved oppositely at significantly less than the speed of sound. Because the imaging is destructive, each image shows a different BEC. The width of the images is $70\mu m$.

understood as arising from the impulse imparted by the optical dipole force, and results in a positive density disturbance that travels at or above the speed of sound. A dark notch is left behind, which is a soliton moving slowly in the $-x$ direction (opposite to the direction of the applied force).

A striking feature of the images is the curvature of the soliton. This curvature is due to the 3-D geometry of the trapped condensate, and occurs for two reasons. First, the speed of sound v_0 is largest at the trap center where the density is greatest, and decreases towards the condensate edge. Second, as the soliton moves into regions of lower condensate density, we find numerically that the density at its center, $n - n_d$, approaches zero, δ approaches π, and v_s decreases to zero before reaching the edge. This is because the soliton depth n_d rather than its phase offset δ appears to be a conserved quantity in a nonuniform medium.

A clear indication that the notches seen in Fig.13 are solitons, rather than simply sound waves, is their subsonic propagation velocity. To determine this velocity, we measure the distance after propagation between the notch and the position of the imprinted phase step along the x direction. Because the position of our condensate varies randomly from shot-to-shot (presumably due to stray, time-varying fields) we cannot always apply the phase step at the center. A marker for the location of the initial phase step is the intersection of the soliton with the condensate edge, because at this point the soliton has zero velocity. Using images taken 5 ms after the imprint, at which time the soliton has not traveled far from the BEC center, we obtain a mean soliton velocity of 1.8(4) mm/s. This speed is significantly less than the mean Bogoliubov speed of sound v_0= 2.8(1) mm/s. From the propagation of the notch in the numerical solutions (Fig.13, lower images) we obtain a mean soliton velocity, $v_s = 1.6$ mm/s, in agreement with the experimental value. The experimental uncertainty is mainly due to the difficulty in determining the position of the initial phase step.

From the lower image of Fig.13 at 5 ms, we can extract the theoretical density and phase profile along the x-axis through the center of the condensate. The dark soliton notch and its phase step are centered at $x = -8$ mm. This phase step, $\delta = 0.58\pi$, is less than the imprinted phase of 1.5π. The difference is caused by the mismatch between the phase imprint and the phase and depth of the soliton solution of the nonlinear Schrdinger equation (Eq.(19)): Our imprinting resolution of 7 μm is larger than the soliton width, which is on the order of the healing length (0.7 μm), and we do not control the amplitude of the wave function.

In order to improve our measurement of the soliton velocity, we avoid the uncertainty in the position of the initial phase step by replacing the razor blade mask with a thin slit. This produces a stripe of light with a Gaussian profile ($1/e^2$ full width $\approx 15\mu$m). With this stripe in the center of the condensate, numerical simulations predict the generation of solitons that propagate symmetrically outwards. We select experimental images with solitons symmetrically located about the middle of the condensate, and measure the distance between them. For a small phase imprint of $\phi \approx 0.5\pi$ (at Gaussian maximum), we observe solitons moving at the Bogoliubov speed of sound, within experimental uncertainty. For a larger phase imprint of $\phi \approx 1.5\pi$, we observe much slower soliton propagation, in agreement with numerical simulations. An even larger phase imprint generates many solitons. The results of these experiments on the creation and propagation of solitons can be found in ref. 133. Solitons in a BEC have also been observed by a group in Hannover [134].

Acknowledgments

Much of the experimental work described here was carried out in the Laser Cooling and Trapping Group in the Atomic Physics Division of the National Institute of Standards and Technology. It is a pleasure to acknowledge the contributions to this work by the other members of the NIST staff of the Laser Cooling Group, Steve Rolston and Paul Lett, as well as former member Chris Westbrook. This work would not have been possible without the contributions of many postdocs, students, and visitors to the group.

While not attempting to be a review of all the work on laser cooling of neutral atoms, these notes have drawn extensively, both explicitly and implicitly, on the work of many groups throughout the world that have advanced the art and science of laser cooling. We are indebted to all of those colleagues, both cited and not, who have contributed so much to our understanding of laser cooling and to the pleasure of working in this field.

The BEC experiments were carried out by present and past members of the Laser Cooling Group, Steve Rolston, Ed Hagley, Lu Deng, Mikio Kozuma, Johannes Denschlag, Jesse Simsarian, Jesse Wen, Yuri Ovchinnikov, Joerg-Helge Muller, Julien Cubizolles, Robert Lutwak, Rob Thompson, Aephraim Steinberg, Mike Gatzke, and Gerhard Birkl. These experiments have also benefited greatly from direct interactions with our theoretical colleagues Paul Julienne, Carl Williams, Eite Tiesinga, Marek Trippenbach, Yehuda Band, Marya Doery, Charles Clark, David Feder, Mark Edwards and Keith Burnett.

We also gratefully acknowledge the financial support not only of NIST, but of the U. S. Office of Naval Research, NASA and NSF.

References

1. *Laser Manipulation of Atoms and Ions, Proc. S.I.F., Course CXVIII*, eds. E. Arimondo, W. D. Phillips. and F. Strumia (North Holland, Amsterdam, 1992).
2. *Fundamental Systems in Quantum Optics*, eds. J. Dalibard, J.-M. Raimond and J. Zinn-Justin (North Holland, Amsterdam, 1992).
3. *Feature issue on laser cooling and trapping of atoms*, eds. S. Chu and C. Wieman, J. Opt. Soc. Am. B **6**, 2020 (1989).
4. *Feature issue on mechanical effects of light*, eds. P. Meystre and S. Stenholm, J. Opt. Soc. Am. B **2**, 1706 (1985).
5. H. Metcalf and P. van der Straten, Phys. Reports **244**, 203 (1994).
6. A. P. Kazantsev, G. I. Surdutovich and V. P. Yakovlev in *Mechanical action of light on atoms* (World Scientific, Singapore, 1990).
7. V. G. Minogin and V. S. Letokhov in *Laser light pressure on atoms* (Gordon and Breach, New York, 1987).
8. S. Chu, Rev. Mod. Phys. **70**, 685 (1998).
9. C. Cohen-Tannoudji, Rev. Mod. Phys. **70**, 707 (1998).
10. W. D. Phillips, Rev. Mod. Phys. **70**, 721 (1998).
11. *Special issue on atom optics and interferometry*, Appl. Phys. B **54**, 321 (1992).

12. *Special issue on atom optics and interferometry*, J. de Physique **4**, 1877 (1994).
13. *Special issue on atom optics and interferometry*, Quant. Semiclass. Optics **8**, 495 (1996).
14. J. Dalilbard and C. Cohen-Tannoudji., J. Opt. Soc. Am. B **2**, 1707 (1985).
15. C. Cohen-Tannoudji, J. Dupont-Roc and G. Grynberg in *Atom-photon interactions: Basic Processes and Applications* (Wiley, New York, 1992).
16. J. P. Gordon and A. Ashkin, Phys. Rev. A **21**, 1606 (1980).
17. R. J. Cook, Phys. Rev. A **22**, 1078 (1980).
18. S. Stenholm, Rev. Mod. Phys. **58**, 699 (1986).
19. C. Cohen-Tannoudji in *Fundamental Systems in Quantum Optics*, eds. J. Dalibard, J.-M. Raimond and J. Zinn-Justin (North Holland, Amsterdam, 1992).
20. J. Dalibard Thése de doctorat d'état en Sciences Physique, Université de Paris VI (1986).
21. P. D. Lett, W. D. Phillips, S. L. Rolston, C. E. Tanner, R. N. Watts and C. I. Westbrook, J. Opt. Soc. Am. B **6**, 2084 (1989).
22. Y. Castin, H. Wallis and J. Dalibard, J. Opt. Soc. Am. B **6**. 2046 (1989).
23. A. Aspect, E. Arimondo, R. Kaiser, N. Vansteenkiste and C. Cohen-Tannoudji, J. Opt. Soc. Am. B **6**, 2112 (1989).
24. A. Aspect, E. Arimondo, R. Kaiser, N. Vansteenkiste and C. Cohen-Tannoudji, Phys. Rev. Lett. **61**, 826 (1988).
25. D. E. Pritchard, K. Helmerson, V. S. Bagnato, G. P. Lafyatis and A. G. Martin in *Laser Spectroscopy VIII* eds. W. Persson and S. Svanberg (Springer-Verlag, Berlin, 1987).
26. H. Wallis and W. Ertmer, J. Opt. Soc. Am B **6**, 2211 (1989).
27. M. Kasevich and S. Chu, Phys. Rev. Lett. **69**, 1741 (1992).
28. W. D. Phillips, J. Prodan and H. Metcalf, J. Opt. Soc. Am. B **2**, 1751 (1985).
29. S. Chu, L. Hollberg, J. Bjorkholm, A. Cable and A. Ashkin, Phys. Rev. Lett. **55**, 48 (1985).
30. T. W. Hodapp, C. Gerz, C. Furtlelehner, C. I. Westbrook and W. D. Phillips, Appl. Phys. B **60**, 135 (1995).
31. W. Phillips and H. Metcalf, Phys. Rev. Lett. **48**, 596 (1982).
32. A. Cable, M. Prentiss and N. P. Bigelow, Opt. Lett. **15**, 507 (1990).
33. C. Monroe, W. Swann, H. Robinson and C. Wieman, Phys. Rev. Lett. **65**, 1571 (1990).
34. S. Andreev, V. Balykin, V. Letokhov and V. Minogin, Pis'ma Zh. Eksp.

Teor. Fiz. **34**, 463 (1981). [JETP Lett. **34**, 442 (1981)].
35. J. Hoffnagle, Opt. Lett. **13**, 102 (1988).
36. W, Ketterle, A. Martin, M. A. Joffe and D. E. Pritchard, Phys. Rev. Lett. **69**, 2483 (1992).
37. R. Gaggl, L. Windholz, C. Umfer and C. Neureiter, Phys. Rev. A **49**, 1119 (1994).
38. M. Prentiss and A. Cable, Phys. Rev. Lett. **62**, 1354 (1989).
39. V. S. Letokhov, V. G. Minogin and B. D. Pavlik, Opt. Commun **19**, 72 (1976).
40. W. D. Phillips and J. V. Prodan in *Laser-Cooled and Trapped Atoms*, ed. W. D. Phillips, Natl. Bur. Stand. (U.S.) Spec. Publ. **653**, 137 (1983); Prog. Quantum Electron. **8**, 231 (1984); in *Coherence and Quantum Optics V*, ed. L. Mandel and E. Wolf (Plenum, New York, 1984); W. Phillips, J. Prodan and H. Metcalf in *Laser Spectroscopy Laser VI*, eds. H. Weber and W. Luthy (Springer-Verlag, Berlin, 1983).
41. W. Ertmer, R. Blatt, J. Hall and M. Zhu, Phys. Rev. Lett. **54**, 996 (1985).
42. C. Salomon and J. Dalibard, C. R. Acad. Sci. Paris **306**, 1319 (1988).
43. J. Prodan, W. Phillips and H. Metcalf, Phys. Rev. Lett. **49**, 1149 (1982).
44. J. Prodan, A. Migdall, W. D. Phillips, I. So, H. Metcalf and J. Dalibard, Phys. Rev. Lett. **54**, 992 (1985).
45. A. Migdall, J. Prodan, W. Phillips, T. Bergeman and H. Metcalf, Phys. Rev. Lett. **54**, 2596 (1985).
46. S. Chu, J. Bjorkholm, A. Ashkin and A. Cable, Phys. Rev. Lett. **57**, 314 (1986).
47. E. Raab, M. Prentiss, A. Cable, S. Chu and D. Pritchard, Phys. Rev. Lett. **59**, 2631 (1987).
48. E. Cornell, C. Monroe and C. Wieman, Phys. Rev. Lett. **67**, 3049 (1991).
49. R. J. C. Spreeuw, C. Gerz, L. Goldner, W. D. Phillips, S. L. Rolston, C. Westbrook, M. W. Reynolds and I. F. Silvera, Phys. Rev. Lett. **72**, 3162 (1994).
50. W. Wing, Phys. Rev. Lett. **45**, 631 (1980).
51. C. G. Aminoff, A. M. Steane, P. Bouyer, P. Desbiolles, J. Dalibard and C. Cohen-Tannoudji, Phys. Rev. Lett. **71**, 3083 (1993).
52. W. Petrich, M. H. Anderson, J. R. Ensher and E. A. Cornell, Phys. Rev. Lett. **74**, 3352 (1995).
53. A. Ashkin, Phys. Rev. Lett. **40**, 729 (1978).
54. A. Ashkin and J. Gordon, Opt. Lett. **4**, 161 (1979).
55. J. Dalibard, S. Reynaud and C. Cohen-Tannoudji, Opt. Comm. **47**, 395 (1983).

56. J. Dalibard, S. Reynaud and C. Cohen-Tannoudji, J. Phys. B **17**, 4577 (1984).
57. P. L. Gould, P. D. Lett, W. D. Phillips, P. S. Julienne, H. R. Thorsheim and J. Weiner in *Advances in Laser Science III*, eds. A. Tam, J. Gole and W. Stwalley (American Institute of Physics, New York, 1987).
58. P. L. Gould, P. D. Lett, P. S. Julienne, W. D. Phillips, H. R. Thorsheim and J. Weiner, Phys. Rev. Lett. **60**, 788 (1988).
59. K. Helmerson in *Interdisciplinary Laser Conference* (Unpublished, Monterey, CA) 1991.
60. S. L. Rolston, C. Gerz, K. Helmerson, P. S. Jessen, P. D. Lett, W. D. Phillips, R. J. Spreeuw and C. I. Westbrook in *1992 Shanghai International Symposium on Quantum Optics*, eds. Yuzhu Wang, Yigui Wang and Zugeng Wang (Proc. SPIE, Shanghai) 1992, Vol. 1726, p. 205.
61. S. Chu, J. E. Bjorkholm, A. Ashkin, J. P. Gordon and L. W. Hollberg, Opt. Lett. **11**, 73 (1986).
62. J. D. Miller, R. A. Cline and D. J. Heinzen, Phys. Rev. A **47**, R4567 (1993).
63. J. D. Miller, R. A. Cline and D. J. Heinzen, Phys. Rev. Lett. **71**, 2204 (1993).
64. C. S. Adams, H. J. Lee, N. Davidson, M. Kasevich and S. Chu, Phys. Rev. Lett. **74**, 3577 (1995).
65. D. M. Stamper-Kurn, M. R. Andrews, A. P. Chikkatur, S. Inouye, H.-J. Miesner, J. Stenger and W. Ketterle, Phys. Rev. Lett. **80**, 2027 (1998).
66. R. Cook and R. Hill, Opt. Comm. **43**, 258 (1982).
67. N. Davidson, H. J. Lee, C. S. Adams, M. Kasevich and S. Chu, Phys. Rev. Lett. **74**, 1311 (1995).
68. H.-J. Lee, C. S. Adams, N. Davidson, B. Young, M. Weitz, M. Kasevich and S. Chu in Atomic Physics 14, eds. D. Wineland, C. Wieman and S. Smith (AIP Press, N.Y., 1995).
69. W. D. Phillips in *Laser Manipulation of Atoms and Ions, Proc. S.I.F., Course CXVIII*, eds. E. Arimondo, W. D. Phillips and F. Strumia (North Holland, Amsterdam, 1992).
70. A. Ashkin and J. Gordon, Opt. Lett. **8**, 511 (1983).
71. J. Dalibard, personal communication, (1986).
72. T. Walker, Laser Physics **4**,, 965 (1994).
73. J. Dalibard and C. Cohen-Tannoudji, J. Opt. Soc. Am. B **6**, 2023 (1989).
74. M. Walhout, J. Dalibard, S. L. Rolston and W. D. Phillips, J. Opt. Soc. Am. B **9**, 1997 (1992).
75. A. Steane and C. Foot, Europhys. Lett. **14**, 231 (1991).

76. J. Werner, H. Wallis and W. Ertmer, Opt. Comm. **94**, 525 (1992).
77. M. Prentiss, A. Cable, J. E. Bjorkholm, S. Chu, E. L. Raab and D. E. Pritchard, Opt. Lett. **13**, 452 (1988).
78. D. Sesko, T. Walker, C. Monroe, A. Gallagher and C. Wieman, Phys. Rev. Lett. **63**, 961 (1989).
79. T. Walker, D. Sesko and C. Wieman, Phys. Rev. Lett. **64**, 408 (1990).
80. W. Ketterle, K. B. Davis, M. A. Joffe, A. Martin and D. E. Pritchard, Phys. Rev. Lett. **70**, 2253 (1993).
81. N. F. Ramsey in *Molecular Beams*, eds. R. J. Elliott, J. A. Krumhansl, W. Marshall and D. H. Wilkinson International Series of Monographs on Physics (Oxford University Press, Oxford, 1956).
82. W. Paul, personal communication, (1985).
83. C. V. Heer, Rev. Sci. Instrum. **34**, 532 (1963).
84. C. V. Heer in Quantum Electronics, ed. C. H. Townes (Columbia University Press, New York, 1960).
85. V. V. Vladimirski, Zh. Eksp. Teor. Fiz. **39**, 1062 (1960).
86. W. Wing, Prog. Quant. Electr. **8**, 181 (1984).
87. W. Ketterle and D. E. Pritchard, Appl. Phys. B **54**, 403 (1992).
88. T. Bergeman, G. Erez and H. J. Metcalf, Phys. Rev. A **35**, 1535 (1987).
89. K. B. Davis, M.-O. Mewes, M. A. Joffe, M. R. Andrews and W. Ketterle, Phys. Rev. Lett. **74**, 5202 (1995).
90. W. Petrich, M. H. Anderson, J. R. Ensher and E. Cornell in *Abstracts of Contributed Papers, ICAP XIV*, 1994, p.1M-7; K. B. Davis, M.-O. Mewes, M. A. Joffe and W. Ketterle, *ibid.*, p.1M-3.
91. D. Sesko, C. G. Fan and C. E. Wieman, J. Opt. Soc. Am B **5**, 1225 (1988).
92. W. D. Phillips, P. L. Gould and P. D. Lett in *Laser Spectroscopy VIII*, eds. W. Persson and S. Svanberg (Springer, Berlin, 1987).
93. P. D. Lett, R. N. Watts, C. I. Westbrook, W. D. Phillips, P. L. Gould and H. J. Metcalf, Phys. Rev. Lett. **61**, 169 (1988).
94. W. D. Phillips, C. I. Westbrook, P. D. Lett, R. N. Watts, P. L. Gould and H. J. Metcalf in *Atomic Physics 11*, eds. S. Haroche, J. C. Gay and G. Grynberg (World Scientific, Singapore, 1989).
95. J. Dalibard, C. Salomon, A. Aspect, E. Arimondo, R. Kaiser, N. Vansteenkiste and C. Cohen-Tannoudji in *Atomic Physics 11*, eds. S. Haroche, J. C. Gay and G. Grynberg (World Scientific, Singapore, 1989).
96. P. J. Ungar, D. S. Weiss, E. Riis and S. Chu, J. Opt. Soc. Am. B **6**, 2058 (1989).
97. B. Sheehy, S.-Q. Shang, R. Watts, S. Hatamian and H. Metcalf, J. Opt. Soc. Am. B **6**, 2165 (1989).

98. D. S. Weiss, E. Riis, Y. Shevy, P. J. Ungar and S. Chu, J. Opt. Soc. Am. B **6**, 2072 (1989).
99. B. Sheehy, S.-Q. Shang, P. van der Straten, S. Hatamian and H. Metcalf, Phys. Rev. Lett. **64**, 858 (1990).
100. G. Nienhuis in *Laser Manipulation of Atoms and Ions, Proc. S.I.F., Course CXVIII*, eds. E. Arimondo, W. D. Phillips and F. Strumia (North Holland, Amsterdam, 1992).
101. O. Emile, R. Kaiser, C. Gerz, H. Wallis, A. Aspect and C. Cohen-Tannoudji, J. Phys. II (Paris) **3**, 1709 (1993).
102. A. Aspect, O. Emile, C. Gerz, R. Kaiser, N. Vansteenkiste, H. Wallis and C. Cohen-Tannoudji in *Laser Manipulation of Atoms and Ions, Proc. S.I.F., Course CXVIII*, eds. E. Arimondo, W. D. Phillips and F. Strumia (North Holland, Amsterdam, 1992).
103. C. Cohen-Tannoudji and W. D. Phillips, Physics Today **43**, 33 (1990).
104. C. Cohen-Tannoudji in *Laser Manipulation of Atoms and Ions, Proc. S.I.F., Course CXVIII*, eds. E. Arimondo, W. D. Phillips and F. Strumia (North Holland, Amsterdam, 1992).
105. Y. Castin, J. Dalibard and C. Cohen-Tannoudji in *Light Induced Kinetic Effects on Atoms, Ions and Molecules*, eds. L. Moi, S. Gozzini, C. Gabbanini, E. Arimondo and F. Strumia (ETS Editrice, Pisa, 1991).
106. Y. Castin, Doctoral Dissertation, Ecole Normale Supérieure (1992).
107. C. Salomon, J. Dalibard, W. D. Phillips, A. Clairon and S. Guellati, Europhys. Lett. **12**, 683 (1990).
108. C. Gerz, T. W. Hodapp, P. S. Jessen, K. M. Jones, W. D. Phillips, C. I. Westbrook and K. Molmer, Europhys. Lett. **21**, 661 (1993).
109. M. H. Anderson, J. R. Ensher, M. R. Matthews, C. E. Wieman and E. A. Cornell, Science **269**, 198 (1995).
110. K. B. Davis, M.-O. Mewes, M. R. Andrews, N. J. van Druten, D. S. Durfee, D. M. Kurn and W. Ketterle, Phys. Rev. Lett. **75**, 3969 (1995).
111. C. A. Sackett, H. T. C. Stoof and R. G. Hulet, Phys. Rev. Lett. **80**. 2031 (1998); C. C. Bradley, C. A. Sackett, J. J. Tollett and R. G. Hulet, Phys. Rev. Lett. **75**, 1687 (1995).
112. F. Dalfovo, S. Giorgini, L. P. Pitaevskii and S. Stringari, Rev. Mod. Phys. **71**, 463 (1999).
113. A. Witte, T. Kisters, F. Riehle and J. Helmcke, J. Opt. Soc. Am. B **9**, 1030 (1992).
114. T. Barrett, S. Dapore-Schwartz, M. Ray and G. Lafyatis, Phys. Rev. Lett. **67**, 3483 (1991).
115. S. G. Miranda, S. R. Muniz, G. D. Telles, L. G. Marcassa, K. Helmerson and V. S. Bagnato, Phys. Rev. A **59**, 882 (1999).

116. H. F. Hess, Phys. Rev. B **34**, 3476 (1986).
117. N. Masuhara, J. M. Doyle, J. C. Sandberg, D. Kleppner, T. J. Greytak, H. F. Hess and G. P. Kochanski, Phys. Rev. Lett. **61**, 935 (1988).
118. P. E. Moskowitz, P. L. Gould, S. R. Atlas and D. E. Pritchard, Phys. Rev. Lett **51**, 370 (1983); P. L. Gould, G. E. Ruff and D. E. Pritchard, Phys. Rev. Lett. **56**, 827 (1986).
119. P. J. Martin, B. G. Oldaker, A. H. Miklich and D. E. Pritchard, Phys. Rev. Lett. **60**, 515 (1988).
120. M. Kozuma, L. Deng, E. W. Hagley, J. Wen, R. Lutwak, K. Helmerson, S. L. Rolston and W. D. Phillips, Phys. Rev. Lett. **82**, 871 (1999).
121. M.-O. Mewes, M. R. Andrews, D. M. Kurn, D. S. Durfee, C. G. Townsend and W. Ketterle, Phys. Rev. Lett. **78**, 582 (1997).
122. E. W. Hagley, L. Deng, M. Kozuma, J. Wen, M. A. Edwards, K. Helmerson, S. L. Rolston and W. D. Phillips, Science **283**, 1706 (1999).
123. E. W. Hagley, L. Deng, M. Kozuma, M. Trippenbach, Y. B. Band, M. A. Edwards, M. Doery, P. S. Julienne, K. Helmerson, S. L. Rolston and W. D. Phillips, Phys. Rev. Lett. **83**, 3112 (1999).
124. G. Lens, P. Meystre, and E. W. Wright, Phys. Rev. Lett. **71**, 3271 (1993).
125. L. Deng, E. W. Hagley, J. Wen, M. Trippenbach, Y. B. Band, P. S. Julienne, J. E. Simsarian, K. Helmerson, S. L. Rolston and W. D. Phillips, Nature **398**, 218 (1999).
126. W. P. Reinhardt and C. W. Clark, J. Phys. B: At. Mol. Opt. Phys. **30**, L785 (1997).
127. L. Dobrek, M. Gajda, M. Lewenstein, K. Sengstock, G. Birkl and W. Ertmer, Phys. Rev. A **60**, R3381 (1999).
128. D. L. Feder, C. W. Clark and B. I. Schneider, Phys. Rev. Lett **82**, 4956 (1996).
129. J. Fujita, M. Morinaga, T. Kishimoto, M. Yasuda, S. Matsui and F. Shimizu, Nature **380**, 691 (1996).
130. A. D. Jackson, G. M. Kavoulakis and C. J. Pethick, Phys. Rev. A **58**, 2417 (1998).
131. Y. Torii, Y. Suzuki, M. Kozuma, T. Kuga, L. Deng and E. W. Hagley, Phys. Rev. A **61**, 041602 (2000).
132. D .M. Giltner, R. W. McGowan and S. A. Lee, Phys. Rev. Lett. **75**, 2638 (1995).
133. J. Denschlag, J. E. Simsarian, D. L. Feder, C .W. Clark, L. A. Collins, J. Cubizolles, L. Deng, E. W. Hagley, K. Helmerson, W. P. Reinhardt, S. L. Rolston, B. I. Schneider and W. D. Phillips, Science **287**, 97 (2000).
134. S. Burger, K. Bongs, S. Dettmer, W. Ertmer, K. Sengstock, A. Sanpera, G. V. Shlyapnikov and M. Lewenstein, Phys. Rev. Lett. **83**, 5198 (1999).

EXPERIMENTAL ASPECTS OF BOSE-EINSTEIN CONDENSATION

ANDREW C. WILSON AND CALLUM R. MCKENZIE
Physics Department, University of Otago, PO Box 56, Dunedin, New Zealand
Email: awilson@physics.otago.ac.nz, Web: http://www.physics.otago.ac.nz/research/bec

An introductory level review of experimental techniques essential for producing and probing Bose condensates formed with dilute alkali vapours is presented. This discussion includes a summary of evaporative cooling techniques, condensate imaging schemes, and a review of current BEC technology.

1 Introduction

The rapid development of experimental research into Bose-Einstein condensation (BEC) is the result of both many years of refinement of laser cooling, magnetic trapping and evaporative cooling by a large number of researchers, and success by JILA and MIT BEC groups in linking these techniques together to achieve the high phase space density required for Bose condensation. The discovery of BEC with dilute alkali vapours in 1995 provided experimentalists with the basic ingredients of a recipe for producing Bose condensates [1,2]. Since then, considerable technical effort by a large number of groups has given rise to variations on the original schemes which in turn have lead to more reliable production of larger condensates, and an array of techniques for manipulating and studying them.

The following discussion leads on from the review of laser cooling and magnetic trapping by Dr William D. Phillips (in this volume), and is meant to provide an introductory review of experimental aspects of evaporative cooling and probing of Bose condensates. To complete the discussion on experimental aspects, a summary of current BEC technology is provided in the final section.

2 Evaporative Cooling in Magnetic Traps

2.1 Introduction to Evaporative Cooling

To date, all dilute atomic Bose condensates have been produced by evaporative cooling of (mostly laser cooled) atoms held in a magnetic trap. Evaporation from a magnetic trap overcomes the density and temperature limits of laser cooling. The important concepts of evaporative cooling of trapped atoms were formulated by Hess in 1986, as a method for cooling atomic hydrogen pre-cooled by cryogenic methods [3]. Evaporative cooling is performed by removing the high-energy component of the thermal distribution of atoms from the magnetic trap. The evaporated atoms carry away more than the average energy, so that when the remaining atoms re-thermalise (by undergoing elastic collisions) the temperature of the vapour decreases. To continue cooling one must continue to remove high-energy atoms. The essential condition for efficient evaporative cooling is a long sample lifetime compared to the collisional re-thermalisation time.

Figure 2.1 Evaporative cooling basics - releasing high-energy atoms from a magnetic trap and subsequent re-thermalisation. We continue cooling by lowering further the rim of the trap.

For a dilute gas confined in a magnetic potential, decreasing the temperature decreases the sample volume. Even though the total number of atoms in the sample decreases during the evaporation process, it is possible to increase the density of the remaining atoms. When this happens, the decreasing temperature leads to an increasing rate of re-thermalisation, despite the loss of atoms. This is called "runaway" evaporation, because the cooling accelerates during the evaporation.

Evaporative cooling is simple to implement and relatively efficient, typically providing six orders of magnitude increase in phase space density at a cost of a factor of 10^3 reduction in the number of atoms. Outside of the laboratory, evaporative cooling is a familiar process - it's how we cool coffee in a cup and why we feel cold after a swim. An excellent review of evaporative cooling of trapped atoms, which contains references to most of the significant models, is provided by Ketterle and van Druten [4]. The discussion here is limited to experimental issues.

2.2 *Radio-Frequency Evaporation*

Pritchard suggested that evaporation could be performed conveniently if the rim of the trap was defined by an RF-resonance condition, rather than by the geometry of the magnetic field [5]. The idea is to use an RF transition (between m_F sub-states) to remove atoms from the magnetic trap (i.e. form a leak) at a distance from the trap centre defined by the Zeeman shift of the transition associated with the trapping field and the frequency of the applied RF field.

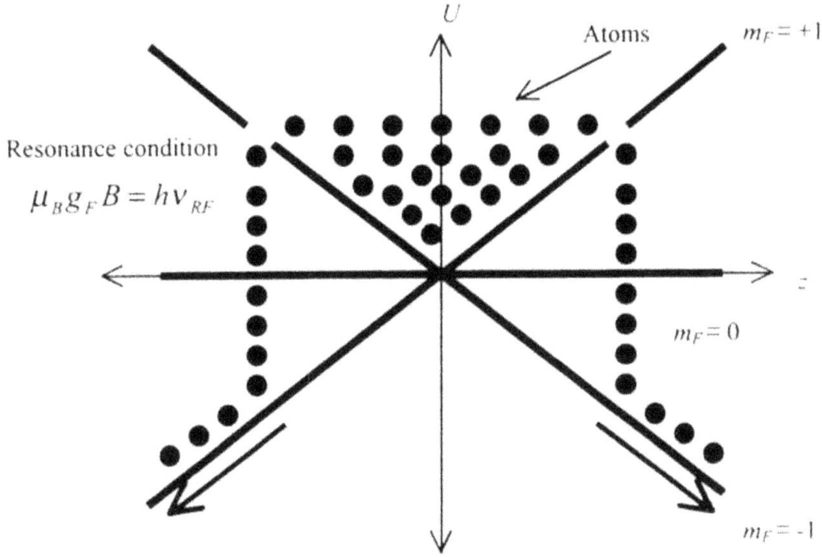

Figure 2.2 Zeeman shift of magnetic sub-levels versus position for a simple quadrupole trap. A leak is formed at the top of the trap using an RF field, allowing high-energy atoms to escape.

In RF-induced evaporation, the RF radiation flips the atomic spin. As a result, the attractive force provided by the magnetic field becomes repulsive and expels the atoms from the trap. This scheme is energy-selective because the resonance frequency is proportional to the magnetic field strength, and therefore to the potential energy of the atoms. In the case of transitions between magnetic sub-levels (i.e. $\Delta m_F = 1$), the resonance condition for the magnetic field strength B is, $\mu_B g_F B = h\nu_{RF}$ where g_F is the Land g-factor and μ_B is the Bohr magneton. For example, the spacing between m_F sub-states is 0.7 MHz per gauss for transitions within the $F = 2$ hyperfine level of ^{87}Rb frequently used to make Bose condensates. RF induced evaporation has the very nice feature that the evaporation process can be separated from the design of the magnetic trapping potential. In particular, there is no need to weaken the trapping potential in order to lower its depth. This makes it easier to reach runaway evaporation.

Ideally, RF evaporation spin-flips atoms on a 3-dimensional resonant shell (of constant magnitude magnetic field) surrounding the trapped atoms. For a static trap, such as an Ioffe-Pritchard style, the shape of the evaporation shell is an ellipsoid centred on the atomic sample (if the influence of gravity can be ignored). However, for weak confinement gravity will displace the atomic sample from the centre of the trap, potentially resulting in (less-efficient) 1D evaporation. For a TOP trap the evaporation geometry is more complex (see below).

The MIT and JILA BEC groups were the first to combine laser cooling with evaporative cooling, and thereby extend the applicability of evaporative cooling to alkali atoms. Increases in phase-space density were first reported in May 1994 [6,7]. Since then numerous groups have implemented RF evaporative cooling of magnetically trapped atomic vapours (see section 3 below). Information on groups who have implemented evaporative cooling and observed Bose condensation is available at the Georgia State University BEC Home Page: http://amo.phy.gasou.edu/bec.html.

In the dressed-atom formalism [8] for RF evaporation one considers the total energy of the atom plus that of the field of N RF photons, and coupling produces a Zeeman pattern with avoided crossing at positions where the RF field is resonant. A slowly moving atom stays on the adiabatic (dressed state) potential curve. When the atom comes close to resonance, the RF field mixes this state with the other m_F states, flattening the potential. Beyond the resonant position, the potential has been adiabatically transformed into an untrapped state. This adiabatic picture is valid if the energy gap due to the avoided crossing is greater than the energy uncertainty associated with the finite time the atom spends in the resonance region. For a two-level system, Landau-Zener theory is used to calculate the transition probability between potential curves [9].

2.3 *Evaporation in a TOP Trap*

Because of the rotating magnetic field, evaporation in a TOP trap is slightly more complicated than in a static trap such as a Ioffe-Pritchard trap. The spatial distribution of an atomic sample confined in a TOP trap is determined by the time-

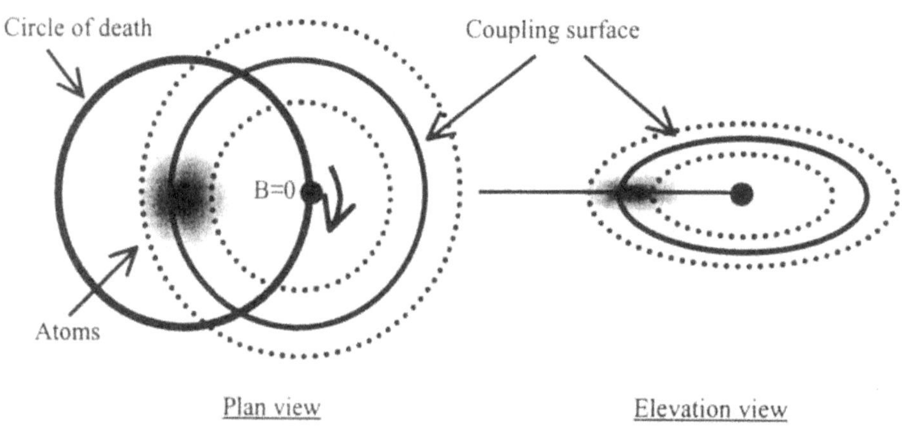

Figure 2.4 Evaporative cooling in a TOP trap.

average of the magnetic potential (which is harmonic), but to understand evaporation in a TOP trap we must consider the instantaneous magnetic field.

The region over which evaporation occurs is an ellipsoid shell with aspect ratio of 2:1 (from the quadrupolar potential), displaced from the centre of the cloud by the rotating bias field. Atoms are removed from the edge of the trap as the evaporation surface rotates about the cloud. The evaporation is essentially 2-dimensional.

With the TOP trap, in addition to RF-induced transitions, one can also induce Majorana spin-flips [10] using the zero magnetic field point of the rotating quadrupole trap. Near the zero magnetic field point of the quadrupole trap the Larmor frequency becomes sufficiently small that the magnetic moment of the atoms in this region can no-longer remain aligned with the local magnetic field direction. In a quadrupole trap ejection occurs at the trap centre. However, as shown in Fig. 2.4, in a TOP trap the zero point orbits the edge of the atomic sample (on a path which Cornell *et al* named the "circle of death") and the resulting spin-flips can be used for evaporative cooling [11]. Because the orbit is fast compared to the motion of the confined atoms the evaporation by Majorana spin flips is essentially 2-dimensional rather than point-like.

The Otago evaporative cooling sequence is a two-step process - a period of evaporation using Majorana spin-flips followed by a period of evaporation using RF-induced transitions [12]. For large thermal clouds, we find that Majorana-based evaporation has two nice features. Firstly, it is easy to implement, since reducing the radius of the circle of death simply involves reducing the strength of the rotating bias field. Secondly, as the bias field is reduced, the trap compression increases, thereby increasing the density (and therefore the elastic collision rate). However, closer to the BEC transition temperature, radio-frequency evaporation appears to give more precise control. The radio-frequency source we use is a GPIB-controlled function synthesiser, and the RF field is applied using an antenna comprised of a single loop (30 mm diam.), oriented to give a magnetic field perpendicular to the plane of the rotating bias field, and driven by a 3 W amplifier. Impedance matching the antenna to maximise the RF power delivered to the sample is difficult to achieve, since evaporative cooling involves ramping the RF frequency over a considerable range (~ 30 MHz to 1 MHz in our case). To set the RF power to an initial sensible value we tune the RF frequency to the atoms at the centre of the trap (i.e. to a value $\mu_B g_F B/h$, where B is the magnitude of the rotating bias field), and increase the RF power until we significantly depopulate the trap. To optimise the RF power for evaporation, we then adjust to give maximum phase space density increase from an evaporative cooling ramp. If the RF power is too high saturation of the transition leads to spatial broadening and poor performance results. If the RF power is set too low, then the evaporation time becomes too long and trap loss from background gas reduces performance (see below). It is worth noting that before we begin the evaporative cooling process proper, we adiabatically compress the cloud by increasing the quadrupole field gradient (to increase the elastic collision rate), and it is likely that some Majorana-based evaporation occurs during this step.

2.4 *Optimisation*

Models of evaporative cooling show that cooling happens on an exponential scale: within a given time interval (naturally measured in units of collision times), number, temperature, and density change by certain factors [4]. In a loss-less trap, in which atoms are removed only by evaporation, the cooling efficiency can be made arbitrarily large by truncating further out on the tail of the thermal distribution. However, in real traps there are loss mechanisms and if the truncation energy is made too large the slow increase in phase space density due to evaporation will be overwhelmed by the decrease due to trap losses. Forced evaporative cooling therefore depends on both the elastic collision rate γ_e, which determines the rate of re-thermalisation, and the trap loss rate γ_{bg}[1]. In general, the larger the ratio of $\gamma_e : \gamma_{bg}$, the more efficient the evaporation. These two rates characterise the truncation process which gives the maximum increase in phase space density for the smallest loss in number.

In practise, we optimise our evaporative cooling process by making moderate steps (removing perhaps a factor of 10 in atom number) and adjusting the time constant for exponential decrease in RF frequency (or radius of circle of death) to find the value which gives the best increase in optical depth (which is proportional to the elastic collision rate). Having optimised each step in the ramp, we obtain a global optimisation [13]. When optimised it is usual for the elastic collision rate to increase during the evaporation — that is for runaway evaporation to occur - so that the re-thermalisation rate increases during the ramp, which in turn means that that the cooling process can proceed more rapidly. If the collision rate increases during the cooling process, the optimum path for decreasing the RF frequency may differ from an exponential. This depends on the initial sample size. A number of groups use a linear ramp.

[1] As discussed in section 2.5, the atomic sample lifetime is limited primarily by collisions with background gas atoms.

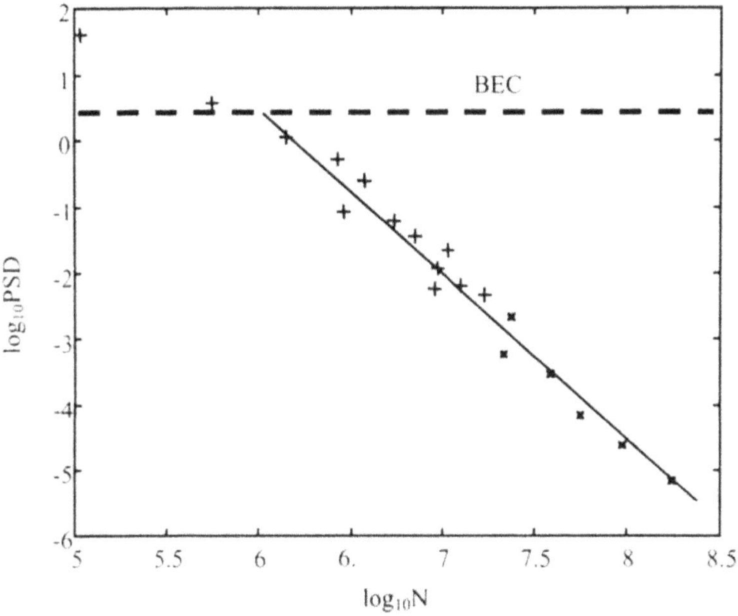

Figure 2.5 Phase space density versus atom number soon after BEC was first observed at Otago. x indicates Majorana evaporation with the circle of death, and + indicates RF evaporation.

2.5 *Collisions: the Good, the Bad, and the Ugly*

Evaporative cooling requires a favourable ratio of (good) elastic collisions which enable re-thermalisation, to (bad) inelastic collisions which lead to trap loss and heating. At the low energies typical of magnetically trapped atoms, the elastic collision rate γ_e is given by $n\sigma v$, where n and v are the density and mean relative velocity of the atoms, respectively, and σ is the s-wave collision cross-section. As evaporation proceeds, the density increases, but the velocity decreases.

Trapped atom clouds are extremely dilute (about ten orders of magnitude less dense than a solid or liquid) and collisional re-thermalisation can take seconds. Typically, the elastic collision rate at the start of evaporative cooling is greater than $1\ s^{-1}$, and to achieve BEC usually requires about 100 - 200 elastic collisions within the trap lifetime. Monte-Carlo simulations indicate that the re-thermalisation time is approximately 3 times the mean period between elastic collisions [14].

In the case of alkali atoms, the dominant inelastic collision process (over a large range of temperatures and densities) is background gas collisions [15]. The loss rate from this mechanism is proportional to the background gas pressure (but essentially independent of the condensate density and temperature). When a residual gas atom collides with an atom in the cold sample, the most likely result is clean ejection with both atoms leaving the trap without further collisions. To overcome background loss, experimentalists go to considerable efforts to optimise the pre-cooling and trapping techniques within an extreme ultra-high vacuum (UHV) environment. The standard technique is to use a combination of an ion pump with a titanium sublimation pump, and to implement differential pumping to isolate the magnetically trapped atoms from the high background gas pressure near the atom reservoir used for pre-cooling. There are two common methods used in BEC labs for achieving a large laser cooled sample in an extreme UHV environment: an atomic beam and Zeeman slower [W.D. Phillips, this volume], or a double MOT and push beam setup [16].

The Otago setup is a double MOT. A collection MOT collects 10^8 atoms in 0.5s from a rubidium sample at room temperature. These atoms are then pushed to a storage MOT formed in a small (16 x 16 x 70 mm) quartz cell, via a narrow (10 mm diameter) tube enabling differential pumping. With multiple transfers from the collection MOT to the storage MOT, we obtain 5×10^9 atoms in the storage MOT in a vacuum with a background pressure in the low 10^{-11} mbar range. At this pressure we obtain a magnetic trap lifetime of approximately 100s. At best we can load about half of the atoms from the storage MOT into a TOP trap.

In addition to loss from collisions with background gas atoms, such collisions are also a source of heat. Experimental efforts to determine the underlying mechanisms and characterise the observed heating effects have not been entirely conclusive [17]. One idea is that glancing collisions may lead to heating but not trap loss. Typically, the trap depth is much greater that the mean energy of the condensate (in terms of temperature: 5 mK compared with 500 nK), so that a glancing collision may cause ejection from the condensate, but not from the trap. JILA have made considerable attempts to analyse this problem [15]. For large dense clouds they considered the possibility of multiple collisions before ejection. Calculations based on this idea show promise, but agreement with experiment is not complete. JILA have proposed that at the end of evaporation, the outer region of the trap is filled with a very dilute, high-energy halo of trapped atoms, which they call the Oort cloud (in analogy with the comets). The large size and low density of the Oort cloud make it difficult to detect with imaging systems designed to look at small dense condensates. However, the JILA BEC group claim direct evidence of the existence of the Oort cloud comes from the reduced heating observed when an RF shield is applied. This is an RF field applied to provide an evaporation shell well beyond the atomic sample, protecting the atomic sample from collisions with high-energy (large orbit) atoms.

In addition to background gas collisions, there is also loss from dipolar relaxation (a 2-body process) and 3-body recombination [18,19]. These inelastic collision processes cause anti-evaporation since they give rise to loss at the most dense (and lowest energy) part of the atomic sample. These effects are more significant in a condensate because the density is much higher than in a thermal cloud. At low energies, the dipolar relaxation rate, γ_{2b}, and 3-body recombination, γ_{3b}, are independent of collisional energy, and scale in the following way: $\gamma_{2b} = \beta n$ and $\gamma_{3b} = \lambda n^2$, where λ and β are the respective rate constants.

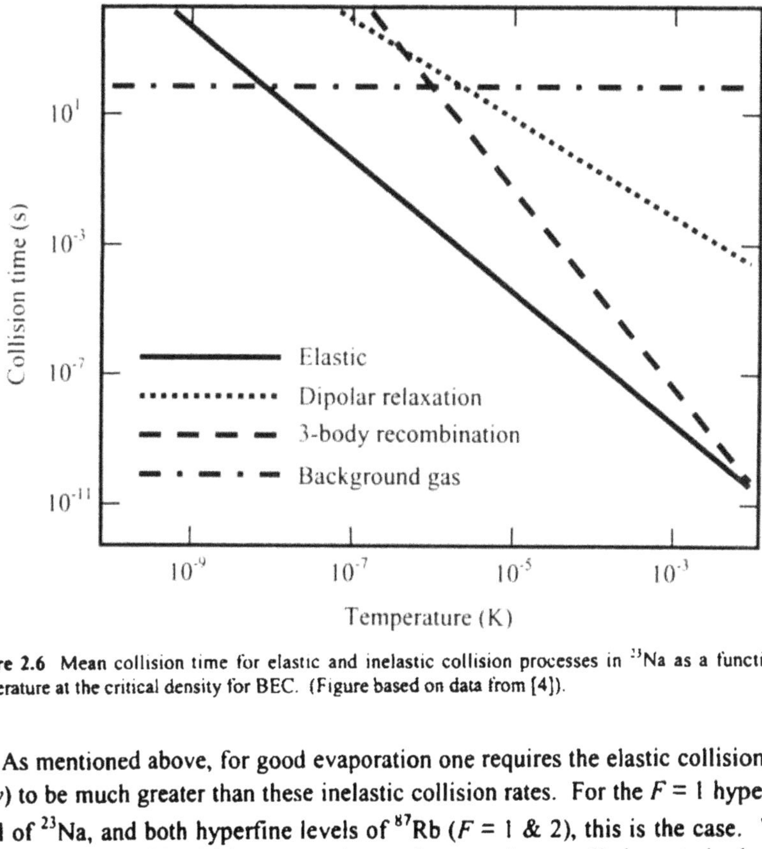

Figure 2.6 Mean collision time for elastic and inelastic collision processes in ^{23}Na as a function of temperature at the critical density for BEC. (Figure based on data from [4]).

As mentioned above, for good evaporation one requires the elastic collision rate ($n\sigma v$) to be much greater than these inelastic collision rates. For the $F = 1$ hyperfine level of ^{23}Na, and both hyperfine levels of ^{87}Rb ($F = 1$ & 2), this is the case. With other atoms, the cold collision properties are in general not well characterised. One can reduce the effect of 3-body recombination by reducing the trap stiffness, since this rate scales as n^2. As for the dipolar relaxation rate, this can be reduced by working at low bias field (say, below 5 G) and in the lowest available hyperfine level, since the hyperfine energy released is minimised [20].

In addition to collisions, heating and trap loss can arise from technical problems. For example, stray RF may lead to spin-flip loss at the trap centre (this may come from the trapping field, or if your bias field gives rise to a Zeeman shift equivalent to a local RF field). Another possible source of heating is magnetic field fluctuations which modulate the trapping potential. For 1% current fluctuations, a simple model predicts a characteristic time for the exponential growth of the energy of a typical cloud of trapped atoms of approximately 100s [21], so that commercial power supplies with a current stability of 10^{-3} to 10^{-4} are more than adequate for achieving BEC. Stability requirements are more severe at the higher trapping frequencies found in optical traps. Another source of heating is mechanical noise, but we are not aware of any groups which have had serious difficulties with this. However, as a precaution we don't use any mechanical vacuum pumps or instruments with noisy cooling fans close to the experiment. Also, stray resonant photons are eliminated with light-tight boxes around the lasers and magnetic trap. Finally, it should be noted that blackbody radiation is not a problem — the number of resonant photons emitted at room temperature and the cross-section for absorption of off-resonant light are both extremely small.

3 Probing Techniques for Bose Condensates

3.1 Introduction to Probing

A substantial amount of quantitative information about Bose condensates formed from a dilute atomic gas can be extracted using a variety of imaging techniques, and this has enabled rapid development of the field. Currently, the two most important techniques for probing Bose condensates are time-of-flight and *in situ* imaging. Both methods yield the density distribution of the atomic sample, from which we can infer properties (such as the temperature and chemical potential) by making comparisons with the results of various models.

Time-of-flight measurement involves switching off the confining potential to allow ballistic expansion of the condensate before imaging. Imaging after time-of-flight has the advantage that the expansion reduces the optical density (and increases the size) of the cloud to enable resonant absorption of probe light. Although relatively simple to implement, ballistic expansion before imaging does create some difficulties. Some traps are difficult to turn off rapidly enough to ensure that no unintentional work is done on the expanding gas. Also, in some circumstances the expansion itself masks the underlying physical behaviour. Finally, multiple images of the same cloud, for (say) dynamical studies, cannot be made using time-of-flight imaging.

The alternative to time-of-flight, imaging *in situ*, avoids these difficulties, but has two main problems. Firstly, the condensate can be small compared to the resolution limit of the imaging system, and secondly, alkali condensates are almost always sufficiently dense so as to be opaque for resonant probe light. Another problem with *in situ* imaging is that the presence of the confining magnetic field can distort images by altering the absorption cross-section across the spatial extent of the cloud. However, with considerable effort, quantitative information has been obtained from images even when the condensate was smaller than the resolution limit of the imaging system (by the Hulet group [22]), when the condensate was extremely optically dense giving rise to refraction of near-resonant probe light (by the Hau group [23]), and in the presence of a full strength quadrupole field (by the Ketterle group [24,25]). The problems of optical density can be circumvented with imaging methods based on far-detuned probe beams, which probe only the real part of the index of refraction: polarisation rotation, dark-field, and phase-contrast imaging.

The interaction of atoms with a beam of light involves two processes of interest to the probing of Bose condensates: absorption of photons and shifting the phase of the transmitted light. These properties are used in absorptive and dispersive imaging methods, respectively. The interaction can be described by the complex index of refraction of the atoms. Following the notation in reference [25], the index of refraction for a two level atom (in the rotating wave approximation and in the limit of low probe intensity) can be expressed as

$$n_{ref} = 1 + \frac{\sigma_0 n \lambda}{4\pi}\left(\frac{i}{1+\delta^2} - \frac{\delta}{1+\delta^2}\right) \tag{2.1}$$

where σ_0 is the resonant cross-section and δ is the detuning (in units of half linewidths). Assuming that light enters and exits the cloud at the same (x, y) co-ordinate (the thin lens approximation), the atoms simply attenuate and phase shift the probe light. The electric field of the probe light can be expressed as

$$E = tE_0 e^{i\phi} \tag{2.2}$$

where the transmission coefficient t and phase shift ϕ depend on the product of the column density $\tilde{n} = \int n\, dz$ and σ_0.

$$t = e^{-\tilde{D}/2} = \exp\left(-\frac{\tilde{n}\sigma_0}{2}\frac{1}{1+\delta^2}\right) \tag{2.3}$$

$$\phi = -\delta\frac{\tilde{D}}{2} = -\frac{\tilde{n}\sigma_0}{2}\frac{\delta}{1+\delta^2} \tag{2.4}$$

where $\tilde{D}$ is the off-resonance optical density. From the column density profile one can obtain the 3-dimensional density profile of the condensate using the symmetry of the trapping potential. Most BEC experiments have relied on spatially imaging the real or imaginary part of the index of refraction. In the following we outline the two most popular imaging methods currently in use.

3.2 Absorptive (or Shadow) Imaging

Absorption imaging is performed by illuminating the atoms with a pulse of resonant laser light and imaging the shadow cast by the atoms onto a CCD camera. The absorption image shows the spatial variation of t^2. This method was used at JILA in the first observation of Bose condensation [1].

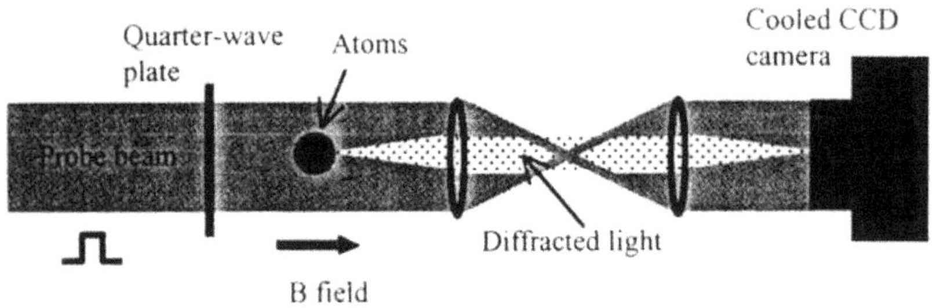

Figure 3.1 A typical shadow imaging configuration, with lenses arranged to form a simple compound microscope.

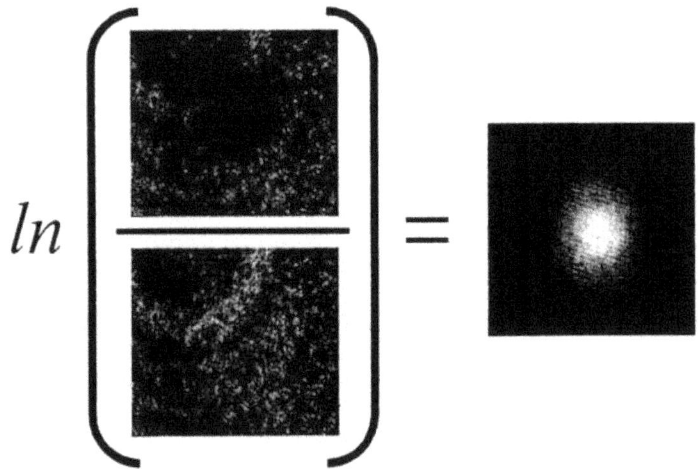

Figure 3.2 Absorptive images before and after processing. The top left image contains absorption profile, the bottom left image is the background (i.e. an image recorded without a atomic sample present), and the processed image is on the right.

The optical depth of a typical confined condensate is > 100, so that essentially all incident resonant light is absorbed. Therefore, resonant absorptive images are almost always performed in time-of-flight. In the Otago experiment, the magnetic trap is switched off in < 1 ms and the cloud is allowed to freely expand for up to 15ms. A short (10 - 50 μs) pulse of light is spatially filtered with a polarisation preserving optical fibre. The probe beam is then circularly polarised and a weak bias magnetic field is applied to the cloud to give well-defined light scattering on a cycling transition. In our case, the bias field is provided by the rotating field of the TOP trap, and the probe pulse is synchronised to the rotation to obtain the appropriate alignment. The shadow formed by the atomic cloud is imaged onto a CCD camera with a pair of doublet lenses (magnification 2 - 10) arranged to form a simple compound microscope. By using large numerical aperture lenses (*f*-number of 2 for the objective) we obtain a spatial resolution of approximately 5 μm, compared with time-of-flight condensate images > 100 μm in the radial direction. The detector is a cooled CCD digital camera which has very low noise, high quantum efficiency (40%) and very high dynamic range (16-Bit). By taking two images - one with a condensate and one without - dividing the intensity values of these images point by point and then taking the natural logarithm we obtain an optical depth profile, from which we extract the column density profile. The quality of our images is probably limited by the optical quality of our quartz vacuum cell (which is not anti-reflection coated).

One important feature of the high performance cooled CCD cameras typical of most BEC experiments is that the pixel readout for a large image can takes seconds to perform, so that taking multiple images in quick succession can be difficult. However, often the image is smaller than the pixel array so that (hopefully, depending on the particular camera) one can choose to readout only part of the array, or alternatively run the camera in a frame-transfer or kinematics mode where multiple images are stored on the array and readout slowly as one large image. The latter method is particularly useful with non-destructive imaging for dynamical studies (see figure 3.5). The intensity of our short single shot probe pulse is approximately the saturation intensity for the transition, and this gives us maximum dynamic range while not appearing to cause significant blurring from recoil-induced motion. For single-shot absorptive probing, if an atom in the sample scatters N photons, it acquires a velocity of $v_{rec}N$ (where v_{rec} is the recoil velocity) along the direction of propagation of the probe beam due to absorption (and a random isotropic velocity $\sqrt{N}v_{rec}$), which may move the condensate beyond the depth of field of the imaging system.

For experiments where high detection sensitivity is paramount, such as for imaging the dilute thermal component of a partially condensed cloud, absorptive imaging performs well. Finally, for (single shot) absorptive imaging, optical pumping can cause calibration problems for imaging samples in non-cycling states, because the photon scattering rate is different for different m_F sub-states.

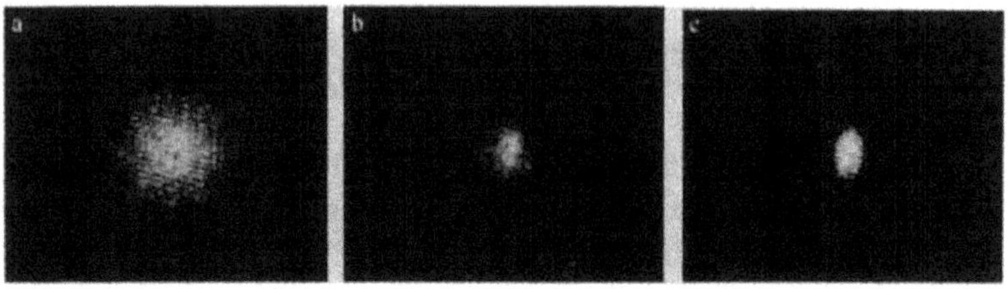

Figure 3.3 Time-of-flight (5 ms) images of Bose-Einstein condensation of ^{87}Rb atoms at Otago, using the absorptive imaging. (a) A thermal vapour. (b) A small Bose condensate within a thermal cloud. (c) An almost pure Bose condensate.

For experiments where high detection sensitivity is paramount, such as for imaging the dilute thermal component of a partially condensed cloud, absorptive imaging performs well. Finally, for (single shot) absorptive imaging, optical pumping can cause calibration problems for imaging samples in non-cycling states, because the photon scattering rate is different for different m_F sub-states.

3.3 *Dispersive Imaging*

As mentioned above, *in situ* resonant imaging does not reveal the density profile, because condensates are nearly always opaque to resonant light. This problem can be circumvented by using far-detuned light. In this case the condensate is transparent, but phase encodes the probe beam. To image the transparent object we must convert the phase shift into an intensity profile for the CCD camera. All dispersive probing methods involve separating the scattered and unscattered components of the probe light and manipulating them independently. This is usually done by spatially filtering in a Fourier plane of the imaging system where the unscattered probe light comes to a focus.

The most common dispersive probing method currently used in BEC labs is phase contrast imaging [26], which is essentially a homodyne detection scheme where the unscattered light acts as the local oscillator and interferes with the scattered radiation. This is accomplished by shifting the phase of the unscattered light by $\pm\pi/2$ with a phase plate, which is an optical flat with a small bump (of height $\lambda/4$) at the centre, placed in the Fourier plane of the objective lens. For small phase shifts, one obtains an intensity profile which is linearly proportional to the phase shift profile, from which the density profile is extracted. This imaging scheme can be made essentially non-destructive in the sense that by using far-detuned probe light, heating by the absorption and spontaneous emission of photons (scattering at large angles) is almost eliminated, and residual heating from the small angle elastic scattering is minimal (but still a limiting factor). In this case, it is possible to record multiple images of the same condensate.

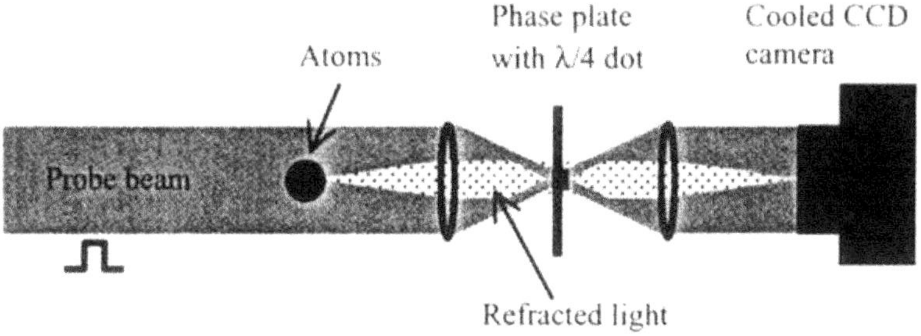

Figure 3.4 A typical phase contrast imaging setup, with a phase plate positioned in the Fourier plane.

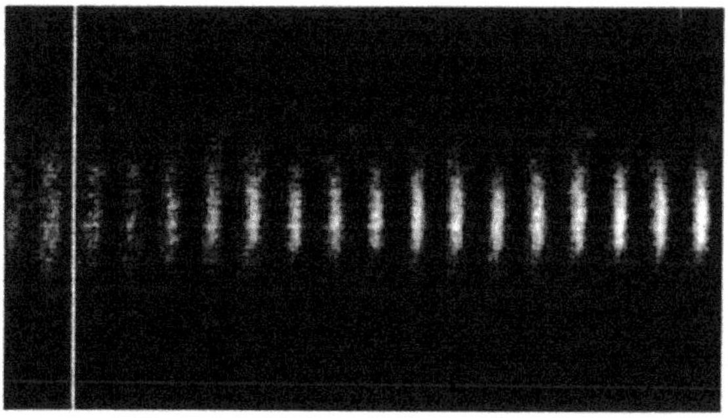

Figure 3.5 The formation of a Bose condensate recorded by the Ketterle group using (non-destructive) phase contrast imaging [27]. Reprinted with permission from Science **279**, 1005 (1998). Copyright 1998 American Association for the Advancement of Science.

3.4 A Brief Mention of Other Imaging Schemes

Early imaging of large dilute magnetically trapped clouds was done by collecting fluorescence, but this does not work well for Bose condensates because of the large optical density and because light scattering can lead to motion of the sample that alters the image. Polarization-contrast imaging is a dispersive scheme which relies on the different phase shifts for orthogonal polarisation components of the probe light [28]. A polarizer placed before the camera can block the unscattered light, resulting in a dark-ground image. Rotating the polarizer leads to interference of the two polarizations, resulting in a phase-contrast image. In applying this method one must consider the effect of magnetic fields on the probing transition. Dark-ground imaging is a simple form of spatial filtering where a small opaque object is placed into the Fourier plane to block the unscattered light. The configuration for this method is similar to phase-contrast imaging except that the small bump on the phase plate is replaced with a similar sized opaque object. For small ϕ the dark-ground signal is quadratic in ϕ. Finally, spectroscopic techniques for measuring condensate parameters have been suggested [29], but so far high resolution spectroscopy has only been applied to hydrogen condensates [30].

4 Survey of Current BEC Experiments

On the next page is a table summarising features of all the functioning BEC experiments (as from 1 January 2000).

4.1 *Summary of Technical Issues*

From the above table one can see both common threads and large variations. The most common feature is that nearly all dilute vapour BEC experiments use laser cooling, followed by magnetic trapping, and finally evaporative cooling. In this section we attempt to review the main issues which have lead to different experimental designs. Related technical details of the Otago BEC experiment are given in reference [12].

4.1.1 Atomic Species

Most BEC groups are currently working with either ^{87}Rb or ^{23}Na. In the two available hyperfine levels of ^{87}Rb, there are three magnetically trapped states, $F = 2$, $m_F = +1, +2$ and $F = 1$, $m_F = -1$. The latter state has a lower heating rate (and therefore a longer sample lifetime). It appears that the heating rate for ^{87}Rb is worse than ^{23}Na. To date, sodium condensates have only been formed in the $F = 1$ state, but optical dipole trapping of condensates (formed by traditional methods) has enabled the confinement of all three available m_F states of the $F = 1$ state of ^{23}Na [31].

There is considerable interest in other atomic species. The Rice group has condensed ^{7}Li, which has a negative scattering length, and considered issues of stability [22,28,32]. At MIT, the spin-polarised hydrogen group have observed condensation [30]. Attempts to Bose condense caesium have so far been unsuccessful, because magnetically confined samples suffer catastrophic inelastic collisional processes [33,34]. However, there is considerable experimental effort being made with optical techniques [35]. The JILA group recently reported condensation with ^{85}Rb [36]. JILA have also been successful in cooling ^{40}K (a fermionic isotope) below the Fermi temperature [37], and groups at Rice and ENS are pursuing a dilute degenerate Fermi gas with ^{6}Li [38]. Numerous other possibilities are being considered for their potential to Bose condense. In future,

Table 4.1 Where, who, with what and how?

Location	Group Leader(s)	Atom	Pre-Cooling	Magnetic Trap
JILA	Cornell/ Wieman	^{87}Rb, ^{85}Rb	Double MOT	TOP & IP (Baseball)
MIT	Ketterle	^{23}Na	Beam + MOT	IP (Cloverleaf)
Rice	Hulet	^{7}Li	Beam + Molasses	IP (Permanent)
Rowland Institute	Hau	^{23}Na	Beam + MOT	IP (4-dee)
Yale	Kasevich	^{87}Rb	Vapour MOT	TOP
Texas	Heinzen	^{87}Rb	Beam + MOT	TOP
Konstanz	Rempe	^{87}Rb	Double MOT	IP (bars)
Max Planck, Munich	Hansch	^{87}Rb	Double MOT	IP (3 coil)
NIST, Gaithersburg	Phillips	^{23}Na	Beam + MOT	TOP
Paris, ENS	Dalibard	^{87}Rb	Double MOT	IP (3 coil)
MIT	Greytak/ Kleppner	H	Cryogenic	IP (Supercond.)
Paris, Orsay	Aspect	^{87}Rb	Beam + MOT	IP (Pole piece)
Hanover	Ertmer	^{87}Rb	Beam + MOT	IP (Cloverleaf)
Otago	Wilson	^{87}Rb	Double MOT	TOP
Sussex	Boshier	^{87}Rb	Double MOT	IP (Baseball)
Tokyo	Takahashi/ Yabuzaki	^{87}Rb	Double MOT	IP (Cloverleaf)
Kyoto	Kuga	^{87}Rb	Double MOT	IP (Cloverleaf)
Oxford	Foot	^{87}Rb	Double MOT (Pyramid)	TOP
Florence	Inguscio	^{87}Rb	Double MOT	IP (3 coil)
Amsterdam	Walraven	^{87}Rb	Double MOT (2D)	IP (bars)
Pisa	Arimondo	^{87}Rb	Double MOT	TOP

Just a numbers game?

Ratio of ^{87}Rb to ^{23}Na condensates	16:3
Ratio of Double MOTs to Zeeman slowers	12:7
Ratio of Ioffe-Pritchard to TOP traps	14:7
Ratio of Northern hemisphere to Southern hemisphere	20:1

Therefore, one might conclude that the best approach is to build a Rb double MOT and Ioffe-Pritchard magnetic trap, somewhere in the northern hemisphere!

buffer-gas loading of molecules and non-alkali atoms [39] and sympathetic cooling [40] may lead to a great many Bose condensed species.

4.1.2 Vacuum Chamber Design

In many experiments Bose condensates are formed in small glass cells with magnetic trapping coils wound outside of the vacuum. This avoids vacuum out-gassing (and cooling) problems with the trapping coils, and difficulties of RF evaporation with a metal chamber. Small glass cells are compact, enabling closely spaced trapping coils, but vacuum conduction may limit very small cells. Accumulation of background rubidium vapour in the glass cell is a common problem, but this can be circumvented by regular warming of the cell, and some groups place activated carbon chunks or non-evaporable getter pumps in the extreme UHV region of the chamber which act as a sorption material for rubidium. This problem is not present in sodium experiments, because the vapour pressure of sodium at room temperature is lower than that for other species typical of a UHV environment (such as H_2 and CO).

4.1.3 MOT Loading

The use of a MOT as a precursor to magnetic trapping is almost universal, but methods for loading the MOT vary amongst experimental groups. The loading method must be optimised to give rapid accumulation of large samples within a UHV environment. The simple approach of loading atoms directly from a background thermal vapour gives either a relatively small number of atoms and loading that is very slow (>100 s), or a magnetic trap lifetime which is severely limited by background collisions with the vapour. The desire for larger condensates has led to the adoption of one of two standard techniques: a Zeeman-slowed beam [41] or a double MOT scheme [16]. It appears that groups with previous experience of atomic beams and Zeeman slowers have tended to adopt these for BEC, and most other groups have chosen the double MOT. A variant on the standard double MOT is to use a pyramid/funnel MOT [42], or LVIS scheme [43], to produce a slow atomic beam which can load directly into a MOT. These methods have the simplification that a Zeeman slower is not required, but the typical atom flux of slow beams generated by these methods is less than in a typical Zeeman slowed thermal beam.

4.1.4 Pre-Cooling

Evaporative cooling can work only if the re-thermalisation time is much shorter than the sample lifetime in the trap. This means maximising the elastic collision rate by loading a cold, dense sample into the magnetic trap. Although the performance of a standard MOT is adequate, a number of groups apply techniques to improve on this just before transfer into the magnetic trap. As mentioned in the discussion on laser cooling techniques [W.D. Phillips, this volume], the density of the MOT can be enhanced by a dark-spot (especially for sodium) [44] or by a brief compression in a MOT (by increasing the magnetic field gradient), and the temperature can be reduced by molasses cooling just before transfer to the magnetic trap [45].

4.1.5 Mode-Matching

In order to maximise the elastic collision rate for efficient evaporative cooling, laser cooled atoms are transferred into a mode-matched magnetic trap, and then adiabatically compressed. The mode-matching condition is satisfied when the phase space density of the laser cooled cloud loaded into the magnetic trap is maximised. After compression in a MOT and molasses cooling, atoms are transferred into a magnetic trap by rapidly switching off the molasses light and then suddenly turning on the magnetic trap, thereby constructing a magnetic potential around the atoms. Ideally, if the potential is mode matched, the volume and temperature of the cloud is maintained during the transfer process. If the trap is too strong, the atoms will be heated by the transfer. If it is too weak, the atoms will expand non-adiabatically. In either case phase-space density will be lost.

To calculate the magnetic trap parameters which give best mode matching we assume the initial total energy is equal to the final total energy. The initial total energy is the kinetic energy due to the temperature in the MOT plus the potential energy due to the size of the cloud when the confining potential is initially switched on. The final total energy is just the sum of the kinetic energy plus the potential energy in the magnetic trap. In the re-thermalised cloud, kinetic energy is equal to the potential energy (the equipartition theorem). From this one can calculate the final and initial phase space densities, and find a curvature for the confining potential which maximises the ratio. In practise, we find that the change in phase space density is fairly insensitive to the curvature.

An obvious complication arises if the shape of the confining potential differs dramatically from the roughly spherical shape of the laser cooled atoms. This is particularly true of Ioffe-Pritchard traps, but this problem can be minimised by careful choice of initial bias field. Finally, it is worth noting that if the trap cannot be switched on fast, the laser cooled cloud will fall under gravity and may undergo centre of mass oscillations when the magnetic potential is applied.

4.1.6 Magnetic Traps

The Ioffe-Pritchard configuration [46,47] is rapidly becoming the "industry standard", but the specific coil designs used by different groups vary considerably (see table 4.1). In general, the trapping coils are based on few turns in order to minimise the switching time. Such coils require high current and water-cooling. Many groups now make their coils with square copper tubing covered with fibreglass or Kapton insulation, but some have opted to flow water around thick copper wire or plates. Often, the coils are wired in series in order to cancel the effect of current fluctuations in the power supply on the stability of the confining potential. Rapid switching is performed using external transistor circuits (based on FETs or IGBTs), and typical switching times are < 1 ms.

4.1.7 Adiabatic Compression

Adiabatic compression plays a crucial role in BEC experiments because although the phase-space density remains constant, the process increases the collision rate before evaporative cooling. Adiabatic compression is performed by increasing the strength of the confining potential at a rate much slower than the trap oscillation frequency and the elastic collision rate - typically over a period of a few seconds (to be on the safe side). If the compression is performed faster than the trap oscillation frequency then the cloud can be set in oscillation and heated. Similarly, if the compression is anisotropic (as is often the case in a Ioffe-Pritchard trap) and performed faster than the elastic collision rate, this will result in an anisotropic temperature and subsequent loss of phase-space density during equilibration. Anisotropic heating is hard to avoid when the initial collision rate is low, but its effect is relatively minor.

Acknowledgements

Our work in this field is supported by the Royal Society of New Zealand Marsden Fund (contract UOO910) and the University of Otago. We would like to thank the other members of the Otago BEC experiment research team - J.L. Martin, N.R. Thomas, R.W. Geursen, and D.M. Warrington - for their assistance with the preparation of this article.

References

1. Anderson M.H., Ensher J.R., Matthews M.R., Wieman C.E. and Cornell E.A., Observation of Bose-Einstein condensation in a dilute atomic vapor. *Science* **269** (1995) pp. 198.
2. Davis K.B., Mewes M.-O., Andrews M.R., van Druten N.J., Durfee D.S., Kurn D.M. and Ketterle W., Bose-Einstein condensation in a gas of sodium atoms. *Phys. Rev. Lett.*, **75** (1995) pp. 3969.
3. Hess H.F., Evaporative cooling of magnetically trapped and compressed spin-polarized hydrogen. *Phys. Rev. B*, **34** (1986) pp. 3476.
4. Ketterle W. and van Druten N.J., Evaporative cooling of trapped atoms. *Adv. At. Mol. Opt. Phys.*, **37** (1996) pp. 181.
5. Pritchard D.E., Helmerson K. and Martin A.G., Atom traps. In *Atomic Physics 11*, ed. by Haroche S., Gay J.C. and Grynberg G. (World Scientific, Singapore 1989) pp. 179-197.

6. Petrich W., Anderson M.H., Ensher J.R. and Cornell E.A. In *Fourteenth International Conference on Atomic Physics (ICAP XIV)*, (Boulder CO, 1994) pp. 1M-7.
7. Davis K.B., Mewes M.-O., Joffe M.A. and Ketterle W. In *Fourteenth International Conference on Atomic Physics (ICAP XIV)*, (Boulder CO, 1994) pp. 1M-3.
8. Cohen-Tannoudji C., Dupont-Roc J. and Grynberg G., *Atom-photon interactions*. John Wiley and Sons, 1992.
9. Rubbmark J.R., Kash M.M., Littman M.G. and Kleppner D., Dynamical effects at avoided level crossings: A study of the Landau-Zener effect using Rydberg atoms. *Phys. Rev. A*, **23** (1981) pp. 3107.
10. Majorana E., Majorana Flops. *Nuovo Cimento*, **9** (1932) pp. 43.
11. Petrich W., Anderson M.H., Ensher J.R. and Cornell E.A., Stable, tightly confining magnetic trap for evaporative cooling of neutral atoms. *Phys. Rev. Lett.*, **74** (1995) pp. 3352.
12. Martin J.L., McKenzie C.R., Thomas N.R., Sharpe J.C., Warrington D.M., Manson P.J., Sandle W.J. and Wilson A.C., Output coupling of a Bose-Einstein condensate formed in a top trap. *J. Phys. B*, **32** (1999) pp. 3065.
13. Sackett C.A., Bradley C.C. and Hulet R.G., Optimization of evaporative cooling. *Phys. Rev. A*, **55** (1997) pp. 3797.
14. Wu H. and Foot C.J., Direct simulation of evaporative cooling. *J. Phys. B*, **29** (1996) pp. L321.
15. Cornell E.A., Ensher J.R., and Wieman C.E., Experiments in dilute atomic bose-einstein condensation. In *Proceedings of the International School of Physics - Enrico Fermi*, ed. by Inguscio M., Stringari S. and Wieman C.E. (IOS Press, 1999) pp. 15.
16. Myatt C.J., Newbury N.R., Ghrist R.W., Loutzenhiser S. and Wieman C.E., Multiply loaded magneto-optical trap. *Opt. Lett.*, **21** (1996) pp. 290.
17. Myatt C.J., Bose-Einstein condensation experiments in a dilute vapor of rubidium. PhD thesis, University Of Colorado (Boulder), 1997.
18. Burt E.A., Ghrist R.W., Myatt C.J., Holland M.J., Cornell E.A. and Wieman C.E., Coherence, correlations, and collisions: What one learns about Bose-Einstein condensates from their decay. *Phys. Rev. Lett.*, **79** (1997) pp. 337.
19. Soding J., Guery-Odelin D., Desbiolles P., Chevy F., Inamori H. and Dalibard J., Three-body decay of a rubidium Bose-Einstein condensate. *Appl. Phys. B*, **69** (1999) pp. 257.
20. Dalibard J., Collisional dynamics of ultra-cold atomic gases. In *Proceedings of the International School of Physics - Enrico Fermi*, ed. By Inguscio M., Stringari S. and Wieman C.E. (IOS Press, 1999) pp. 321.
21. Gehm M.E., O Hara K.M., Savard T.A. and Thomas J.E., Dynamics of noise-induced heating in atom traps. *Phys. Rev. A*, **58** (1998) pp. 3914.
22. Bradley C.C., Sackett C.A. and Hulet R.G., Analysis of in situ images of Bose-Einstein condensates of lithium. *Phys. Rev. A*, **55** (1997) pp. 3951.

23. Hau L.V., Busch B.D., Liu C., Dutton Z., Burns M.M. and Golovchenko J.A., Near-resonant spatial images of confined Bose-Einstein condensates in a 4-Dee magnetic bottle. *Phys. Rev. A*, **58** (1998) pp. R54.
24. Andrews M.R., Bose condensates and the atom laser. PhD thesis, Massachusetts Institute of Technology, 1998.
25. Ketterle W., Durfee D.S. and Stamper-Kurn D.M., Making, probing and understanding Bose-Einstein condensates. In *Proceedings of the International School of Physics - Enrico Fermi*, ed. by Inguscio M., Stringari S., and Wieman C.E. (IOS Press, 1999) pp. 67.
26. Andrews M.R., Mewes M.-O., van Druten N.J., Durfee D.S., Kurn D.M. and Ketterle W., Direct, nondestructive observation of a Bose condensate. *Science*, **273** (1996) pp. 84.
27. Miesner H.-J., Stamper-Kurn D.M., Andrews M.R., Durfee D.S., Inouye S. and Ketterle W., Bosonic stimulation in the formation of a Bose-Einstein condensate. *Science*, **279** (1998) pp. 1005.
28. Bradley C.C., Sackett C.A. and Hulet R.G., Bose-Einstein condensation of lithium: Observation of limited condensate number. *Phys. Rev. Lett.*, **78** (1997) pp. 985.
29. Lye J.E., Cuthertson B.D., Bachor H-A. and Close J.D., Phase modulation spectroscopy: a nondestructive probe of Bose-Einstein condensates. J. Opt. B: Quantum Semiclas. Opt., **1** (1999) pp. 402.
30. Fried D.G., Killian T.C., Willmann L., Landhuis D., Moss S.C., Kleppner D. and Greytak T.J., Bose-Einstein condensation of atomic hydrogen. *Phys. Rev. Lett.*, **81**(1998) pp. 3811.
31. Stenger J., Inouye S., Stamper-Kurn D.M., Miesner H.-J., Chikkatur A.P. and Ketterle W., Spin domains in ground-state Bose-Einstein condensates. *Nature*, **396** (1999) pp. 345.
32. Bradley C.C., Sackett C.A., Tollett J.J. and Hulet R.G., Evidence of Bose-Einstein condensation in an atomic gas with attractive interactions. *Phys. Rev. Lett.*, **75** (1995) pp. 1687.
33. Hopkins S.A., Webster S., Arlt J., Bance P., Cornish S., Marago O. and Foot C.J., Measurement of elastic cross section for cold cesium collisions. *Phys. Rev. A*, **61** (2000) pp. 2707.
34. Guery-Odelin D., Soding J., Desbiolles P. and Dalibard J., Is Bose-Einstein condensation of atomic cesium possible? *Europhys. Lett.*, **44** (1998) pp. 26.
35. Kerman A.J., Vuletic V., Chin C. and Chu S., Beyond optical molasses: 3D raman sideband cooling of atomic cesium to high phase-space density. *Phys. Rev. Lett.*, **84** (2000) pp. 439.
36. Cornish S.L., Claussen N.R., Roberts J.L., Cornell E.A. and Wieman C.E., Stable ^{85}Rb Bose-Einstein condensation with widely tunable interactions. cond-mat/0004290, 2000.
37. DeMarco B. and Jin D.S., Onset of Fermi degeneracy in a trapped atomic gas. *Science*, **285** (1999) pp.1703.

38. Mewes M.-O., Ferrari G., Schreck F., Sinatra A. and Salomon C., Simultaneous magneto-optical trapping of two lithium isotopes. *Phys. Rev. A*, **60** (2000) pp. 1403.
39. deCarvalho R., Doyle J.M., Guillet T., Kim J., Weinstein J.D. and Patterson D., Buffer-gas loaded magnetic traps for atoms and molecules: A primer. *Euro. Phys. J. D*, **7** (1999) pp. 289.
40. Myatt C.J., Burt E.A., Ghrist R.W., Cornell E.A. and Wieman C.E. Production of two overlapping Bose-Einstein condensates by sympathetic cooling. *Phys. Rev. Lett.*, **78** (1997) pp. 586.
41. Prodan J.V., Phillips W.D. and Metcalf H.J., Laser production of a very slow monoenergetic atomic beam. *Phys. Rev. Lett.*, **49** (1982) pp. 1149.
42. Lee K.I., Kim J.A., Noh H.R. and Jhe W., Single-beam atom trap in a pyramidal and conical hollow mirror. *Optics Lett.*, **21** (1996) pp. 1177.
43. Lu Z.T., Corwin K.L., Renn M.J., Anderson M.H., Cornell E.A. and Wieman C.E., Low-velocity intense source of atoms from a magneto-optical trap. *Phys. Rev. Lett.*, **77** (1996) pp. 3331.
44. Ketterle W., Davis K.B., Joffe M.A., Martin A. and Pritchard D.E., High densities of cold atoms in a dark spontaneous-force optical trap. *Phys. Rev. Lett.*, **70** (1993) pp. 2253.
45. Lett P.D., Watts R.N., Westbrook C.I., Phillips W.D., Gould P.L. and Metcalf H.J., Observation of atoms laser cooled below the Doppler limit. *Phys. Rev. Lett.*, **61** (1988) pp. 169.
46. Pritchard D.E., Cooling neutral atoms in a magnetic trap for precision spectroscopy. *Phys. Rev. Lett.*, **51** (1983) pp. 1336.
47. Bagnato V.S., Lafyatis G.P., Martin A.G., Raab E.L., Ahmad-Bitar R.N. and Pritchard D.E., Continuous stopping and trapping of neutral atoms. *Phys. Rev. Lett.*, **58** (1987) pp. 2194.

www.ingramcontent.com/pod-product-compliance
Ingram Content Group UK Ltd.
Pitfield, Milton Keynes, MK11 3LW, UK
UKHW020359150926
8624IPUK00004B/235

9 789810 244842